ICM-90 Satellite
Conference Proceedings

A. Fujiki · K. Kato · T. Katsura
Y. Kawamata · Y. Miyaoka (Eds.)

Algebraic Geometry and Analytic Geometry

Proceedings of a Conference held in Tokyo, Japan
August 13-17, 1990

Springer-Verlag
Tokyo Berlin Heidelberg
New York London Paris
Hong Kong Barcelona

Editors
Akira Fujiki
College of General Education, Kyoto University
Yoshida-nihonmatsu, Sakyo-ku, Kyoto, 606 Japan

Kazuya Kato and Yujiro Kawamata
Department of Mathematics, University of Tokyo
Hongo, Bunkyo-ku, Tokyo, 113 Japan

Toshiyuki Katsura
Department of Mathematics, Ochanomizu University
Otsuka, Bunkyo-ku, Tokyo, 112 Japan

Yoichi Miyaoka
Department of Mathematics, Rikkyo University
Nishi-ikebukuro, Toshima-ku, Tokyo, 171 Japan

Mathematics Subject Classification (1980): 14-06, 14F99, 14J28, 14K99, 14L99

ISBN 4-431-70086-2 Springer-Verlag Tokyo Berlin Heidelberg New York
ISBN 3-540-70086-2 Springer-Verlag Berlin Heidelberg New York Tokyo
ISBN 0-387-70086-2 Springer-Verlag New York Berlin Heidelberg Tokyo

Printing and binding: Permanent Typesetting and Printing Co., Ltd., Hong Kong

Preface

The International Conference "Algebraic Geometry and Analytic Geometry, Tokyo 1990" was held at Tokyo Metropolitan University and the Tokyo Training Center of Daihyaku Mutual Life Insurance Co., from August 13 through August 17, 1990, under the co-sponsorship of the Mathematical Society of Japan. It was one of the satellite conferences of ICM90, Kyoto, and approximately 300 participants, including more than 100 from overseas, attended the conference.

The academic program was divided into two parts, the morning sessions and the afternoon sessions. The morning sessions were held at Tokyo Metropolitan University, and two one-hour plenary lectures were delivered every day. The afternoon sessions at the Tokyo Training Center, intended for a more specialized audience, consisted of four separate subsessions: Arithemetic Geometry, Algebraic Geometry, Analytic Geometry I and Analytic Geometry II. This book contains papers which grew out of the talks at the conference.

The committee in charge of the organization and program consisted of A. Fujiki, K. Kato, T. Katsura, Y. Kawamata, Y. Miyaoka, S. Mori, K. Saito, N. Sasakura, T. Suwa and K. Watanabe. We would like to take this opportunity to thank the many mathematicians and students who cooperated to make the conference possible, especially Professors T. Fukui, S. Ishii, Y. Kitaoka, M. Miyanishi, Y. Namikawa, T. Oda, F. Sakai and T. Shioda for their valuable advice and assistance in organizing this conference.

Financial support was mainly provided by personal contributions from Professors M. Nagata, T. Shioda and S. Iitaka as well as grants and fellowships from the following institutions:

The Inamori Foundation,
Inoue Foundation for Science,
Japan Association for Mathematical Sciences,
The Nikko Securities Co., Ltd.
Tokyo Metropolitan Univ..

We would like to thank Tokyo Metropolitan University for making available the lecture hall for the morning sessions. Last, but not least, we would like to express our hearty gratitude to Daihyaku Mutual Life Insurance Co. for generously providing facilities and accommodations.

The editors

Tokyo, April 1991

CONTENTS

POINCARE POLYNOMIALS OF SOME MODULI VARIETIES

V.BALAJI[1] AND C.S.SESHADRI

§1 Introduction:

For smooth projective varieties $V/\mathbb{C}$, the Weil conjectures as established by Deligne, tell us that the number of rational points of the corresponding variety V over $\mathbb{F}_{q^n}$, for all n, determine the Betti numbers of V (for precise details cf.,§5). This theme has been taken up by Harder and Narasimhan in [H-N] and by Desale and Ramanan in [D-R] to compute the Poincaré polynomial of the moduli space M(n,d) of semi-stable vector bundles of rank n and degree d, where n and d are coprime. More recently, Atiyah and Bott [A-B] following a geometric approach compute the Poincaré polynomial of the moduli space M(n,d) when (n,d) = 1,and also show that there is no torsion in the cohomology in this case. Let N be the smooth compactification of $M(2,0)^s_{\mathcal{O}_X}$ (the stable bundles with $\det E \cong \mathcal{O}_X$) constructed in [S]. In [B-S], an approach modelled on [A-B] was studied and this gave only partial success in the computation of the cohomology of N.

In this paper, we base ourselves on the Harder-Narasimhan approach; more precisely, we use the strata obtained in [B-S] to compute the number of $\mathbb{F}_q$-rational points of N and we deduce the Poincaré polynomial of N in a compact form.

---

[1]Supported by N.B.H.M.

The layout of the paper is as follows: in §2 we recall some facts about N and fix some notations. The heart of the work is in §3. Here we analyse the Siegel formula (cf., [H-N],[D-R],[Bi])to obtain an expression for the number of rational points of $M(2,0)^s_{\mathcal{O}_X}$. In §4 we compute the rational points of the remaining strata and this provides the ingredients for obtaining the number of rational points of N. We end by obtaining the expression for the Poincaré polynomial of N making use of a lemma of Kirwan (cf., [K] pp., 186).

It is easy to see that the computations in this paper give the intersection Poincaré polynomial of $M(2,0)_{\mathcal{O}_X}$. It is very likely that these methods generalize to give the intersection Poincaré polynomial for all M(n,d), obtained by Kirwan by other methods (cf., [K-2]).

Acknowledgements: We thank D.Indumati for helping us *reduce* the polynomial expressions and the referee for his meticulous reading of the paper.

§2 **Preliminaries.**

We first recall the basic information on the desingularization model N defined in [S].

X is a smooth projective curve of genus $g \geq 2$. Let M denote the normal projective variety of equivalence classes of semi-stable vector bundles of rank 2 and trivial determinant, under the equivalence relation $V \sim V'$ iff $gr(V) \cong gr(V')$. Let M^s be the smooth open subset of M consisting of the stable bundles.

For a vector bundle V on X, the notion of a parabolic structure has been defined in [M-S] (for the purposes of the present paper (cf., [B-1],[B-2]).

Let PV_4 denote the category of semi-stable vector bundles (V,Δ) with parabolic structure Δ, as in [B-2], where V is a vector bundle of rank 4 on X with det $V \cong \mathcal{O}_X$, and Δ is the parabolic structure on V given at a point $P \in X$ with suitable weights (cf.,[S]) . Then we have

Definition :(cf.,[S],[B-S]) N is the set of isomorphism classes of $(V,\Delta) \in PV_4$, such that End V is a specialisation of $\mathcal{M}_2$ - the 2 x 2 matrix algebra.

We recall for convenience the definition of the Kummer variety associated to the Jacobian J of X. The Kummer variety for our purposes can be defined as follows:

K = { the isomorphism classes of vector bundles of the form $\{L \oplus L^{-1}\}$ where L is a line bundle of degree 0 } .

It is known that K has only 2^{2g} nodal singularities which we denote by K_o, and we have the canonical morphism $\phi : J \longrightarrow K$, which is defined by mapping a line bundle L to $\phi(L) = L \oplus L^{-1}$. We have the following description of K :

$$K - K_o = \{L \oplus L^{-1} \mid L^2 \not\cong \mathcal{O}_X,\ L \in J\}$$

$$K_o = \{L \oplus L \mid L^2 \cong \mathcal{O}_X,\ L \in J\}$$

All that is needed for our purposes is concentrated in the following theorem. (cf., [B-1],[B-2],[B-S])

Theorem 2.1. There is a natural structure of a smooth projective variety on N and there exists a canonical morphism $\pi : N \longrightarrow M$, which is an isomorphism over M^s. Further, there is a natural stratification of N by locally closed subvarieties $\{N_i\}$ $i = 1,2,3,$ which can be described as follows:

a) $\pi : N - N_1 \longrightarrow M^s$, is an isomorphism.

b) $N_1 - N_2$ which we denote by Y is a $\mathbb{P}^{g-2} \times \mathbb{P}^{g-2}$ bundle over $K - K_o$, where K is the Kummer variety defined above which has been identified with the singular locus of M ; indeed $K = M - M^s$. (cf., [N-R]).

c) Let $p \in K_o$, then the fibre $\pi^{-1}(p)$ is the disjoint union of two closed subschemes R and S where

(i) R is a vector bundle of rank $(g - 2)$ over $G(2,g)$, the Grassmanian of 2-planes in g-space.

(ii) S is isomorphic to $G(3,g)$.

Thus

$$N_2 - N_3 = \coprod R \text{ (the union is over points of } K_o)$$

and

$$N_3 = \coprod S \text{ (union over points of } K_o).$$

§3. The Siegel Formula.

Let X be a curve of genus g defined over a finite field $\mathbb{F}_q$. Then the ζ-function of X takes the form

$$\zeta_X(s) = \frac{\prod_{i=1}^{2g} (1-\omega_i q^{-s})}{(1-q^{-s})\ (1-q.q^{-s})}$$

where ω_i are algebraic integers (depending on X) with $|\omega_i| = q^{1/2}$. Let E be a vector bundle of rank n over X defined over $\mathbb{F}_q$, having fixed determinant, $\det E = L$, of degree d. The Siegel formula is the following:

$$\sum \frac{1}{N_q(\mathrm{Aut}E)} = \frac{1}{(q-1)} \; q^{(n^2-1)(g-1)} \zeta_X(2) \ldots \zeta_X(n).$$

where the summation is over <u>isomorphism</u> classes of bundles E of rank n defined over $\mathbb{F}_q$, such that detE = L, and $N_q(\mathrm{Aut}E)$ = the cardinality of the set of $\mathbb{F}_q$-valued points of AutE, the group <u>scheme</u> of automorphisms of E (defined over $\mathbb{F}_q$). (cf., [A-B],[H-N],[D-R]).

We shall restrict ourselves to the case of bundles E of rank 2 with detE = $\mathcal{O}_X$. Then the Siegel formula takes the form:

$$\sum \frac{1}{N_q(\mathrm{Aut}E)} = \frac{1}{(q-1)} \; q^{3(g-1)} \zeta_X(2). \qquad (*)$$

the summation being over isomorphism classes of bundles E of rank 2 defined over $\mathbb{F}_q$ with detE = $\mathcal{O}_X$.

Since the stable bundles admit only scalar automorphisms these contribute

$$\frac{N_q(M(n,d)^s_{\mathcal{O}_X})}{(q-1)}$$

to the left hand side of the Siegel formula, where the numerator denotes the number of rational points of $M(n,d)^s_{\mathcal{O}_X}$, the open subset of stable bundles of $M(n,d)_{\mathcal{O}_X}$.

Following [D-R] we introduce

$$\beta(n,d) = \sum \frac{1}{N_q(\mathrm{Aut}E)} \qquad (1)$$

where the summation is over the semi-stable bundles, and

$$\beta'(n,d) = \sum \frac{1}{N_q(\mathrm{Aut}E)} \qquad (2)$$

summed over the remaining bundles (the non-semi-stable ones).

For the purposes of the present work , we assume that $n = 2$ and $d = 0$, with $\det E = \mathcal{O}_X$, $\forall$ E.

In (2) we have the contribution of the non-semi-stable bundles, which have a canonical (Harder-Narasimhan) filtration defined over $\mathbb{F}_q$:

$$0 \longrightarrow L \longrightarrow E \longrightarrow L^{-1} \longrightarrow 0 \qquad (3)$$

where the degree of the bundles L (defined over $\mathbb{F}_q$) ranges from $r = 1,2,3,\ldots,$ (one must observe that $\det E \cong \mathcal{O}_X$). Note that $r = 0$ is not included here since the corresponding extensions define <u>semi-stable</u> bundles. To compute $N_q(\mathrm{Aut}E)$, we consider (following [A-B])the following two cases:

(I) <u>the split extensions</u>: For $E = L \oplus L^{-1}$ the automorphisms group is easily seen to be isomorphic as a <u>scheme</u> to the product $\mathbb{G}_m \times \mathbb{G}_m \times \mathbb{A}$ (all over $\mathbb{F}_q$) where $\mathbb{A}$ is the affine space of dimension = $\dim H^0(X,L^2) = \dim \mathrm{Hom}(L^{-1},L)$

Hence we have :

$$N_q(\mathrm{Aut}E) = (q-1)^2 h_o$$

where $h_o = N_q(H^0(X,L^2))$.

(II) <u>the non-split extensions</u>: Here because of the non-split nature the automorphism group is isomorphic to the product $\mathbb{G}_m \times \mathbb{A}$ (over $\mathbb{F}_q$) where $\mathbb{A}$ is as in (I) above. Hence we have in this case :

$$N_q(\mathrm{Aut}E) = (q-1)\, h_o$$

The number of isomorphism classes of bundles E which come from non-split extensions with a fixed L as in (3) above , is equal to $N_q(\mathbb{P}(H^1(X,L^2)) = (h_1 - 1)/(q-1)$.

where $h_1 = N_q(H^1(X,L^2))$.

Thus the contribution to $\beta'(2,0)$ from a <u>fixed</u> <u>L</u> is given by

$$\frac{1}{(q-1)^2 h_o} + \frac{h_1 - 1}{(q-1)^2 h_o} = \frac{h_1}{h_o} \frac{1}{(q-1)^2}$$

By Riemann-Roch , we get

$$\frac{h_o}{h_1} = q^{2r-g+1}$$

Therefore we have

$$\beta'(2,0) = \frac{N_q(J)}{(q-1)^2} \cdot \sum_{r=1}^{\infty} \frac{1}{q^{2r-g+1}} \tag{4}$$

(cf.,pp., 596 [A-B]).

Let us now examine in detail the term $\beta(2,0)$. As we have observed

$$\beta(2,0) = \beta_K + \frac{N_q(M^s)}{(q-1)}$$

where the second term is the contribution of the stable bundles and β_K the contribution of the <u>semi-stable</u> <u>bundles</u> <u>which</u> <u>are</u> <u>not</u> <u>stable</u> (all defined over $\mathbb{F}_q$); the K stands for the Kummer variety since $M - M^s \cong K$. (cf., [B-1]).

We now set

$$\beta_1 = \sum \frac{1}{N_q(\text{AutE})}$$

where the summation is over bundles E (defined over $\mathbb{F}_q$) which are extensions of the type

$$0 \longrightarrow L \longrightarrow E \longrightarrow L^{-1} \longrightarrow 0$$

with $L^2 \not\cong \mathcal{O}_X$, deg L = 0.

We now define :

$$\beta_2 = \sum \frac{1}{N_q(\mathrm{Aut}E)}$$

where the summation is over the bundles E which are extensions of the type

$$0 \longrightarrow L \longrightarrow E \longrightarrow L \longrightarrow 0$$

with $L^2 \cong \mathcal{O}_X$. The line bundles such that $L^2 \cong \mathcal{O}_X$ are 2^{2g} in number and therefore, without loss of generality we can suppose that L and the extensions are defined over $\mathbb{F}_q$ (by taking $\mathbb{F}_q$ to be the field over which X and all these objects are defined).

We see that

$$\beta_K = \beta_1 + \beta_2 .$$

Proposition 3.1. With β_2 as defined above we have

$$\beta_2 = \frac{2^{2g}}{N_q(GL(2))} + \frac{2^{2g}}{q(q-1)} N_q(\mathbb{P}^{g-1})$$

Proof : Consider the bundles which contribute to the sum β_2 . We examine two cases :

1) the split ones : let $E = L \oplus L$, $L^2 \cong \mathcal{O}_X$. Then

$$\mathrm{Aut}E \cong GL(2)$$

and therefore the contribution to β_2 arising from a given L is

$$\frac{1}{N_q(GL(2))}$$

Since the line bundles L such that $L^2 \cong \mathcal{O}_X$ are 2^{2g} in number, the contribution to β_2 by the split extensions is

$$\frac{2^{2g}}{N_q(GL(2))} \qquad \text{(i)}$$

2) <u>The non-split ones</u> : let E be given by

$$0 \longrightarrow L \longrightarrow E \longrightarrow L \longrightarrow 0$$

it is easy to see that $\mathrm{Aut}E \cong \mathbb{G}_m \times \mathbb{G}_a$. Hence

$$N_q(\mathrm{Aut}E) = q(q-1)$$

Thus a given L contributes $1/q(q-1)$ to β_2.

The non-split extensions correspond to non-zero elements of $H^1(X,\mathcal{O})$ modulo $\mathbb{G}_m$ i.e $\mathbb{P}(H^1(X,\mathcal{O}))$. Hence the contribution to β_2 of the non-split E is given by

$$\frac{2^{2g}\, N_q(\mathbb{P}(H^1(X,\mathcal{O})))}{q(q-1)} = \frac{2^{2g}\, N_q(\mathbb{P}^{g-1})}{q(q-1)} \qquad \text{(ii)}$$

The expressions (i) and (ii) together prove Prop 1.

Before we consider the term β_1 , we need the following remarks.

Remark 3.2:

Let E be a vector bundle on X which is semi-stable but not stable (defined over $\mathbb{F}_q$) such that $\det E \cong \mathcal{O}_X$. Then $E \longrightarrow gr(E)$ defines an isomorphism $\theta : M - M^s \longrightarrow K$, (which without loss of generality can be assumed to be defined over $\mathbb{F}_q$). Observe that this bundle E can be given as an extension

$$0 \longrightarrow L^{-1} \longrightarrow E \longrightarrow L \longrightarrow 0 \qquad (*)$$

where the extension and the line bundles L and L^{-1} are defined over $\bar{\mathbb{F}}_q$ (the algebraic closure of $\mathbb{F}_q$). We can now break up the $\mathbb{F}_q$ rational points of $K - K_o$ into the following subsets:

(i) The points $\{k \mid \theta(E) = k\}$ such that (*) is non-split.

(ii) The points $\{k \mid \theta(E) = k\}$ such that the line bundle L is defined over $\mathbb{F}_q$ and the extension (*) splits over $\mathbb{F}_q$.

(iii) The remaining ones.

Denote the subsets (i),(ii) and (iii) above by A_1, A_2 and B respectively. We observe that for bundles E corresponding to a point in A_1 , L^{-1} is the unique line sub-bundle of degree zero. Hence by descent theory the sub-bundle L^{-1} as well as L and the extension (*) are defined over $\mathbb{F}_q$.

Let E correspond to a point of B. Then $E \cong L \oplus L^{-1}$ over $\bar{\mathbb{F}}_q$. Since E is $\mathbb{F}_q$-rational we see that $\phi(E)$ is in $(K - K_o)(\mathbb{F}_q)$, where ϕ is the canonical 2-sheeted covering $\phi : J \longrightarrow K$ (without loss of generality , ϕ can be assumed to be defined over $\mathbb{F}_q$). We see then that L and L^{-1} are in $(J - J_o)(\mathbb{F}_{q^2})$, where $J_o = \phi^{-1}(K_o)$. Thus E is given by an extension over $\mathbb{F}_{q^2}$ as in (*) above. We claim that this extension splits over $\mathbb{F}_{q^2}$. To see this we note that

$$\dim_{\bar{\mathbb{F}}_q}(\mathrm{Hom}_{\bar{\mathbb{F}}_q}(L,E)) = \dim_{\bar{\mathbb{F}}_q}(\mathrm{Hom}_{\bar{\mathbb{F}}_q}(L,L\oplus L^{-1})) = 1$$

(since $L \not\cong L^{-1}$)

We have

$$\mathrm{Hom}_{\bar{\mathbb{F}}_q}(L,E) = (\mathrm{Hom}_{\mathbb{F}_{q^2}}(L,E)) \otimes \bar{\mathbb{F}}_q$$

so that

$$\dim_{\mathbb{F}_{q^2}}(\mathrm{Hom}_{\mathbb{F}_{q^2}}(L,E)) = 1$$

which implies easily that (*) splits over $\mathbb{F}_{q^2}$.

We observe that $L \notin (J-J_O)(\mathbb{F}_q)$, for otherwise the extension is defined over $\mathbb{F}_q$ and by a similar argument as above this extension would split over $\mathbb{F}_q$.

Thus we could club together the subsets A_1 and A_2 into the set A which can be described as follows:

$$A = \left\{ \begin{array}{l} \text{Isomorphism classes of bundles E defined by extensions} \\ 0 \longrightarrow L^{-1} \longrightarrow E \longrightarrow L \longrightarrow 0, \text{ defined over } \mathbb{F}_q \text{ with } L \in (J-J_o)(\mathbb{F}_q) \end{array} \right\}$$

and the set B can described as follows:

$$B = \left\{ \begin{array}{l} \text{Isomorphism classes of bundles } E \text{ defined over } \mathbb{F}_q \text{ such} \\ \text{that } E = L \oplus L^{-1} \text{ over } \mathbb{F}_{q^2} \text{ with } L \in (J-J_o)(\mathbb{F}_{q^2}) - (J-J_o)(\mathbb{F}_q). \end{array} \right\}$$

Lemma 3.3. Let $E \in A$. Then the automorphisms of E can be given as follows:

(a) If the extension $0 \longrightarrow L^{-1} \longrightarrow E \longrightarrow L \longrightarrow 0$, is non-split then

$$\mathrm{Aut}E \cong \mathbb{G}_m \quad (\text{over } \mathbb{F}_q)$$

(b) If $E = L \oplus L^{-1}$, then

$$\mathrm{Aut}E \cong \mathbb{G}_m \times G_m \ (\text{over } \mathbb{F}_q)$$

Proof. (a) In this case the bundles are in fact simple and hence AutE is $\mathbb{G}_m$.

(b) Since $L^2 \not\cong \mathcal{O}_X$, in this case $\mathrm{Aut}E \cong \mathrm{Aut}L \times \mathrm{Aut}L^{-1}$, and the result follows.

Digression on Restriction of scalars.

Let K/k be a finite Galois extension and V a K-scheme . Let $\mathcal{G} = \mathrm{Gal}(K/k)$, be the Galois group. Assume that V is quasi-projective. Given $\sigma \in \mathcal{G}$, define the K-scheme V^σ as follows:

If the K-scheme structure on V is given by the morphism

$$V \longrightarrow \mathrm{Spec}(K)$$

define V^σ by the morphism (cf.,[Se]).

$$V \longrightarrow \mathrm{Spec}(K) \xrightarrow{\sigma} \mathrm{Spec}(K)$$

Let V be a K-scheme and set

$$W = \prod_{\sigma \in \mathcal{G}} V^\sigma$$

(products as K-schemes). Then there is a natural action of $\mathcal{G}$ on W lifting the $\mathcal{G}$ action on K since $W^{\tau} \cong W$ (as K-schemes) $\forall \tau \in \mathcal{G}$.

Since V is quasi-projective so is W and therefore by general theory of Galois descent we see that there exists a k-scheme W_1 such that (cf.,[Se]).

$$W = W_1 \otimes_k K \quad \text{(base change of } W_1 \text{ to K)}$$

Indeed $W_1 \cong W/\mathcal{G}$.

Definition. The k-scheme W_1 defined above is called <u>the k-scheme obtained from V by restriction of scalars to k</u>, and we write it as

$$W_1 = \mathrm{Res}_{K/k}(V)$$

Then we have the following standard fact

Proposition 3.4. For any k-algebra B, we have

$$\mathrm{Res}_{K/k}(V)(B) = V(B \otimes_k K)$$

Digression ends here.

Lemma 3.5. Let $E \in B$, B being as defined in Remark 3.2. Then we have

$$N_q(\mathrm{Aut}E) = N_{q^2}(\mathbb{G}_m) = (q^2-1)$$

Proof. By Remark 3.2 we can assume that the bundle E is in fact a split extension over $\mathbb{F}_{q^2}$, i.e

$$E = L \oplus L^{-1}$$

where L^{-1}, $L \in (J - J_o)(\mathbb{F}_{q^2})$. Let σ be the generator of $\mathrm{Gal}(\mathbb{F}_{q^2}/\mathbb{F}_q)$. Since E is defined over $\mathbb{F}_q$ we have $E^{\sigma} \cong E$, so that $\sigma(L) \cong L$ or L^{-1}. If $\sigma(L) \cong L$, then $L \in J(\mathbb{F}_q)$, which is <u>not</u> the case. Therefore

$$\sigma(L) \cong L^{-1}$$

Since the group scheme AutE behaves well under base change (being defined by a representable functor), we have :

$$\mathrm{Aut}\, E \otimes_{\mathbb{F}_q} \mathbb{F}_{q^2} \cong \mathrm{Aut}L \times \mathrm{Aut}L^{-1} \; (\cong \mathbb{G}_m \times \mathbb{G}_m)$$

Now

$$\mathrm{Aut}L^{-1} \cong \mathrm{Aut}\sigma(L) \cong \mathrm{Aut}L^{\sigma} \qquad (\text{over } \mathbb{F}_{q^2})$$

$\mathrm{Aut}L^{\sigma}$ being defined as in the digression above.

Therefore by the Proposition 3.4 on restriction of scalars it is clear that

$$\mathrm{Aut}E = \mathrm{Res}_{\mathbb{F}_{q^2}/\mathbb{F}_q}(\mathrm{Aut}L) = \mathrm{Res}_{\mathbb{F}_{q^2}/\mathbb{F}_q}(\mathbb{G}_m)$$

and thus

$$\mathrm{Aut}E\, (\mathbb{F}_q) = \mathrm{Aut}L\, (\mathbb{F}_{q^2}) = \mathbb{G}_m(\mathbb{F}_{q^2})$$

Hence

$$N_q(\mathrm{Aut}E) = N_{q^2}(\mathbb{G}_m) = (q^2-1)$$

This proves the lemma.

We summarise the contribution of the terms arising from $K - K_o$ to the Siegel formula in the following proposition.

Proposition 3.6. The term β_1 defined in this section can be given as follows:

$$\beta_1 = \frac{N_q(A)}{(q-1)^2} + \frac{N_q(B)}{(q^2-1)} + \frac{N_q(J) - 2^{2g}}{(q-1)}$$

Proof. The first two terms follow immediately from Lemmas 3.3,and 3.5. The third term arises from the non-split extensions lying in the class A. The number of rational points in A is given by

$$N_q(A) = \frac{1}{2}(N_q(J) - 2^{2g})$$

but extensions of L by L^{-1} and the extensions of L^{-1} by L give rise to distinct isomorphism classes of bundles. Therefore one has $2\,N_q(A)$ distinct isomorphism classes and all of them being simple bundles (by Lemma 3.3) we obtain the third term as well.

Note that $N_q(B)$ is given by

$$N_q(B) = N_q(K - K_o) - N_q(A).$$

Proposition 3.7. The number of $\mathbb{F}_q$-rational points of the moduli space of stable bundles M^s is given by the following expression:

$$N_q(M^s) = N_q(M_L(2,1)) + \frac{q^{g-1}N_q(J)}{(q^2 - 1)} - (\beta_1 + \beta_2)(q-1)$$

where $M_L(n,d)$ is the moduli space of semi-stable bundles of rank n and degree d with determinant L and where $\beta_1$and β_2 are as given in Propositions 3.1 and 3.6.

Proof. Recall the Siegel formula (*) as given in the beginning of this section. It can be conveniently expressed as

$$\beta(2,0) + \beta'(2,0) = \frac{q^{3g-3}}{(q-1)}\,\zeta_X(2).$$

By the expression (4) of this chapter the term $\beta'(2,0)$ is given by

$$\beta'(2,0) = \frac{q^{g-3}}{(1- q^{-2})}\;\frac{N_q(J)}{(q-1)^2}$$

The term $\beta(2,0)$ is given by

$$\beta(2,0) = \frac{N_q(M^s)}{(q-1)} + \beta_1 + \beta_2$$

Thus we have

$$N_q(M^s) = \left\{ \frac{q^{3g-3}}{(q-1)} \zeta_X(2) - \beta'(2,0) - \beta_1 - \beta_2 \right\}(q-1)$$

Adding and subtracting the expression

$$\frac{q^{g-2}N_q(J)}{(q-1)^2(1-q^{-2})}$$

inside the flowery brackets and simplifying the expression, keeping in mind the fact that $N_q(M_L(2,1))$ is given by (cf., [H-N],[D-R],[A-B])

$$N_q(M_L(2,1)) = \left\{\frac{q^{3g-3}}{(q-1)} \zeta_X(2) - \frac{q^{g-2}N_q(J)}{(q-1)^2(1-q^{-2})}\right\} (q-1)$$

we get the required expression for $N_q(M^s)$.

§4. Rational Points of N.

Proposition 4.1. Let Y be the variety $N_1 - N_2$ defined as a $\mathbb{P}^{g-2}\times \mathbb{P}^{g-2}$-bundle over $K - K_o$ (cf., Thm 2.1 §2). Then

$$N_q(Y) = N_q(A)\ N_q(\mathbb{P}^{g-2}\times \mathbb{P}^{g-2}) + N_q(B)\ N_{q^2}(\mathbb{P}^{g-2})$$

where A and B are subsets of $(K - K_o)(\mathbb{F}_q)$ as defined in Remark 3.2.

Proof. We observe that if

$$Y \xrightarrow{\pi} K - K_o$$

is the fibration, then the fibre $\pi^{-1}(L \oplus L^{-1})$ can be described as follows:

$$\pi^{-1}(L \oplus L^{-1}) = \mathbb{P}(H^1(X,L^2)) \times \mathbb{P}(H^1(X,L^{-2})) \qquad (*)$$

(cf., [B-1],[B-S],[B-2]). Without loss of generality π can be assumed to be defined over $\mathbb{F}_q$.

To compute the number of $\mathbb{F}_q$ rational points of Y, it suffices to compute the rational points on the fibres $\pi^{-1}(p)$, $p \in (K - K_0)(\mathbb{F}_q)$ and then sum them up.

Let $E = L \oplus L^{-1} \in A$; in other words , L and $L^{-1} \in J(\mathbb{F}_q)$. Hence by (*) above , $\pi^{-1}(L \oplus L^{-1})$ is defined over $\mathbb{F}_q$ and it is easy to see that

$$N_q(\pi^{-1}(L \oplus L^{-1})) = N_q(\mathbb{P}^{g-2})^2$$

$\forall\ L \oplus L^{-1} \in A$.

Now let $E \in B$. Then $E \cong L \oplus L^{-1}$ over $\mathbb{F}_{q^2}$, where L and $L^{-1} \in J(\mathbb{F}_{q^2}) - J(\mathbb{F}_q)$.

Let σ be the generator of $Gal(\mathbb{F}_{q^2}/\mathbb{F}_q)$, then as we have noted in §3, $\sigma(L) = L^{-1}$, for bundles E and L as above. Hence one sees easily that

$$\mathbb{P}(H^1(X,L^2))^{\sigma} \cong \mathbb{P}(H^1(X,L^{-2}))$$

(cf., *Digression on page* 11)

By the discussion above

$$\pi^{-1}(E) \cong \mathbb{P}(H^1(X,L^2)) \times \mathbb{P}(H^1(X,L^2))^{\sigma} \qquad (\text{over } \mathbb{F}_{q^2})$$

implying by remarks in §3

$$\pi^{-1}(E) \cong \mathrm{Res}_{\mathbb{F}_{q^2}/\mathbb{F}_q}(\mathbb{P}(H^1(X,L^2))) \qquad (\text{over } \mathbb{F}_{q^2})$$

Therefore by Prop 3.4

$$\pi^{-1}(E)(\mathbb{F}_q) = \mathbb{P}^{g-2}(\mathbb{F}_{q^2})$$

i.e

$$N_q(\pi^{-1}(E)) = N_{q^2}(\mathbb{P}^{g-2})$$

This is true of every $E \in B$. Thus the proof of the proposition is complete.

Remark 4.2

The strata $N_2 - N_3$ and N_3 do not afford any difficulty in the computation of the $\mathbb{F}_q$ rational points since $N_2 - N_2 = \amalg R$, $N_3 = \amalg S$, where R is a vector bundle of rank $(g-2)$ over $G(2,g)$ and $S \cong G(3,g)$ (cf., Thm 2.1 §2)

Now that we have all the necessary ingredients we can compute the number of rational points of N.

Theorem 4.2. The number of rational points of N over $\mathbb{F}_q$ is given by the following expression:

$$N_q(N) = N_q(M^s) + N_q(Y) + N_q(N_2 - N_3) + N_q(N_3)$$

where

(i) $N_q(M^s) = N_q(M_L(2,1)) + \dfrac{q^{g-1}N_q(J)}{(q^2-1)} - (\beta_1 + \beta_2)(q-1)$

(ii) $N_q(Y) = N_q(A)\ N_q(\mathbb{P}^{g-2} \times \mathbb{P}^{g-2}) + N_q(B)\ N_{q^2}(\mathbb{P}^{g-2})$

(iii) $N_q(N_2 - N_3) = 2^{2g}\ N_q(G(2,g))$

(iv) $N_q(N_3) = 2^{2g}\ N_q(G(3,g))$.

The β_1 and β_2 above are defined by:

(a) $\beta_1 = \dfrac{N_q(A)}{(q-1)^2} + \dfrac{N_q(B)}{(q^2-1)} + \dfrac{N_q(J) - 2^{2g}}{(q-1)}$

(b) $\beta_2 = \dfrac{2^{2g}}{N_q(GL(2))} + \dfrac{2^{2g}}{q(q-1)}\ N_q(\mathbb{P}^{g-1})$

(c) $N_q(A) = \frac{1}{2}(N_q(J)-2^{2g})$

(d) $N_q(B) = N_q(K-K_o)-N_q(A)$

Proof: This is immediate from Propositions 3.1, 3.6, 3.8 and 4.1.

Remark 4.4

We have a canonical desingularisation $\bar{K}$ of K. This is obtained by blowing up the 2-torsion points J_o of the Jacobian and then taking the $\mathbb{Z}/(2)$ quotient of the blow-up. We then see easily that the cardinality of the $\mathbb{F}_q$-rational points of $(K-K_o)$ (required in (d) above) can be expressed as

$$N_q(K-K_o) = N_q(\bar{K}) - 2^{2g}\, N_q(\mathbb{P}^{g-1})$$

Remark 4.5.

We note that in all our computations we could have replaced $\mathbb{F}_q$ by $\mathbb{F}_{q^r}$ $\forall\, r \geq 1$.

§5. The Poincare Polynomial of N.

Let V be a smooth projective variety over $\mathbb{F}_q$ obtained from a smooth variety $V_{\mathbb{C}}$, defined over a ring of algebraic integers R, by reduction modulo a maximal ideal $\mathfrak{p}$ of R.

Define the zeta function of V as

$$Z(V,t) = \exp\left(\sum_{r=1}^{\infty} N_{q^r}(V)\,\frac{t^r}{r}\right) \in \mathbb{Q}[[t]]$$

Then the Weil conjectures state that we can express Z(V,t) in the form

$$Z(V,t) = \frac{P_1(t)\ldots P_{2n-1}(t)}{P_0(t)\ldots P_{2n}(t)}$$

where $P_0(t) = (1-t)$, $P_{2n}(t) = (1 - q^n t)$, and $\forall\ 1 \le i \le 2n-1$, $P_i(t)$ is a polynomial over $\mathbb{Z}$ given by

$$P_i(t) = \prod_j (1 - \gamma_{ij} t)$$

where the γ_{ij} 's are algebraic integers with $|\gamma_{ij}| = q^{1/2}$.

Furthermore the Betti numbers of $V_{\mathbb{C}}$ are given by

$$B_i(V_{\mathbb{C}}) = \deg(P_i)$$

Definition.

For $r \ge 1$, and V a quasi-projective variety over $\mathbb{F}_q$ of dimension n, we set

$$\bar{N}_{q^r}(V) = q^{-rn} N_{q^r}(V)$$

(cf.,Kirwan[K-1]pp.,178)

Lemma 5.1 Suppose that $Y_1,\ldots,Y_k$ are smooth projective varieties over $\mathbb{F}_q$ obtained as reduction mod p of varieties defined in characteristic 0. Suppose that f is a rational function of (k+1) variables with integer coefficients such that

$$f(q^{-r},\ \bar{N}_{q^r}(Y_1),\ldots,\bar{N}_{q^r}(Y_k)) = 0$$

$\forall\ r \ge 1$. Then

$$f(t^2, P_t(Y_1),\ldots,P_t(Y_k)) = 0$$

where $P_t(Y_i)$ is the Poincaré polynomial of the corresponding variety over $\mathbb{C}$.

Proof.(cf., Kirwan [K-1] pp.,186).

We write below the *'expression* $\bar{N}_q$*'* of various terms to obtain the final expression of the Poincaré polynomial:

$$\bar{N}_q(M^s) = \bar{N}_q(M_L(2,1)) + \bar{N}_q(J)\left\{\frac{1}{q^{g-2}(q^2-1)}\right\} - \bar{\beta}_1 - \bar{\beta}_2 .$$

where $\bar{\beta}_1$ and $\bar{\beta}_2$ are given as follows:

$$\bar{\beta}_1 = \left\{ \bar{N}_q(\bar{K}) \left\{ \frac{1}{q^{2g-3}(q+1)} \right\} + \bar{N}_q(J) \left\{ \frac{1}{q^{2g-5}(q^2-1)} \right\} - \bar{N}_q(\mathbb{P}^{g-1}) \left\{ \frac{2^{2g}}{q^{2g-2}(q+1)} \right\} \right.$$

$$\left. - \left\{ \frac{2^{2g}}{q^{3g-5}(q^2-1)} \right\} \right\}$$

(where $\bar{K}$ is as in Remark 4.4).

$$\bar{\beta}_2 = \left\{ \frac{2^{2g}}{q^{3g-2}(q^2-1)} \right\} + \bar{N}_q(\mathbb{P}^{g-1}) \left\{ \frac{2^{2g}}{q^{2g-1}} \right\} .$$

$$\frac{1}{q} \bar{N}_q(Y) = S_q(B) \left\{ \frac{(q^{2g-2}-1)}{(q^2-1)} \right\} + \bar{N}_q(J)\ \bar{N}_q(\mathbb{P}^{g-2})^2 \left\{ \frac{1}{2q} \right\}$$

$$- \bar{N}_q(\mathbb{P}^{g-2})^2 \left\{ \frac{2^{2g-1}}{q^{g+1}} \right\}$$

where $S_q(B)$ is given as follows:

$$S_q(B) = \bar{N}_q(\bar{K})\left\{\frac{1}{q^{2g-3}}\right\} - \bar{N}_q(\mathbb{P}^{g-1})\left\{\frac{2^{2g}}{q^{2g-2}}\right\} - \bar{N}_q(J)\left\{\frac{1}{2q^{2g-3}}\right\}$$

$$+\left\{\frac{2^{2g-1}}{q^{3g-3}}\right\}$$

$$\bar{N}_q(N_2 - N_3)\left\{\frac{1}{q^3}\right\} = 2^{2g}\,\bar{N}_q(G(2,g))\left\{\frac{1}{q^3}\right\}$$

$$\bar{N}_q(N_3)\left\{\frac{1}{q^6}\right\} = 2^{2g}\,\bar{N}_q(G(3,g))\left\{\frac{1}{q^6}\right\}.$$

Thus we have the following expression for $\bar{N}_q(N)$:

$$\bar{N}_q(N) = \bar{N}_q(M^s) + \bar{N}_q(Y) + \bar{N}_q(N_2 - N_3)\left\{\frac{1}{q^3}\right\} + \bar{N}_q(N_3)\left\{\frac{1}{q^6}\right\}$$

the individual terms as given above.

By virtue of the Lemma 4.5 we have the following expression for the Poincaré polynomial :

Consider the following expressions:

$$\bar{\beta}_1(t) = P_t(\bar{K})\left\{\frac{t^{4g-4}}{(1+t^2)}\right\} + P_t(J)\left\{\frac{t^{4g-6}}{(1-t^4)}\right\} - P_t(\mathbb{P}^{g-1})\left\{\frac{2^{2g}\;t^{4g-2}}{(1+t^2)}\right\}$$

$$-$$

$$-\;2^{2g}\left\{\frac{t^{6g-6}}{(1-t^4)}\right\}.$$

$$\bar{\beta}_2(t) = 2^{2g}\left\{\frac{t^{6g}}{(1-t^4)}\right\} + \left\{P_t(\mathbb{P}^{g-1})\;2^{2g}\;t^{4g-2}\right\}$$

$$P_t(Y) = S_t(B)\left\{\frac{(1 - t^{4g-4})}{t^{4g-8}\ (1-t^4)}\right\} + P_t(J)\ P_t(\mathbb{P}^{g-2})^2\ t^2/2$$

$$- P_t(\mathbb{P}^{g-2})^2\ 2^{2g-1}t^{2g+2}.$$

where $S_t(B)$ is given by :

$$S_t(B)=P_t(\bar{K})t^{4g-6}-P_t(\mathbb{P}^{g-1})2^{2g}t^{4g-4}-1/2P_t(J)t^{4g-6} + 2^{2g-1}t^{6g-6}$$

Note that $\bar{K}$ is the desingularisation of the Kummer variety and its Poincaré polynomial can be given as follows:

$$P_t(\bar{K}) = \sum_{i=0}^{2g} b_i(\bar{K})\ t^i$$

where the $b_i(\bar{K})$ are given as follows:

$$b_i(\bar{K}) = \begin{cases} 0 & \text{if } i \text{ is odd} \\ 1 & \text{if } i = 0,2g \\ \binom{2g}{2} + 2^{2g}, & \text{if } i \text{ is even and } 0 < i < 2g. \end{cases}$$

where the $\binom{n}{r}$ are the usual binomial coefficients.(cf., [Sp]).

Theorem 5.2 The Poincaré polynomial of N can be given as follows:

$$P_t(N) = P_t(M_L(2,1)) + P_t(J)\left\{\frac{t^{2g}}{(1 - t^4)}\right\} - (\bar{\beta}_1(t)+\bar{\beta}_2(t)) + P_t(Y)$$

$$+ 2^{2g}\ t^6\ P_t(G(2,g)) + 2^{2g}\ t^{12}\ P_t(G(3,g)).$$

where $P_t(Y)$, $\bar{\beta}_1(t),\bar{\beta}_2(t)$ are as given above and the other terms are as given below:

$$P_t(M_L(2,1)) = \frac{(1+t^3)^{2g} - t^{2g}(1+t)^{2g}}{(1-t^2)(1-t^4)}$$

$$P_t(J) = (1 + t)^{2g}$$

and the terms $P_t(G(2,g))$ and $P_t(G(3,g))$ are the standard Poincaré polynomials of Grassmanian varieties.

We give below the Betti numbers of the variety N for some low genus curves.

BETTI NUMBERS FOR LOW GENUS

	g = 3	g = 4	g = 5
B_1	0	0	0
B_2	2	2	2
B_3	6	8	10
B_4	17	31	48
B_5	6	16	20
B_6	96	385	1328
B_7	6	72	150
B_8	17	413	1583
B_9	6	128	522
B_{10}	96	413	2862
B_{11}	6	72	642
B_{12}	17	385	3072

References:

[A-B] M.F.ATIYAH AND R.BOTT, *The Yang-Mills equations over Riemann surfaces,* Phil.Trans.Royal.Soc.London, A 308, (1982), 523-615

[B-1] V.BALAJI, *Cohomology of certain moduli spaces of vector bundles* Proc.Ind.Acad.Sci.(Math.Sci) Vol 98 (1988), 1-24.

[B-2] V.BALAJI, *Intermediate jacobian of some moduli spaces of vector bundles on curves,* Amer.J.Math.,Vol 112, (1990),611-630.

[B-S] V.BALAJI AND C.S.SESHADRI, *Cohomology of a moduli space of vector bundles,* "The Grothendieck Festschrift " Volume I,(1990), 87-120

[Bi] E.BIFET, *Sur les points fixes du schema* $Quot_{\mathcal{O}^r_{X}/X/k}$ *sous l'action du tore* $\mathbb{G}^r_m$, C.R.Acad.Paris,t. 309,Série I, (1982), 609-612.

[D-R] U.V.DESALE AND S.RAMANAN, *Poincaré polynomials of the variety of stable bundles,* Math.Annalen, 216, (1975), 233-244.

[H-N] G.HARDER AND M.S.NARASIMHAN, *On the cohomology of moduli spaces of vector bundles on curves,* Math.Annalen 212, (1975) 299-316.

[K-1] F.C.KIRWAN, *Cohomology of quotients in symplectic and algebraic geometry,* Mathematical Notes 31, Princeton.Univ.Press (1984).

[K-2] F.C.KIRWAN, *On the homology of compactifications of moduli spaces of vector bundles over a Riemann surface,* Proc.Lond.Math.Soc.53, (1986), 237-267.

[S] C.S.SESHADRI, *Desingularisation of moduli varieties of vector bundles on curves,* International Symposium on Algebraic Geometry, Kyoto,(1977), 155-184, Kinokunia (Tokyo).

[Se] J.-P.Serre, *Algebraic Groups and Class Fields*, Graduate Texts in Mathematics 117, Springer-Verlag.

[Sp] E.H.SPANIER, *The Homology of Kummer manifolds*, Proc.Amer.Math.Soc. 7 (1956), 155-160.

SCHOOL OF MATHEMATICS, SPIC SCIENCE FOUNDATION,
92, G.N. CHETTY ROAD, T. NAGAR,
MADRAS-600 017, INDIA.

On Two Conjectures in Birational Algebraic Geometry

Fedor A. Bogomolov

Steklov Institute of Mathematics, Academy of Science of USSR
ul Vavilova 42, GSP-I, II7966 Moscow, USSR

In this article I want to formulate and prove a synthetic version of two well known conjectures. One of them is the so-called Bloch-Kato conjecture. It provides a description of the torsion cohomology groups for any field in terms of Milnor K-functor. Another one was formulated by A. Grothendieck and concerns only the fields of rational functions on algebraic varieties over number fields. Namely, it claims that the Galois group of the algebraic closure of such field considered as an abstract profinite group defines the field in a functorial way. In fact, the Bloch-Kato conjecture can also be reformulated in terms of some quotient of the Galois group above.

DEFINITION. *Assuming that the field K has characteristic different from p and contains all roots of unity we define $PGal^cK$ as a maximal pro-p-quotient of the group $GalK/[[GalK, GalK], GalK]$. Here $GalK$ denotes the Galois group of the algebraic (separable) closure of K over K.*

The main statement can now be reformulated as follows:

THEOREM. *Let K be a field of rational functions on an algebraic variety over a nontrivial algebraically closed field k, char $K \neq p$. Then $PGal^cK$ considered as an abstract group defines K, modulo purely inseparable extensions, if the dimension of K over k is greater than one.*

REMARK 1: It is easy to see that the groups $PGal^c$ are naturally isomorphic for fields which are purely inseparable extension of one another. Therefore the result is maximally exact. Also $PGal^c$ is a free central extension of its abelian quotient if the dimension of K is one; thus giving no extra information about the field.

REMARK 2: In order to omit set-theoretic complications we assume countability of the ground field k.

In the main body of the proof I also assume the absence of nonramified abelian extensions of the field K. It essentially simplifies the proof from the technical point of view, but does not change the answer. Actually we prove the theorem, restoring the field by means of the $PGal^c$. The main ingredient of the restoration process is the description of all commuting pairs of the elements in $PGal^c K$ in terms of the valuation subgroups of this group. The complete proof of the corresponding statement can be found in [I] but I give the main elements of it in paragraphs 2 and 3. I present in this article just the proof, leaving aside its consequences and further speculations around it.

The article corresponds to the talk given by the author at the Tokyo Conference on Algebraic Geometry, which was held in August, 1990. I want to thank the organizers of the Conference for the invitation and very stimulating atmosphere. I am also very grateful to V. Voevodsky for describing to me the Grothendieck conjecture.

§1

The Bloch-Kato conjecture can be formulated for any field, but I'll describe it here for the field K containing all p^n-roots of unity with characteristic non equal to p. Denote by $GalK$ the Galois group of its algebraic (separable) closure over K. By K_i^M denote i-th Milnor K-group of the field K. Bloch-Kato conjecture claims the existence of the natural isomorphism: $K_i^M/p^n = H^i(GalK, Z/p^n)$.

Consider now the group $PGal^cK$, defined in the introduction as a maximal pro-p-quotient of the group $GalK/[[GalKGalK], GalK]$. This group is a central pro-p-extension of the maximal abelian pro-p-quotient of $GalK$, denoted in standard way as $PGal^{ab}K$. In fact, $PGal^cK$ is a universal group with ambiguous property. It is torsion-free under mild assumptions on the field K, which are satisfied by the fields of rational functions. There is then an exact sequence of groups, in which all groups but $PGal^cK$ are torsion-free abelian pro-p-groups:

$$1 \longrightarrow S \longrightarrow \Lambda^2 PGal^{ab}K \longrightarrow PGal^cK \longrightarrow PGal^{ab}K \longrightarrow 1.$$

The group $\Lambda^2 PGal^{ab}K$ is a quotient of the tensor product $PGal^{ab}K \otimes PGal^{ab}K$ over the ring Z_p of integer p-adic numbers, factorized by the relation: $a \otimes a$ is equal to zero. This group is thus generated by the commutators of the pairs x, y from $PGal^{ab}K$ and S is a subspace of nontrivial relations among the commutators x, y in $PGal^cK$. It is naturally dual to the image of the group $H^2(PGal^{ab}K,\ Q_p/Z_p)$ in $H^2(PGal^cK, Q_p/Z_p)$. This duality is a consequence of the analogous duality for the finite quotients of $PGal^cK$ established in [2]. Since $PGal^{ab}K$ is torsion-free we know $H^2(PGal^{ab}K, Q_p/Z_p)$ is generated by the products of one-dimensional elements. By Kummer theory $H^I(PGal^{ab}K, Q_p/Z_p)$ is naturally isomorphic to the group $\check{K}^*$, obtained from K^* as an inductive limit of the groups $K^*/p^n \xrightarrow{h} K^*/p^{n+I}$ with $h(x) = x^p$.

As it follows from Merkuriev-Suslin theorem (see for example [3]) the image of $H^2(PGal^{ab}K, Q_p/Z_p)$ in $H^2(PGal^cK, Q_p/Z_p)$ coincides with the group $\check{K}_2$ defined by K_2 of the field K in the same way as $\check{K}^*$ above. Indeed the elements of the second cohomology group describe the central extensions of it and symbols $(a, b)_{p^n}$ describe the central extensions of $PGal^{ab}K$ after lifting on $GalK$. On the other hand, an extension which becomes trivial on $GalK$ is trivial already on the level of the group $PGal^cK$. This follows from the fact that any map of $GalK$ into central extension of abelian group passes through $PGal^cK$.

It makes possible to give a geometrical description of the group S. Namely all relations in K_2 are generated by the elements of type $(x, x-a)_{p^n}$ thus coming from the fields of transcendence degree one in K over k. If L is such a field, then the group S_L is trivial subgroup of $PGal^cL$ and the map of pro-p-groups $\Lambda^2 PGal^{ab}K \to \Lambda^2 PGal^{ab}L$ induced by imbedding L into K is trivial on S. The fact that the kernel of cohomology map is generated by such symbols is equivalent S coincides with the intersection of the above type projections for $\Lambda^2 PGal^{ab}$.

REMARK: So the statement above proves also that S is a primitive subspace in $\Lambda^2 PGal^{ab}K$, which means that $x \in S$ if there is $m \in Z_p - 0, with\ mx \in S$. It means also that the system of relations in $H^2(PGal^cK, Q_p/Z_p)$ corresponds to a system of linear subspaces in K^*-multiplicative groups of the subfields of transcendence degree one in K.

By the above construction we can construct for $PGal^cK$ a fundamental system of quotients, which are torsion-free groups and have a presentation as central extension of finite dimensional space over Z_p by another finite dimensional space and center coinciding with the commutator of the group. Let R be such a quotient and R^c is a center; R^c equal to $[R, R]$ and R^{ab} is the maximal abelian quotient of R. The following lemma describes then the image of the cohomology ring $H^*(R^{ab}, Q_p/Z_p)$ in $H^*(R, Q_p/Z_p)$.

LEMMA 1.1. *The kernel of the natural map $H^*(R^{ab}, Q_p/Z_p)$ into $H^*(R, Q_p/Z_p)$ coincides with an ideal generated by the kernel of the corresponding map on the second cohomology group $I \subset H^2(R^{ab}, Q_p/Z_p)$.*

COROLLARY. *The image of $H^i(PGal^{ab}K, Q_p/Z_p)$ in $H^i(PGal^cK, Q_p/Z_p)$ coincides with $\check{K}_i^M$, where $\check{K}_i^M$ is the inductive limit of the quotient groups K_i^M/p^n of i-th Milnor K-groups of the field K, by the system of maps $h : K_i^M/p^n \to K_i^M/p^{n+1}$, $h(x) = x^p$. In order to prove the corollary just notice that the symbolic relation defining K_i^M is coming from the symbols of degree two.*

PROOF OF THE LEMMA: Consider first the case of a one-dimensional Z_p-space R^c. The spectral sequence computing torsion cohomology of R in this case has only one nontrivial differential, which coincides with multiplication by the class w in $H^2(R^{ab}, Q_p/Z_p)$ defining the extension R of R^{ab}. For the general case we have to consider the dual map of homology groups $H_*(R, Z_p) \to H_*(R^{ab,}, Z_p)$, and the dual statement for homology groups will be: the image of $H_*(R, Z_p)$ in $H_*(R^{ab,}, Z_p)$ coincides with the intersection of the images of $H_*(R_j, Z_p)$, where j runs through all quotient groups of R, which are one-dimensional extensions of R^{ab}. Let us use induction on the dimension of R^c over Z_p. Assume that the statement is true for $R' = R/Z_p$, where Z_p is a subgroup of R^c. If we take a character $r : R^c \to Z_p$ supplementary to the projection of R^c to R/Z_p then we define a cyclic Z_p extension of R^{ab} as $R/\ker r$ and the product $R' \times_{R^{ab}} R/\ker r$ will be isomorphic to R. If s is an element of $H_*(R^{ab}, Z_p)$ and s', s^r are its liftings to $R', R/\ker r$, respectively, then the product $s' \times_s S^r$ is defined in the homology group $H_*(R, Z_p)$ and this element lifts s to R. Thus the image of $H_*(R, Z_p)$ in $H_*(R^{ab}, Z_p)$ coincides with the intersection of the images for R' and $R/\ker r$. This proves the lemma for $H_*(R, Z_p)$ and by duality for the cohomology groups. Q.E.D.

As it follows from the above lemma the Bloch-Kato conjecture can be formulated as a statement, comparing the properties of $GalK$ and $PGal^cK$ (if K contains all p^n-roots of unity and has characteristic different from p). The theorem formulated in the introduction makes very plausible that it is possible to deduce this conjecture from the theorem using purely categorical arguments. More precisely, the formulation of the theorem in terms of the equivalence between category of the fields of rational functions and dual category of the $PGal^c$-groups category (we actually have to be more accurate defining it) presumably leads to an equivalence of the properly defined cohomological functors. Remark that among all fields containing nontrivial closed subfield the fields of rational functions constitute a dense subset in inductive topology and the statement of Bloch-Kato conjecture is continuous in respect with it. Thus to prove it for the fields of rational functions will be sufficient for extending it on all inductive limits of such fields.

§2

Here I start a description of the commuting pairs of elements in $PGal^c$. This property of the elements x, y in $PGal^c K$ depends only on their images in the quotient group $PGal^{ab}K$. Thus I shall speak about commuting pairs in $PGal^{ab}K$, which will mean preimages of the corresponding elements in $PGal^c K$ commute. The theorem describing such pairs is contained in [1]. I present here the sketch of the proof along the same line as in [I], but with several improvements. I also omit some technical lemmas. Let us start with the description of the elements of $PGal^{ab}K$. Any of them defines a character of the multiplicative group K^* of K with values in Z_p and vice versa. In the case when K does not have nonramified abelian extensions the group $PGal^{ab}K$ has a more geometric description. Namely in this case we can consider a normal projective variety X with $k(X)$ equal to K and finitely generated and torsion-free Picard group. K^*/k^* in this case naturally identifies with the subspace of functions having finite support on the set of all irreducible divisors. This space has finite codimension in the space of all functions with finite support on this set, which we shall denote by S. The quotient space is isomorphic to $Pic(X)$. Then by a standard argument from the functional analysis, the dual space $PGal^{ab}K$ identifies with the quotient of all functions $Z_p(S)$ by $Pic(X) \otimes Z_p$, where the elements of the latter space are described by functions with value depending only on the class of the element of S in $Pic(X)$.

Any nonarchimedean valuation ν of the field K defines the completion K_ν of this field and natural maps: $j^{ab} : PGal^{ab}K_\nu \longrightarrow PGal^{ab}K$ and $j^c : PGal^c K_\nu \to PGal^c K$. Both maps are imbeddings and moreover the image of $PGal^{ab}K_\nu$ is a direct summand in $PGal^{ab}K$. If A is a valuation ring of ν on K and m is the corresponding ideal in A then the elements of $PGal^{ab}K_\nu$ correspond to the functions trivial on the subset $1 + m$ in A. Actually this property demands the restriction that the characteristic of the residue field of the valuation not to be equal to p. But instead of imposing such a restriction on A_ν we shall better change the definition of the valuation group, thus extending the theory for the latter case.

DEFINITION. *Consider for the valuation ν of K the ideal m in A such that the elements of* *type* $1+m$ *are invertible in* A *and define the valuation subgroup of* $PGal^{ab}K$ *as a subgroup of* Z_p *characters on*

K^*, *trivial on* $1+m$.

In the same way the notion of inertia element can be defined.

DEFINITION. *We call* f *to be an inertia element of the valuation group if it is trivial on a multiplicative group* A^*.

As I already have pointed out, this notion differs from the standard one in the case if the characteristic of the residue field of A is equal to p. If Γ is an abelian group K^*/A^* then any inertia element of $PGal^{ab}K_\nu$ corresponds to a character of this group, $f(x) = \chi(\nu(x))$, $\chi : \Gamma \to Z_p$. It implies a special functional property on f, considered as homogeneous function with values in Z_p on linear k-space K. In the case when the initial valuation is trivial on k it can be axiomatized as follows.

DEFINTION F. *We say that function* f *on a linear space* L *satisfies the flag property (or is a flag function) if there exists a flag of* k*-subspaces in* $L : 0 = L_1 \subset L_2 \subset \ldots \subset L_n \subset L$ *such that* f *is constant on any* $L_{j+1} \setminus L_j$.

We respectively call an element of $PGal^{ab}K$ to be a flag element if it is a flag-function on any finite-dimensional linear k-subspace of K. The flag property for the inertia elements above follows from the nonarchimedean property of the valuation, trivial on k.

Inertia elements of general nonarchimedean valuations are described in a more complicated way and their properties are axiomatized in the following definition.

DEFINITION GF. *We call* k*-invariant function* f *on a finite-dimensional linear space over* k *to be a generalized flag function (or* GF*-function) if one of the following assumptions holds:*

1) f *is a flag function on* L.

2) There exists a flag of k*-subspaces* $O = L_1 \subset L_2 \subset \ldots \subset L_n = L$ *such that for any* $x \in L_{i+1} \setminus L_i$, $y \in L_i$ $f(x+y) = f(y)$ *and the function* f *on* L_{i+1}/L_i *defined as* $f(x(\text{mod } L_i) = f(x)$ *for* $x \notin L_i$ *and for*

$x \in L_i$, $f(x(\text{mod } L_i) = 0$ *is again a* GF*-function.*

3) *In the case if the filtration from 2) is trivial, there exists a valuation on* k, *with valuation subring* B, *maximal ideal* $m \in B$, *residue field* k *and* B *sublattice* M *in* L *with the properties:*

a) $f(x+y) = f(x)$ *for any* $x \in M,\ y \in mM, x \notin mM$.

b) *the function* f *defined on the* k*-space* M/mM *as* $f(x(\text{mod } mM)) = f(x),\ x \notin mM$ *and* $f(0) = 0$ *is a generalized flag function on* M/mM.

REMARK: If the field K has finite characteristic then by reducing the ground field k to the algebraic closure of a finite field we reduce the notion of GF-function to the previous notion of flag function. In the case of characteristic zero it can in the same way be reduced to the case 3) of the definition with k isomorphic to Q (field of rational numbers) and B to be the ring of l-integral rational numbers for prime $l \neq p$.

THEOREM 2.1. *Any* GF*-element of* $PGal^{ab}K$ *is an inertia element for some valuation subgroup.*

REMARK: If the field K has a presentation as an inductive limit of the subfields K_i in K, then the statement of the theorem for K follows from the one for all K_i. Indeed, the order defined by the valuation on the greater field has to be compatible with the ordering on the smaller field. Thus the function f which is constant on the classes of the valuation of the smaller field will be automatically constant on the intersection of the classes and therefore has the property to be constant on classes of elements in K. Thus we can actually consider only the case of the finite transcendence extension of a finite field or Q. In the first case our element will be a flag element and in the second one it will be either flag element or the p-valuation on Q will be involved.

PROOF: Let f first be a flag element. For any linear subspace $L \subset K$ over k we can define then the notion of generic point, which is just the point of the maximal stratum of L defined by f (see definition F above). Let us consider all two dimensional subspaces V_i in K containing fixed element x of the field K and such that x is generic in V_i and f is not constant on $V_i - 0$.The sum of any finite number of such spaces has a property that x is generic in it and the codimension of the next stratum is one.Linear envelope of all subspaces V_i will be the universal subspace in K with this property.

Indeed the sum of one dimensional strata in V_i is a nongeneric stratum in the sum of V_i ,since it is a linear envelope of nongeneric points.But the sum is generated by the set of one dimensional strata and x itself.Thus x is generic in the sum of V_i.Any linear space with x being generic and codimension one nongeneric strata is a sum of the two dimensioinal subspaces as V_i described above.

Define R_x to be the linear envelope of V_i.This space cannot coinside with the field because f is multiplicative and therefore K contains two dimensional subspaces where x is not generic.R_x can also be defined as a minimal subspace containing x and all points in K which appear in the smaller strara than x in linear subspaces of K .

Indeed let x and y lie in a linear space L with x being in a more generic stratum than x.We can find then a linear subspace L' in L with x to be generic in L' and, y to be in a codimension one nongeneric stratum of L'.

Consider now the special stratum of R_x and denote it as m_x.The set of spaces m_x defines a multiplicative scale on K^* since ym_x is equal m_{xy} by multiplicativity of f and for different points x ,y either $m_x \subseteq m_y$ or $m_y \subseteq m_x$.If m_x and m_y coincide then x,y lie in the same stratum of any linear subspace containing both of them.The opposite is also evidently true.If we consider then the property of the point in K to be in the same or smaller stratum than x for any linear subspace in K then it defines a linear subspace. Therefore we can define a maximal linear subspace A_x in K consisting of such points. Since f is multiplicative on K, $yA_x = A_{xy}$ and A_x does not coincide with K for any x. A_1 is also a subring of K.

In order to obtain a valuation from this I remark, that for any $x, y \in K$ either $A_x \subseteq A_y$ or $A_y \subseteq A_x$. Therefore the set of A_x defines an equivalence relation on K^*, where $x \sim y$ if A_x and A_y coincide. The set $x \sim 1$ constitute then a subgroup of invertible elements in A and the relation above defines an ordering on the quotient group K^*/A^*. Therefore we constructed a valuation on K. It remains to remark that $x \in A^*$ if for any linear k-subspace in K it lies in the same stratum as 1. Thus $f(x) = f(1) = 0$ and f is a function induced from the quotient K^*/A^*, or equivalently f is an inertia element of valuation above.

Let K be now a finite transcendence degree extension of Q. Consider the Q-flag structure on the linear subspaces defined according to 2) in the definition GF. If it is nontrivial for some L in K, then the system of Q-subspaces A_x in K is defined as above. Therefore, A_1 will be a valuation ring and the function f on the residue field A/m will be defined by the same rule as in3)of the definition GF. If f corresponds then to the intertia element on A/m with valuation ring B and B^* a set of invertible elements in B, f will be induced from the character of K^*/B^* thus being an inertia element of the combined valuation.

We have to consider the case when Q-filtration (or speaking more generally, flag filtration from 2)) is trivial for any subspace L. By 3) we can define for any x in L a maximal sublattice m over B in L with x (mod mB) to be generic in M/mM. This lattice will be unique if f is not constant on L and L has dimension two. Now we have to repeat the same construction as we have used in the flag case.

Let us take the sum of the two dimensional sublattices containing 1 as generic point and with non constant function f on these sublattices.The resulting space will be the some of B and the ideal of a composite valuation on K. In order to prove that let us consider the sublattices over B in a fixed linear space with the properties that f is induced from the quotient of the sublattices above by maximal ideal.If then there exists one sublattice M such that f is induced from the flag function on M/mM then it is true for all sublattices with the above property.

It follows from the fact that it is true for two dimensional spaces and the property if function to be flag if it is flag on two dimensional subspaces(see the next paragraph).Now by induction on dimension we can prove that the group of linear transformations stabilizing f transforms the induced function on the quotient linear space as follows: any set of the transformations can be decomposed into a sequence of the elementary ones which correspond to multiplying of the chosen strata $L_{i+1} - L_i$ on the normal subspace and contracting all higher dimensional strata by L_{i+1}/L_i to smaller ones if L_i is not empty. Therefore any space L has a unique sublattice M with f a flag-function induced from M/mM and x generic on M/mM. I recall that we have to take into account only the case of $B = \mathbb{Z}_\ell$ for some ℓ prime. Therefore f is a flag function on the residue space. It is easy to see now that M has a property of the composite valuation ring if we started with $x = 1$.This follows from the fact that a change of lattice with induction property does not change the cyclic order of strata (on the projective space $P(M/mM)$). Using now the same arguments as in the flag case we obtain a valuation ring A_1 from this data. Q.E.D.

§3

Let now x, y be a pair of the elements in $PGal^{ab}K$ independent over Z_p, which commute , being lifted in $PGal^cK$. I want to prove that for any such pair there exists a nontrivial valuation v of the field K such that both x and y lie in the valuation subgroup $PGal_v^{ab}$ of $PGal^{ab}K$. The main point in obtaining this result is to prove the following statement.

THEOREM 3.1. *Z_p space spanned by two independent elements over Z_p which commute in $PGal^c$ contains also some nontrivial inertia element of $PGal^{ab}$.*

REMARK: Since the kernel of the projection of $PGal^c$ on $PGal^{ab}$ coincides with the center of the first group ,any liftings of x and y commute if some of them do. That is why we consider pairs of elements in $PGal^{ab}$ but the commutation property in $PGal^c$. It has to be taken into account to prevent a confusion when the commuting property is considered for the elements of abelian group $PGal^{ab}$.

The proof of 3.1 contains a lot of intermediate lemmas and all this paragraph will be devoted to it. I start with the following geometric observation: for any one-dimensional subfield K' of K the group $PGal^cK'$ is a free central extension of the abelian group $PGal^{ab}K'$ and therefore the images of commuting elements x, y have to be proportional in $PGal^{ab}K'$. That means x, y are in fact proportional Z_p-functions being restricted on K' as a subspace of K. In particular, if s is a generator of the subfield $k(s)$ in K then the functions x, y have to be proportional on the two-dimensional k-subspace $k + ks$ in K and this is equivalent for them to be proportional on $k(s)$. For a general two-dimensional subspace in K we have a slightly different condition for x, y.

LEMMA 3.2. *If x, y commute in $PGal^c$ then on any two-dimensional k-subspace V in K the space of Z_p-functions generated by x, y and 1 on V has dimension strictly less than 3.*

PROOF: If s is a nontrivial element of V then $s^{-1}V$ contains subfield k and therefore $x - x(s)$ and $y - y(s)$ are linearly dependent Z_p functions on $s^{-1}V$ but this implies linear dependence between $x, y, 1$ on V. This is equivalent to the statement of the lemma.

NOTATION: We denote the dimension of space spanned by $x, y, 1$ as $rk/x, y, 1/$ and the space itself as $/x, y, 1/$.

Now the statement of Theorem 3.1 follows from the following series of statements below.

THEOREM 3.3. *If x, y are two functions on the projective plane P^2 such that $rk/x, y, 1/$ on it is equal to 3, but for any projective line P^1 in P^2 $rk/x, y, 1/$ is less than 3 the space $/x, y, 1/$ then contains a nontrivial GF-function, if the base field k for P^2 contains more than three elements.*

LEMMA 3.4. *If x is GF-function being restricted on any two-dimensional subspace in a linear space L then x is also GF-function on L unless some residue field for GF-structure of the restriction of x coincides with the field of two elements.*

LEMMA 3.5. *If x, y commute in $PGal^c$, but $/x, y/$ in $PGal^{ab}$ does not contain GF-element (i.e., the element which is a GF-function on K) then we can find three-dimensional linear k-subspace L in K, such that $x, y, 1$ are linearly independent as functions on $P^2 = P(L)$ and $/x, y/$ as a space of functions on L does not contain nontrivial GF-function either.*

Let us deduce first 3.5 from 3.4.

Assume on the contrary that $/x, y/$ does not contain a GF-function. Then by 3.4 we can find a two-dimensional subspace V_x in K such that x is not a GF-function on V_x. Since $rk/x, y, 1/$ is less than 3 on V_x by the assumption of Theorem 3.3 we can find an element y' in $/x, y/$, such that $/x, y/ = /x, y'/$ and y' is constant on V_x. Let us define an analogous space $V_{y'}$. For some s in K the intersection sV_x and $V_{y'}$ is nontrivial. Denote then by L the direct sum $sV_x + V_{y'}$. This space has dimension 3 over k and $/x, y/$ does not contain a GF-function, as a space of functions on L.

Indeed, since x is not a GF-function on V_x it cannot be a GF-function on L.The function y' is constant on V_x and thus on sV_x by multiplicativity of y' on K.Therefore if $/x, y/$ contains a

GF-function on L then it has to be proportional to y'. But it is not GF on the subspace $V_{y'}$. Thus $/x, y/$ on L does not contain a GF-function if it is true for K.

Now I want to turn to the proof of Lemma 3.4 in the case of flag functions. The general case easily reduces to the case of Q-spaces and valuations corresponding to prime numbers for GF-structures on two-dimensional subspaces. In fact, the proof follows the same steps as the proof of Lemma 3.4′ below, but is much more technical. I refer for my article [1] for complete proof of 3.4.

LEMMA 3.4′. *If x is a flag function on any two-dimensional subspace of a linear space L over a field k and k contains more than three elements then x is a flag function on L too.*

PROOF: For any linear space V with a flag function x on it we can define the notion of principal value $\bar{x}(V)$ which is equal to the value of x on the principal stratum. Suppose that the lemma is proved already for the spaces of dimension $n-1$ and prove it for L of dimension n. The function $\bar{x}(V)$ for codimension one linear subspaces in L is thus defined on projective space $P(L^*)$.

SUBLEMMA. *The function $\bar{x}$ on $P(L^*)$ is either a constant or a delta-function of one point (x) in $P(L^*)$.*

PROOF OF SUBLEMMA: Let us take any projective line P^1 in $P(L^*)$ and consider the restriction of $\bar{x}$ on it. For this P^1 we can find a two-dimensional subspace V^2 in L which intersects all spaces V_t, t in P^1. Suppose that $s1, s2, s3, s4$ are the points on P^1 with $\bar{x}(s1)$ different from both $\bar{x}(s3)$ and $\bar{x}(s4)$ and $\bar{x}(s2)$ also different from $\bar{x}(s3), \bar{x}(s4)$. It can't happen because we can find V^2 intersecting all the spaces V_{si} in general stratum. Thus for any four points on P^1 the function x takes the same values at three of them. Therefore x on P^1 is either constant or delta-function. If it is constant for any line P^1 in $P(L^*)$ then it is constant also on $P(L^*)$. If it is a delta-function on P^1 of the point (x) then it will be the delta-function of the same one point on $P(L^*)$. Indeed for any other projective line P^1, we can find two-dimensional subspace intersecting all V_s, s in $P^{1\prime}$. Using generality condition for the intersection of this space with V_{si}, si in P^1 we obtain that generic value on $P^{1\prime}$ coincides with generic value of $\bar{x}$ on P^1. Then (x) is the point of singular value of x for any line P^1 containing (x). Thus x is delta-function of (x) on $P(L^*)$. It finishes the proof of sublemma.

Consider now the case when $\bar{x}$ is constant on $P(L^*)$. Linear envelope of the special strata for flag function is also a special stratum. Thus any subset of points which lie in the union of special strata for codimension one subspaces in L lie also in some codimension two subspace if it fits in some codimension one subspace. But then it is also true without the latter condition. It means that all special strata for x in codimension one subspaces constitute a proper linear subspace of codimension two in L. The function x is then constant outside it. Thus in the case $\bar{x}$ is constant, we prove the lemma.

If $\bar{x}$ is a delta-function of (x) then from the similar considerations we conclude that x is constant outside the subspace $V_{(x)}$. It proves the inductive statement and the lemma.

Now we turn to the proof of Theorem 3.3.

Let us consider the map $(x, y) : P^2 \to Z_p + Z_p$ which can be considered as an affine plane over Z_p. The conditions imposed on (x, y) by the theorem are equivalent to the property that the map above transforms projective lines into affine lines, or in other words can be considered as a map between geometries. The following lemma states that such maps have a simple structure: there exists a projective line on P^2 and a map of this projective line into affine line such that the map of projective plane is actually a cone over this map.

LEMMA 3.6. *The image of P^2 under the map (x, y) in $Z_p + Z_p$ is contained in a union of line and a point.*

Instead of proving this lemma I shall prove a similar statement for the map of P^2 into affine plane over finite field, $A^2 = Z/p + Z/p$.

LEMMA 3.6'. *The image of P^2 in A^2 as above under a map which transforms projective lines into affine lines is contained in a union of a line and a point.*

PROOF: If $p = 2$, then the statement says that the image does not coincide with A^2. If it coincides on the contrary then there are two parallel nonintersecting lines in A^2, but any two lines intersect in P^2. This easily yields a contradiction. We have to take on P^2 two pairs of points, such that points of the first pair are transformed into different points of the first affine line and another pair is transformed into different points on a parallel line on A^2. If we take then projective lines corresponding to these pairs their intersection must map into the intersection of the affine lines, but it is trivial.

For general prime p we have to prove that the image of P^2 cannot contain five points such that there are two triples among them lying on the lines and corresponding two lines are different.

If such configuration of points exists then we take a projective imbedding of A^2 and create a coordinate system, such that the five points will obtain the following projective coordinate: $(1,0,0),(1,a,0),(1,0,b),(0,1,0),(0,0,1)$. The line joining a point in finite part of the plane with $(0,0,1)$ consists of points with the same second coordinate. Thus if $(1,d,f)$ lies in the image of P^2 then the intersection points of lines joining this point with the points of the image also lie there. In particular, $(1,d,0)$ and $(1,0,f)$ is there, but then also $(1,\frac{a+d}{2},\frac{b+f}{2})$. Therefore, the image includes a subset $(1,x,0)$ where together with x,y it contains $\frac{x+y}{2}$. Since multiplication by $1/2$ is invertible in Z/p we obtain that the image of P^2 contains all points $(1,x,0)$, x in Z/p. But then it contains a projective line in projective completion of A^2. This yields a contradiction in the case of general prime p. This proves Lemma 3.6′.

REMARK: The difference of 3.6′ with 3.6 has a technical nature since the notion of line is more easy to use in the case of field than in the case of ring. The proof in the case of Z_p can be obtained by induction on the quotient rings Z/p^n of Z_p [1].

I want to remark now that after linear transformation of the functions x,y we can assume that x takes only two values: $0,1$ and $y=0$ if $x=1$. Let h be any function with values $0,1$ on Z_p. Then $x,h(y)$ define a map of P^2 in Z_p+Z_p with the image consisting of three points: $(1,0),(0,0),(0,1)$. The map $(x,h(y))$ also satisfies the properties above for (x,y), thus mapping projective lines in P^2 to affine lines in Z_p+Z_p. Moreover the following statement is true.

LEMMA 3.7. *If y takes more than two values and x is not GF-function on P^2 then we can find h such that $/x,h(y)/$ does not contain GF-function.*

PROOF: Remark that any projective line, where y takes at least two values,contains points, where y takes any value from the set $y(P^2)$. Indeed for any fixed value y' we have just to take the intersection of the line above with projective line, where y takes value y' and x takes value 1 at some point. Now we fix a line P^1 with y non constant and I want to find

h such that $h(y)$ is not GF-function on P^1. We can reduce the set $y(P^2)$ to three values $0, 1, 2$ and denote the corresponding level sets in P^1 as $S1, S0, S2$. The GF-property of the function $h(y)$ on P^1 can be expressed also in the following: P^1 is decomposed into a union of level sets $A0$ and $A1$, if we fix infinite point inside one of them and zero at another then the latter one will have a structure of an additive subgroup of the field k inside P^1 (see $B1$). Now the union of two sets in k is an additive group, when both of them are affinely equivalent to the additive groups only if char $k = 2$. Thus we reduce our lemma to that case. I only have to remark that the union of two subsets isomorphic to valuation ideals will also be like one if k contains exactly two elements. Since this case was excluded in Theorem 3.1, we proved the lemma.

Thus we have two functions x, y (we use it instead $h(y)$ further) with values $0, 1$ and the following properties: P^2 consists of the three level sets for x, y denoted as S_{00}, S_{10}, S_{01}, respectively, and any projective line lies in the union of no more than two of them. This is just a translation of Theorem 3.3 conditions for the case of two valued functions. If we fix an infinite line P_b^1 in such a way that it lies in a union of S_{10} and S_{01} then S_{00} will obtain the direct product structure with respect to any coordinate system on the complementary affine plane such that the axes of it correspond to the points in different level subsets of P_b^1. It provides with an action of affine group stabilizing the level sets. We can find then a level set S (one of S_{00}, S_{10}, S_{01}) with the property: for any P^1 with nontrivial intersection A of P^1 and S, A and $\bar{A} = P^1 - A$ are affinely equivalent to the additive subgroup of k in P^1 if we fix infinite point in the complementary set. As it was indicated before, it leads to a conclusion that the function levels are like the levels of GF-function for corresponding combination of x and y. This proves Theorems 3.3.and 3.1.

Now I shall describe the commuting elements for inertia elements. We have to consider them as functions on the corresponding valuation subring A_v.

For any two-dimensional subspace V in A_v containing k and the element h of the maximal ideal m_v of the ring A_v the function x commuting with inertia element d_v of $PGal_v^{ab}K$ will be proportional to d_v if the latter one is nontrivial on V. Thus x is constant on the elements of type $1 + h$. Since the elements h with the above property constitute the complementary subset in m_v to a proper ideal m_v', they multiplicatively generate all the subset of the elements of type $1 + m_v$ in K^*. Thus the element x is constant on $1 + m_v$. This is equivalent to the fact that x is an element of $PGal_v^{ab}$ in a refined definition of valuation subgroup provided in the first paragraph.

§4

The procedure of the restoration of K by means of the group $PGal^c K$ includes several steps.

First we define for any element x of $PGal^{ab} K$ its centralizing subgroup in $PGal^{ab} K$ with respect to the group operation in $PGal^c K$. This group will be denoted C_x. It is a primitive subgroup of $PGal^{ab} K$ and is isomorphic to Z_p for generic x. As it follows from Theorem 3.1 and the description of the commutating elements for inertia elements the group C_x is greater than Z_p only if $\dim_k K$ is greater than one and x lies in some valuation group $PGal_v^{ab} K$.

From the theorem proved in the previous paragraph we can also describe closed abelian subgroups A of $PGal^{ab} K$ which can be lifted into abelian subgroups of $PGal^c K$. Namely if the group A has rank r over Z_p then A contains a subgroup of rank at least $r - 1$ consisting of the inertia elements. It is evident from the above description that the inertia elements commute iff they correspond to the same composite valuation. It follows immediately from the compatibility of the corresponding flag structures on the linear subspaces in the field K.

Therefore if the dimension of the field K over k is n we obtain that the dimension of the abelian subgroup A above is not greater than n.It can be equal to n then only if corresponding valuation has the field $k_v(((x_1)...)x_n)$ as a residue field.Here k_v is the residue field of the restriction of our valuation on the ground field k.

If k is isomorphic to the algebraic closure of finite field k_v coincides with k and we can define divisorial valuation subgroups of $PGal^{ab} K$ in terms of $PGal^c K$ as follows.Let us take the system of closed abelian subgroups in $PGal^{ab}$ with liftings in $PGal^c K$ and of maximal rank over Z_p.Consider for any closed cyclic subgroup in these groups it's centraliser in $PGal^c$ and then pick up the ones with maximal centralisers.The refined version of this recipe also works in the case of any algebraically closed field k,as shows the following lemma.

LEMMA 4.1. *Consider abelian closed subgroups of the group $PGal^{ab}K$ which can be lifted into $PGal^{c}K$ and have maximal rank over Z_p among all such groups.Define for any cyclic over Z_p subgroup of such group it's centraliser in $PGal^{c}$ .Let us take maximal groups among these centralisers in respect with imbeddings.They coincide then with divisorial valuation groups $PGal^{ab}_{v}$ after projection into $PGal^{ab}$ and the inertia subgroup of the divisorial valuation v coincides with the centraliser of the whole $PGal^{ab}_{v}$ in $PGal^{c}K$.*

PROOF: According to the preceeding arguments we have to prove only that any divisorial valuation on the residue field K^v of K can be lifted into divisorial valuation on K.Here K^v has the same dimension over residue field k_v of k as K has over k.

Geometrically it corresponds to the existing of the model for the field K which has smooth reduction at the general point of the divisor describing the divisorial valuation of the residue field. It suffices to find such model on the formal level, but then the question reduces to the well examined one-dimensional case.

If irreducible divisor D of the field K reduces to the divisor D^v of K^v then the residue field of the latter one valuation on K^v can be considered as the residue field of the function field on divisor D. Thus we obtain an imbedding of the corresponding valuation groups: $PGal^{ab}_{D^v} \in PGal^{ab}_{D}$.Therefore the latter one is greater and we finished the proof of the lemma.

REMARK: We have used onedimensinality of K when assumed that maximal abelian liftable subgroups have rank strictly greater than 1 over Z_p

COROLLARY. *Thus from the structure of $PGal^{c}$ we deduce the system of inertia subgroups Z_v in $PGal^{ab}K$ and the groups $PGal^{ab}_{v}$ for the whole set of divisorial valuations of K*

This structure of one-dimensional subgroups in Z_p-space $PGal^{ab}$ very much reminds the one defining compact toric variety with the difference that our space is infinite-dimensional and we have a system of lines instead of the system of vectors.

I recall that every subgroup of valuation elements for a divisorial valuation v is a primitive subgroup of $PGal^{ab}$ isomorphic to Z_p and will be further denoted by Z_v.

The group $PGal^{ab}K$ is naturally dual to $\check{K}^*$. It is an abelian group obtained from the multiplicative group K^* by taking an inductive limit: $\check{K}^* = \text{ind } \lim K^*/p^i$, where the map K^*/p^i into K^*/p^{i+1} is a multiplication by p.

Let us take any projective normal variety X over k with $k(X) = K$ and consider the natural exact sequence

$$0 \longrightarrow (K^*/k^* \approx Div^0 X) \longrightarrow DivX \longrightarrow PicX \longrightarrow 0.$$

Here $Div^0 X$ is a group of principal divisors.

For any n we obtain a dual sequence of groups:

$$0 \longrightarrow Hom(PicX, Z/p^n) \longrightarrow Hom(DivX, Z/p^n) \longrightarrow$$

$$Hom(Div^0 X, Z/p^n) \longrightarrow Ext^1(PicX, Z/p^n) \longrightarrow Ext^1(DivX, Z/p^n).$$

Since $DivX$ is free of torsion $Ext^1(PicX, Z/p^n) = 0$ if $tors_p PicX = 0$.

COROLLARY. *If $PicX$ is torsion-free, then $PGal^{ab}K$ contains a Z-sublattice, which is dual to K^*/k^*.*

Namely,the group $DivX$ coincides with the group of integer functions with finite support over the set S of irreducible divisors on X (this presentation depends on X). Thus the dual object corresponds to the space of arbitrary functions on S. For example, $Hom(DivX, Z/p^n) = Z/p^n(S)$. Since the group $PicX$ is free abelian and finitely generated, $PGal^{ab} \approx \mathbb{Z}_p(S)/PicX \otimes \mathbb{Z}_p$ contains the image $Z(S)/PicX$ as a natural Z-sublattice. The subgroup $Z(S)/PicX \subset PGal^{ab}$ is dual to the multiplicative group K^*/k^* imbedded into $DivX \otimes_Z \mathbb{Z}_p$ in a natural way. If $(\alpha, f) \in Z$, for $\alpha \in PGal^{ab}$ and any $f \in K^*$ then $\alpha \in Z(S)/PicX$ and vice versa.This duality corresponds to the Risz-type duality between the spaces of functions with finite support and all functions on a given set.

The condition above on K is equivalent to the triviality of the group of nonramified abelian extensions of K, or $\pi_1^{ab}(X) = 0$. We shall assume this condition on K in all further considerations though Propositions 4.1, 4.2, 4.3, 4.4 are valid without this assumption.

S1: (Simplifying assumption)The field K does not have abelian nonramified extensions.

REMARK: In the general case $Hom(DivX, Z/p^n)$ maps non surjectively on $PGal^{ab}/p^n$ and the image $Z(S)/PicX$ in $PGal^{ab}$ dual to the extension of the multiplicative group K^* by the torsion group. The latter one is generated by the elements of type $f^{1/m}$, where f in m-th power locally everywhere.

We denote the subgroup $Z(S)/PicX$ by G_Z. This notation assumes that G_Z does not depend on X, but it is clear from the above duality argument. Now I want to restore G_Z in terms of the $PGal^c$, but first I would like to define also $G_{Z(p)} = G_Z \otimes Z(p)$, where $Z(p) \subset Q$ is a subring of p-integer rational numbers. The group $G_{Z(p)}$ is $Z(p)$ sublattice of $PGal^{ab}$.

REMARK: By assumption on K, ($PicX$ is free and finitely generated abelian group), I identify $PGal^{ab}$ with a quotient of $\mathbb{Z}_p(S)$ by $PicX \otimes \mathbb{Z}_p$-finite dimensional $\mathbb{Z}_p$-subspace with a lattice. Thus $PGal^{ab}$ has the same principal properties as $\mathbb{Z}_p(S)$. Consider any line $\langle v \rangle$ and an intersection $\langle v \rangle \cap G_Z$. This intersection is isomorphic to cyclic group Z_v and $Z_v \otimes_Z Z_p \approx \langle v \rangle$. It follows from the fact that any divisorial valuation is realized as a divisor, smooth in general point on some model of K.

REMARK: The amount of divisorial valuations in K is much greater than the amount of irreducible divisors on a fixed model X. Namely, irreducible divisors on X constitute a subset in $PGal^{ab}$ which generates this group with finite dimensional space of relations (in $S1$ case). Divisorial valuations correspond to a bigger set of lines $\langle v \rangle$ in $PGal^{ab}$.

It happens that G_Z is uniquely defined by the property of $G_Z \cap \langle v \rangle$ described above,if the field K has characteristic zero.

WARNING: Working with infinitely generated profinite objects forces me to use a modified version of tensor product.Namely tensor product for that kind of objects will actually mean the projective limit of tensor products.If the spaces considered are the spaces of functions on some set then the extension of the group of values (from Z to $Z(p)$) for example will be denoted as tensor product on $Z(p)$ over Z.

$Z(p)$ lattice $G_{Z(p)} = Z(p)(S)/PicX \otimes_Z Z(p)$ is dual to the group $K^*/k^* \otimes Z(p)$ which consists of the characters $\chi : PGal^{ab} \to \mathbb{Z}_p$ such that $\chi(G_{Z(p)}) \subset Z(p) \subset \mathbb{Z}_p$. We call such characters to be rational. They correspond to the elements of type f^α, $\alpha \in Z(p)$, $f \in K^*/k^*$.It follows from duality between spaces of all functions and functions with finite support having values in $Z(p)$ and defined on the set of all irreducible divisors of a fixed model variety X. Generalizing this notion I want to define the notion of rational projection.

Namely, if V_Z is a projective system of free abelian groups and $V = V_Z \otimes \mathbb{Z}_p$, then we call surjective linear projection $\pi : V \to L$ to be rational if there is projective system L_Z and surjective map $\pi_Z : V_Z \to L_Z$ such that $L_Z \otimes_Z \mathbb{Z}_p = L$ and π coincides with $\pi_Z \otimes \mathbb{Z}_p$. In the same way we define $Z(p)$-rational maps.

EXAMPLE: Let $k(x) \overset{i}{\subset} K$ be an imbedding of the fields, where $i(k(x))$ is algebraically closed in K. Then the map $PGal^{ab}K \to PGal^{ab}k(x)$ is rational with respect to $Z(p)$ $((Z)$ structure induced by subgroups $G_{Z(p)}(G_Z)$. The proof is evident, since the image $k(x)^* \overset{i}{\subset} K^*$ is dual to the map $i^* : G_Z(K) \to G_Z(k(x))$ and the last one is surjective. It follows from the primitivity of $k(X)^*$ as a sublattice in K^*.

THEOREM 4.1. *Let $H_{Z(p)}$ be any $Z(p)$-sublattice in $PGal^{ab}$ with the property that $H_{Z(p)} \cap \langle v \rangle = Z(p)$ and this group generates $\langle v \rangle$ as $\mathbb{Z}_p$-module. Then $H_{Z(p)} \approx \lambda G_{Z(p)}$ for some $\lambda \in \mathbb{Z}_p$.*

PROOF: The group $H_{Z(p)}$ defines for every $\langle v \rangle$ an elements $\lambda_v \in \mathbb{Z}_p^*(\text{mod } Z(p)^*)$ such that $\lambda_v H_{Z(p)} \cap \langle v \rangle = G_{Z(p)} \cap \langle v \rangle$. If $\lambda_v \sim 1$ for all v, then $H_{Z(p)}$ contains $G_{Z(p)}$ and thus coincides with $G_{Z(p)}$.

Let us consider first the case $K = k(x)$. The subgroups $\langle v \rangle$ are parametrized by the set of closed points of P^1 and δ_v is a natural set of generators for G_Z, $\delta_v \in \langle v \rangle$ then there is only one relation $\sum_{v \in P^1} \delta_v = 0$, or $\delta_\infty = -\sum_{v \in P^1 \setminus \infty} \delta_v$. Respectively the relation for elements δ'_v generating $H_{Z(p)} \cap \langle v \rangle$ will be $-\sum_{v \in P' \setminus \infty} \lambda_v \delta'_v = \lambda_\infty \delta'_\infty$,but by assumption that $H_{Z(p)} \cap \langle \infty \rangle$ is rational means $\lambda_v / \lambda_\infty \in Z_{(p)}$, or $\lambda_\infty G_{Z(p)} = H_{Z(p)}$.

In order to understand the general case we consider the subfields of K which are isomorphic to $k(x)$ and algebraically closed in K. We have then the projection $\pi : PGal^{ab}K \to PGal^{ab}k(x)$. The group $\ker \pi$ is isomorphic to $PGal^{ab}K/k(x)$ and the subgroups $\langle v \rangle$ contained in $PGal^{ab}K/k(x)$ generate a subgroup of finite index. It follows from the finiteness of torsion of $PicX/P^1_x$ which lies in the image of finitely generated abelian group $PicX$ in $PicX/P^1$ (X is normal projective model of K). That means $H_{Z(p)} \cap PGal^{ab}K/k(x) \otimes \mathbb{Z}_p$ is a subgroup of finite index in $PGal^{ab}K/k(x)$ and as a consequence that the image of $H_{Z(p)}$ in the quotient $PGal^{ab}/k(x)$ is $Z(p)$-lattice. The previous considerations yield that the image of $H_Z(p)$ in $PGal^{ab}k(x)$ coincides with $\lambda_{k(x)} \cdot ImG_{Z(p)}$, or $\lambda_v (\text{mod } Z(p)^*)$ is constant on the set of divisorial valuation groups $\langle v \rangle$ with nontrivial projections on $k(x)$.

For any two divisorial valuations v, v' we can find such function $y \in K^*$ that $\langle v \rangle$ and $\langle v' \rangle$ will both have nontrivial projections in $PGal^{ab}k(y)$. That means λ_v is constant and the theorem is proved

In the above proof we have used the assumption $S1$ that K does not have nontrivial unramified abelian extensions.In the general case we shall need the fact that infinite part of the torsion subgroup for relative model X/P^1_x corresponds to the torsion in Pic^0X over k

Since the subgroup $G_{Z(p)}$ is already defined by $PGal^c$ group structure (modulo multiplication by $\lambda \in \mathbb{Z}_p^*$) I want to define a subgroup $G_Z \subset G_{Z(p)}$ in the same terms. Consider first a pair of rational characters $f_1^{\alpha_1}, f_2^{\alpha_2} : PGal^{ab} \to \mathbb{Z}_p$. I recall that they naturally correspond to the elements of $K^* \otimes Z(p)$. If $f_1^{\alpha_1}, f_2^{\alpha_2}$ are algebraically dependent elements of K then the map $f_1^{\alpha_1}, f_2^{\alpha_2} : PGal^{ab} \to \mathbb{Z}_p \times \mathbb{Z}_p$ can be lifted into a map of $PGal^c$ into the group which is central $\mathbb{Z}_p$-extension of $\mathbb{Z}_p \times \mathbb{Z}_p$. It happens that the opposite is also true.

LEMMA 4.2. *If* $(f_1^{\alpha_1}, f_2^{\alpha_2})_{p^n}$, $\alpha_i \in Z(p)$, $t_i \in K^*$ *and any* $n > 0$ *defines trivial element in* $H^2(PGal^c, \mathbb{Q}_p/\mathbb{Z}_p)$ *then* f_1, f_2 *are algebraically dependent.*

PROOF: Remark first that $(f_1^{\alpha_1}, f_2^{\alpha_2})_{p^n} = 0$ for any n if and only if $(f_1, f_2)_{p^N} = 0$ for any N.

For any component of the divisor of f_1 the restriction of f_2 is constant because in the opposite case $(f_1, f_2)_{p^n}$ is not trivial for some n already under restriction on the local subgroup of this divisor. If we consider the model X_1, where f_1 defines regular map on P^1, then the divisor of f_2 on this model will also be vertical. Thus f_1, f_2 are algebraically depended.

REMARK 1: On the complex variety X the lemma is equivalent to the statement that $\frac{dt_1}{t_1} \wedge \frac{dt_2}{t_2}$ define nontrivial element in $H^2(\widetilde{X}, Z)$ as soon as it is not identically zero here $\widetilde{X}$ is an open subvariety of X where both f_1, f_2 are regular and invertible. Triviality $\frac{dt_1}{t_1} \wedge \frac{dt_2}{t_2}$ is equivalent to algebraic dependence of them, because that means f_2 is constant on the level divisors of f_1.

REMARK 2.6: I don't know if this lemma is true for any two $\mathbf{Z}_p$ characters χ_1, χ_2 of $PGal^{ab}$ without rationality assumption.

COROLLARY. *Using Lemma 4.1 we can characterize in terms of $PGal^c$ all projections of $PGal^{ab}$ on $PGal^{ab}k(x)$ dual to the algebraically closed imbeddings of $k(x)$ into K. They are exactly maximal rational projections onto the quotient groups V of dimension more than one, with the property that corresponding map of $H^2(V, \mathbf{Q}_p/\mathbf{Z}_p)$ into $H^2(PGal^c, Q_p/Z_p)$ is trivial.*

The image of G_Z in $PGal^{ab}k(x)$ is a lattice with the property that there is a set of generators $\delta_v \in \langle v \rangle$ with only one relation $\sum_{v \in P^1} \delta_v = 0$. Let $H_Z \subset G_{Z(P)}k(x)$ be a Z-sublattice with the property that H_Z is generated by $H_X \cap \langle v \rangle$,and it generates $G_{Z(p)} \cap \langle v \rangle$ by tensoring with $Z(p)$. Suppose also that for some choice of generating elements $\delta_{v'}$ of $H_Z \cap \langle v \rangle \sum \delta_{v'} = 0$. In this case $H_Z = \lambda G_Z k(x)$, by the same argument as in the proof of Theorem 4.1, $\lambda \in Z(p)^*$. That means for a dual group that $H_Z \subset PGal^{ab}k(x)$ is Z-dual to $\lambda^{-1}k(x)^*$. Now I formulate the general statement.

THEOREM 4.3. *Let H_Z be any Z-sublattice in the group $PGal^{ab}K$ (as usual, $\pi_1^{ab}(X) = 0$ for normal projective model) with the following properties: 1) $H_Z \subset G_{Z(p)}$, 2) H_Z is generated by the intersections with subgroups $\langle v \rangle$, , and $H_Z \cap \langle v \rangle$ generates $\langle v \rangle$ over $Z(p)$, 3)the image of H_Z under natural projection into $PGal^{ab}k(x)$ coincides with a subgroup $\lambda G_Z(k(x))$ for any algebraically closed in K subfield $k(x) \subset K$ and some λ depending on $k(x)$. Then H_Z is isomorphic to $\lambda G_Z(K')$ for some $\lambda \in Z(p)^*$ and some subfield $K' \subset K$, such that $K : K'$ is purely inseparable extension.*

If H_Z is a lattice as above, then $H' = H + G_Z$ also satisfies 1), 2), 3).Then H' is dual to a submodule $M \subset K^*$.This module satisfies the properties dual to the ones described above.Namely, for any algebraically closed subfield $k(x)$ in K , the intersection of M with $k(x)*$ coinsides with a subgroup $k(x)*^r$ for some r.In fact we can assume that for some y in K the module M contains $k(y)$.In order to obtain it we have just to multiply H_Z by $\lambda^{-1}k(y)$.

Let us prove that for any submodule $M' \subset K^*$ with the properties:

a) if $x \in K^* \setminus \bigcup_{q>1} K^{*q}$ then $x \in M' \iff k^*(x) \subset M'$,

b) for any x, $x^r \in M$ for some r,

then M' corresponds to a subfield M in K, $[K : M]$-purely inseparable.

Prove first if $x_1, x_2 \in M' \Longrightarrow x_1 + \alpha x_2 \in M$.

If $x_1/x_2 \in K^* \setminus \bigcup_{q>1} K^{*q}$, then $x_1/x_2 + \alpha \in k(x_1/x_2)$ and thus $x_1 + \alpha x_2 \in M$.

If $x_1/x_2 \in K^{*q}$, $x_1 \in K^* \setminus \bigcup_{q>1} K^{*q}$ then we can find β such that none of the elements: $x_1 + \beta$, $\frac{x_1+\beta}{x_2}$, $x_1 + \alpha x_2 + \beta$ lies in $\bigcup_{q>1} K^{*q}$. In this case $x_1 + \beta \in k(x_1)^* \subset M$ and $\frac{x_1+\beta}{x_2} + \alpha \in k(x_1 + \beta/x_2)^* \subset M$ and therefore

$$x_1 + \alpha x_2 \in k(x_1 + \beta - \alpha x_2)^* \subset M.$$

If we then find $r \in M$ such that neither rx_1 nor $rx_2 \notin \bigcup_{q>1} K^{*q}$ then $rx_1 + \alpha r x_2 \in M$ or $x_1 + \alpha x_2 \in M$. Such r exists because by assumption M contains the multiplicative group of some subfield $k(y)$ in K. That means M is a subfield. Since M contains some power of any element the extension $K : M$ has to be purely inseparable. For $charK = 0$ the module M then is equal to K^{*q}.

In the case of finite characteristic any subfield $M \subset K$ such that $K : M$ is purely inseparable defines a subgroup $M_{Z'}$,and the module obtained above corresponds to the subfield K' of K such that K is finite purely inseparable extension of K'.

Corollary. *In the case of* $charK = \ell \neq p$, $G_Z \otimes Z[\frac{1}{p}]$ *is uniquely defined by* $PGal^c$. *If* $K' \supset K$ *is purely inseparable, then* $K^{[l^n]}$ *contains* K *for some* n, $\{x \in K^{[l^n]} \iff x = h^{l^n}, h \in K\}$ *and* $PGal^{ab}K^{[l^n]} \approx PGal^{ab}K$ *and* $\check{K}_2[K^{[l^n]}] \approx \check{K}_2[K]$. *Therefore* $PGal^cK = PGalK^{[l^n]}$.

Thus, by $PGal^c$ we restored the sublattice G_Z in $PGal^{ab}$ and by duality the multiplicative group K^*/k^* (or one of the groups $K^{[F]*}/k^*$ for purely inseparable extensions of K) and divisorial valuations.

I want to restore now the additive structure of K. The set K^*/k^* has a structure of k-rational points of the projective space $P_k(K)$ and I want to reconstruct the projective space structure on this set from the data described above. In order to do this let us return back to the maps of G_Z to G_Z $k(x)$. Choosing the generator $+\delta_\infty$ or $-\delta_\infty$ at one point $\infty \in P^1$ makes it possible to fix the sign for all the rest of δ_v, $v \in P^1$ by the relation $\Sigma\delta_v = 0$. For any two divisorial valuations v, v' we can find a function $x \in K^*$ such that $k(x)$ is algebraically closed in K and $x(\delta_v) - x(\delta_{v'}) \neq 0$ in $PGal^{ab}k(x)$,and both $x(\delta_v)$ and $x(\delta_{v'})$ are nonzero. Because of that all signs for δ_v have to be compatible.

The set of such subfields $k(x) \subset K$ defines a system of projective lines P^1 in K^*/k^*.Three elements f, g, h lie on the same line if $f/h, g/h$ are algebraically dependent ,but the lines we can restore satisfy additional condition.Namely we can define for the closure of the field generated by $f/h, g/h$ in K the set of valuation subgroups in corresponding Galois group.Since the field obtained that way is isomorphic to $k(y)$ for some y we can consider the support of any element in this field over the set of all valuations.The field generators are described then by the property that their support consists exactly of two valuations(points in the corresponding projvtive line).Among them we can distinguish the subsets which constitute two dimensional linear subspaces.They contain elements with the same pole,or in our terms the generators of the field $k(y)$ which have negative value on the same valuation.Since the generators δ_v were already fixed we thus obtain partial projective structure on K^*/k^* .

In terms of functions f, g, h above that means we have to demand in addition to algebraic dependence of f/h and g/h also that both of them will be generators of the corresponding subfield and have the same pole.

THEOREM 4.4. *There is only one projective structure on K^*/k^* which corresponds to some field structure and is compatible with partial projective structure described above if $dim(K : k) \geq 2$.*

REMARK: It is evidently not true for curves, since the $PGal^{ab}$ structure with sublattice G_Z is invariant under any permutation of the elements $v \in P^1$.

REMARK: This partial projective structure already restores the weaker relation of algebraic dependence between pairs of the elements in K^*/k^*

Let f, g,h be algebraically dependent elements in K^*/k^* in respect with the above structure and assume that they lie on a projective line .Let y be an element with the property that y is algebraically independent with any of them.Suppose also that the elements y,$(y+\alpha h)/f$,$(y+\beta h)/g$ generate algebraically closed subfields of K,for any α and β in k . Consider now projective plane in K^*/k^* which contains f,g,h,y.The projective lines corresponding to the pairs $(y+\alpha h, f)$ and $(y+\beta h, g)$ intersect nontrivially in the projective plane.

If on the contrary f, g, h don't lie in the same projective line being nevertheless algebraically dependent then the projective envelope of f, g, h, y will be three dimensional projective space in K^*/k^*.Therefore the lines described above will not intersect unless α is equal to β.Thus we can describe the property of f, g, h to be contained in one projective line in terms of partial projective structure on K^*/k^*.

The only problem is to find an element y with the properties demanded above,but by classical Bertini theorem this is a property satisfied by generic element of any big enough linear subspace in K.Therefore we can reconstruct the projective structure on K^*/k^* by the partial projective structure given by $PGal^c$ and the fact that the one is obtained from the field.

LEMMA.4.5. *The projective structure on K^*/k^* uniquely restores the structure of the field k and the additive structure on K*

REMARK: We prove this statement assuming that k is infinite as we usually assumed before,but the proof demands only that corresponding projective line over k will contain more than four points.

Consider the elements $1, g, g+\alpha$ in K^*/k^*, α is an element of k.They lie in the same projective line.The element $g+\alpha$ in this line can be characterized by the property that $(g+\alpha)(g-\alpha)$ lies in the projective line generated by 1 and g^2.Thus for any two points on any projective line we can define projective involution stabilizing them.It is defined purely in terms of projective structure on K^*/k^*.

This involutions reconstruct the field k.Namely fixing one point as ∞ we define a reflection S_a^∞ with respect to any point. Then we define the summation by $S_0^\infty S_\alpha^\infty x = x - 2\alpha$ in case $char k$ is not equal to two. Fixing points $(1, -1)$ in P^1 we are able to define multiplicative operation in the field k,or more exactly in the set $P^1 \setminus pt$.

For $char k = 2$ we need slightly more complicated products of S_β^∞, (β is a root of unity in characteristic two field with more than two elements.

COROLLARY. *If h is an isomorphism of abstract groups $PGal^c K$ ($dim K$ is more than 1) and $PGal^c K'$ then the imbedding of fields $\widetilde{h} : K' \to K$ can be defined with K being finite purely inseparable extension of the field K'.The group isomorphism is then different from the initial one by multiplication by constant on the abelian quotient.*

The degree $K : K'$ can be bounded in the following way: h induces an isomorphism

$$G_Z K \subset G_Z K \otimes Z[\frac{1}{p}] \approx G_Z K' \otimes Z[\frac{1}{p}]'.$$

Fixing $\lambda \subset Z(p)^*$ we consider a finite set $x_1, \dots, x_n \in K$ such that $k(x_1)$ is algebraically closed in K and x_i generate K as a field over subfield generated by k,and a set $y_1, \dots, y_n \in K'$ with the same property. Then $y_i^{p^{m_i}} \in h(G_Z)$ and $x_i^{p^{n_i}} \in h^{-1}(G_{Z'})$ for some choice m_i, n_i but that means $K' \supset k(x_i^{[n_i]}) \supset K^{p^{max\ m_i}}$, $k \supset K'^{[p^{max\ n_i}]}$ and thus the map defines finite extension of fields. It finishes the proof of the main statement.

The result obtained above has many different applications ,but I plan to put them into a separate publication.I am very thankful for a support of Harvard Department of Mathematics during in preparing the above manuscript.I also want to thank referee for useful remarks.

References

1: F.A.Bogomolov "On the structure of abelian subgroups of the Galois groups"Izvestiya of the Academy of Science of the USSR(in russian) N1,1991 .

2: A.S.Merkuriev,A.A.Suslin "K-cohomology of Severi-Brauer varieties and norm residue homomorphism"Izvestiya of the Academy of Science of USSR 1982 v.46 N.5 p.1011-1061.

Ample sheaves on moduli schemes

Hélène Esnault
Eckart Viehweg

Universität - GH - Essen, Fachbereich 6, Mathematik
Universitätsstraße 3, D-4300 Essen 1 , Germany

In this note we take up methods from [3] and [14], III, to give some criteria for certain direct image sheaves to be ample. To give a flavour of the result obtained let us state:

Theorem 0.1 *Let $f : X \to Y$ be a flat projective Gorenstein morphism of complex reduced schemes whose fibres are irreducible normal varieties with at worst rational singularities. Assume that for some $N > 0$ the map $f^* f_* \omega^N_{X/Y} \to \omega^N_{X/Y}$ is surjective and that for some $\mu > 0$ the sheaf $det(f_* \omega^\mu_{X/Y})$ is ample on Y. Then for all $\eta \geq 2$ the sheaf $f_* \omega^\eta_{X/Y}$ is ample, whenever it is non zero.*

Similar statements, replacing "ample" by "maximal Kodaira-dimension" and allowing degenerate fibres, played some role in "Iitaka's program" (see [1] or [9] and the references given there).

The proof of (0.1) is more or less parallel to the proof of a similar result in [3], where we assumed Y to be a curve and allowed degenerate fibres. Therefore it is not surprising that (0.1) is effective again, i.e. that one can measure the ampleness of $f_* \omega^\eta_{X/Y}$ with $det(f_* \omega^\mu_{X/Y})$ and invariants of the fibres.

The interest - if ever - in results like (0.1) comes from the theory of moduli spaces. Let $\mathcal{C}_h$ denote the functor of families of compact complex canonically polarized manifolds with Hilbert-polynomial h and C_h the coarse moduli scheme. If $\nu > 0$ is choosen such that ω^ν_X is very ample for all $X \in \mathcal{C}_h(Spec(\mathbb{C}))$, and if $\lambda_\eta \in Pic(C_h) \otimes \mathbb{Q}$ denotes the "sheaf" corresponding to $det(f_* \omega^\eta_{X/Y})$ for $f : X \to Y \in \mathcal{C}_h(Y)$, then the ample sheaf on C_h obtained by the second author in [14], II was of the form $\lambda^a_\nu \otimes \lambda^b_{\mu \cdot \nu}$ for $\mu, a, b >> 0$. Adding (0.1) to the methods employed in [14], II, we will show in §4 that in fact λ_ν itself is ample.

A quite similar improvement of the description of ample sheaves on moduli spaces M_h will be obtained in §5 for the moduli functor $\mathcal{M}_h$ of pairs $(f : X \to Y, \mathcal{H})$, where f is smooth,

$\omega_{X/Y}$ numerically effective along the fibres and $\mathcal{H}$ a polarization with Hilbert polynomial h, up to isomorphism. As we explained in [14], III, §1, this is not the moduli functor $\mathcal{P}_h$ of polarized manifolds usually considered, at least if one allows the irregularity of the manifolds to be positive. $\mathcal{P}_h$ is a quotient of $\mathcal{M}_h$. In [16] the second author constructed coarse quasi-projective moduli spaces P_h for $\mathcal{P}_h$. As explained in (5.11) the natural morphism $M_h \to P_h$ is finite, if for all $(F, \mathcal{H}) \in \mathcal{M}_h(Spec(\mathbb{C}))$ one knows that $\omega_F^\delta = \mathcal{O}_F$ for some δ. Under this assumption (5.10) implies:

The coarse moduli space P_h of polarised manifolds F with $\omega_F^\delta = \mathcal{O}_F$ exists as a quasi-projective scheme. If $\gamma_\delta \in Pic(P_h) \otimes \mathbb{Q}$ denotes the "sheaf" corresponding to $f_*\omega_{X/Y}^\delta$ for

$$f : X \to Y \in \mathcal{P}_h(Y),$$

then γ_δ is ample.

The same result, for moduli of K3 surfaces, is due to Pjatetskij-Šapiro and Šafarevich [12].

The final version of [16] will contain a discussion of ample sheaves on P_h in the general case, building up on the results explained here.

This note does not really contain any substantially new ideas. We use the proof of (0.1) and of the ampleness of λ_ν to recall and clarify some of methods from [3] and [14], III.

In §1 we consider the relation between weak positivity and ampleness. §2 contains a discussion of the invariant $e(\Gamma)$ introduced in [3] to measure the singularities of a divisor Γ on some manifold X. As sketched in [14], III, we extend the properties of $e(\Gamma)$ to varieties X with rational Gorenstein singularities. In fact, the arguments in [14], III, were a little bit too sketchy at this point and one statement has to be corrected (see 2.12).

In §3 we prove (0.1) and some similar result needed for the polarized case. We include a short discussion about the "effectivity" of (0.1).

After the discussion of ample sheaves on the moduli spaces C_h and M_h in §4 and §5 we will sketch in §6 some generalizations of (0.1) to fibre spaces with degenerate fibres. However, the arguments in §6 should be considered as a guide line to the possible proofs, far from being complete.

We keep the conventions from [14], II. Especially one should have in mind, that all schemes are supposed to be seperated and of finite type over $\mathbb{C}$, that points should be $\mathbb{C}$-valued and that locally free sheaves should be of finite rank, which is the same for different components of the base.

1 Weak positivity and ampleness

Whereas in [14] we had to use the notation "weakly positive over some open set and with respect to a desingularization of a compactification", we will get along in this note with a much simpler set up, related to ampleness. Using this, the properties of weakly positive sheaves needed in the sequel can be deduced in a quite simple way.

Let Y be a reduced quasi-projective scheme, $\mathcal{H}$ be an ample invertible sheaf and $\mathcal{F}$ be a locally free sheaf on Y.

Definition 1.1 $\mathcal{F}$ is called *weakly positive over* Y if for all $a > 0$ one can find some $b > 0$ such that the natural map $H^\circ(Y, S^{a\cdot b}(\mathcal{F}) \otimes \mathcal{H}^b) \otimes_{\mathbb{C}} \mathcal{O}_Y \to S^{a\cdot b}(\mathcal{F}) \otimes \mathcal{H}^b$ is surjective.

If $\mathcal{F}$ is invertible and Y is compact then "weakly positive over Y" is equivalent to "numerically effective". More generally we have:

Lemma 1.2 *$\mathcal{F}$ is weakly positive over Y if and only if for all $\eta > 0$ the sheaf $S^\eta(\mathcal{F}) \otimes \mathcal{H}$ is ample.*

Before proving (1.2) let us recall some simple properties of ample sheaves.

Lemma 1.3 *The following conditions are equivalent:*
a) $\mathcal{F}$ is ample.
b) For some $\eta > 0$ the sheaf $S^\eta(\mathcal{F}) \otimes \mathcal{H}^{-1}$ is generated by global sections.
c) For some $\eta > 0$ the sheaf $S^\eta(\mathcal{F}) \otimes \mathcal{H}^{-1}$ is weakly positive over Y.

Proof. The equivalence of a) and b) is shown in [5], 2.5, and obviously b) implies c). If c) holds true, then

$$S^{2\cdot b} S^\eta(\mathcal{F}) \otimes \mathcal{H}^{-2\cdot b + b}$$

is generated by global sections, as well as the quotient sheaf

$$S^{2\cdot b\cdot\eta}(\mathcal{F}) \otimes \mathcal{H}^{-b}.$$

Hence $S^{2\cdot b\cdot\eta}(\mathcal{F})$ is ample as a quotient of an ample sheaf and by [5], 2.4, we are done.

The last condition in (1.3) motivates the following definition, which will be used in §3.

Definition 1.4 Let $\mathcal{F}$ and $\mathcal{A}$ be locally free sheaves on Y, $\mathcal{A}$ of rank 1. We write $\mathcal{F} \succeq \frac{b}{\eta} \cdot \mathcal{A}$ if $S^\eta(\mathcal{F}) \otimes \mathcal{A}^{-b}$ is weakly positive over Y.

If $\mathcal{A}$ is ample in (1.4), then the statement $\mathcal{F} \succeq \frac{1}{\eta} \cdot \mathcal{A}$ implies that $\mathcal{F}$ is ample and measures "how ample" $\mathcal{F}$ is compared to $\mathcal{A}$.

Lemma 1.5 *Let $\tau : Y' \to Y$ be a finite morphism such that $\mathcal{O}_Y \to \tau_* \mathcal{O}_{Y'}$ splits. Then $\mathcal{F}$ is ample on Y if and only if $\tau^* \mathcal{F}$ is ample on Y'.*

Proof. It follows directly from the definition of ampleness that $\tau^*\mathcal{F}$ is ample if $\mathcal{F}$ is ample.

Assume that $\tau^*\mathcal{F}$ is ample. Let us choose b such that $\tau_*\mathcal{O}_{Y'} \otimes \mathcal{H}^b$ is generated by global sections. For some $\eta >> 0$ the sheaf $S^\eta(\tau^*\mathcal{F}) \otimes \tau^*\mathcal{H}^{-b-1}$ is generated by global sections. Since τ is finite we have surjections

$$\oplus \tau_*\mathcal{O}_{Y'} \to \tau_*\mathcal{O}_{Y'} \otimes S^\eta(\mathcal{F}) \otimes \mathcal{H}^{-b-1} \to S^\eta(\mathcal{F}) \otimes \mathcal{H}^{-b-1}$$

and

$$\oplus(\tau_*\mathcal{O}_{Y'}) \otimes \mathcal{H}^b \to S^\eta(\mathcal{F}) \otimes \mathcal{H}^{-1}.$$

By the choice of b and by (1.3), $\mathcal{F}$ must be ample.

Of course the assumption made for τ in (1.5) holds true if Y is normal and τ is finite. Other examples frequently used here are:

Lemma 1.6 *Let $\mathcal{L}$ be an invertible sheaf on Y.*
a) If for some $N > 0$ and some effective divisor D one has $\mathcal{L}^N = \mathcal{O}_Y(D)$, then one can find some scheme Y', a Cartier divisor D' on Y' and a finite flat morphism $\tau : Y' \to Y$ such that $\tau^*D = N \cdot D'$ and such that $\mathcal{O}_Y \to \tau_*\mathcal{O}_{Y'}$ splits.
b) For all $N > 0$ one can find a finite flat morphsim $\tau : Y' \to Y$ of schemes and an invertible sheaf $\mathcal{L}'$ on Y' such that $\tau^*\mathcal{L} = \mathcal{L}'^N$ and such that $\mathcal{O}_Y \to \tau_*\mathcal{O}_{Y'}$ splits.

Proof. In a) we may take

$$Y' = Spec(\bigoplus_{i \geq 0} \mathcal{L}^{-i}/(\mathcal{L}^{-N} \hookrightarrow \mathcal{O}_Y)).$$

Since $\mathcal{L} = \mathcal{O}_Y(B - C)$ for effective divisors B and C it is enough to consider b) for $\mathcal{L} = \mathcal{O}_Y(B)$. For $\mathcal{H}$ ample and $\mu >> 0$, $\mathcal{H}^{\mu \cdot N}(-B)$ is generated by global sections. If H is the zero set of a general section we can apply a) to $(\mathcal{H}^\mu)^N = \mathcal{O}_Y(H + B)$.

Proof of 1.2 Let $\mathcal{F}$ be weakly positive and $\eta > 0$. We find some $\tau : Y' \to Y$, by (1.6,b), satisfying the assumption made in (1.5) with $\tau^*\mathcal{H} = \mathcal{H}'^\eta$ for some ample sheaf $\mathcal{H}'$. By definition weak positivity is compatible with pullback and $S^{2 \cdot b}(\tau^*\mathcal{F}) \otimes \mathcal{H}'^b$ will be generated by global sections for some $b > 0$. Then $S^{2 \cdot b}(\tau^*\mathcal{F}) \otimes \mathcal{H}'^{2 \cdot b}$ is ample as well as $\tau^*\mathcal{F} \otimes \mathcal{H}'$ and $S^\eta(\tau^*\mathcal{F}) \otimes \tau^*\mathcal{H}$. By (1.5) we are done.

On the other hand, if $S^\eta(\mathcal{F}) \otimes \mathcal{H}$ is ample we can find some $b > 0$ such that $S^b S^\eta(\mathcal{F}) \otimes \mathcal{H}^b$ is globally generated as well as its quotient $S^{b \cdot \eta}(\mathcal{F}) \otimes \mathcal{H}^b$.

The close connection between ample and weakly positive sheaves can also be expressed using coverings.

Lemma 1.7 *The following conditions are equivalent:*
a) $\mathcal{F}$ *is weakly positive over* Y.
b) There exists some finite morphism $\tau : Y' \to Y$ *such that* $\tau^*\mathcal{F}$ *is weakly positive over* Y' *and such that* $\mathcal{O}_Y \to \tau_*\mathcal{O}_{Y'}$ *splits.*
c) There exists some $\mu > 0$ *such that for all flat finite morphisms* $\tau : Y' \to Y$ *and all ample sheaves* $\mathcal{A}'$ *on* Y' *the sheaf* $\tau^*\mathcal{F} \otimes \mathcal{A}'^{\mu}$ *is weakly positive over* Y'.
d) There exists some $\mu > 0$ *such that for all flat finite morphisms* $\tau : Y' \to Y$ *and all ample sheaves* $\mathcal{A}'$ *on* Y' *the sheaf* $\tau^*\mathcal{F} \otimes \mathcal{A}'^{\mu}$ *is ample.*

Proof. The equivalence of a) and b) follows from (1.2) and (1.5). Obviously a) implies d) and d) implies c) for given $\mu > 0$. Assume that c) holds true and let $\eta > 0$ be given. Let $\tau : Y' \to Y$ be the finite morphism constructed in (1.6, b) with $\tau^*\mathcal{H} = \mathcal{H}'^{\eta\cdot\mu+\eta}$. Then $\tau^*\mathcal{F} \otimes \mathcal{H}'^{(\mu+1)}$ as well as $S^\eta(\tau^*\mathcal{F}) \otimes \mathcal{H}'^{\eta\cdot(\mu+1)}$ is ample and (1.5) implies the ampleness of $S^\eta(\mathcal{F}) \otimes \mathcal{H}$.

(1.7) allows to carry over properties of ample sheaves to weakly positive sheaves. Translating the corresponding statements in [5] one obtains as in [14], III, 2.4.:

Lemma 1.8 *a) Let* $\mathcal{F}$ *and* $\mathcal{G}$ *be locally free and weakly positive over* Y. *Then* $\mathcal{F} \otimes \mathcal{G}$ *and* $\mathcal{F} \oplus \mathcal{G}$ *are weakly positive over* Y.
b) Let $\mathcal{F}$ *be locally free. Then* $\mathcal{F}$ *is weakly positive over* Y*, if and only if for some* $r > 0$, $\otimes^r \mathcal{F}$ *is weakly positive over* Y.
c) Positive tensor bundles of sheaves, weakly positive over Y *are weakly positive over* Y.

The more general notation of weakly positive sheaves one has to use in order to prove the existence of quasi-projective moduli schemes will only appear in §6:

Definition 1.9 Let Y be a reduced scheme, $j : Y_o \to Y$ be an open dense subscheme and $\sigma : Y' \to Y$ be a desingularization. Let $\mathcal{H}$ be ample and invertible on Y, let $\mathcal{F}$ be a coherent sheaf on Y and $\mathcal{F}'$ be a coherent sheaf on Y'. Then we call $\mathcal{F}_0 = j^*\mathcal{F}$ *weakly positive over* Y_0 *with respect to* $(Y', \mathcal{F})$ if the hollowing holds true:
i) $\mathcal{F}_0$ is locally free
ii) For all $\nu > 0$ one has inclusions

$$S^\nu(\mathcal{F}_0) \to j^*\delta_* S^\nu(\mathcal{F}')$$

iii) For all $a > 0$ one finds some $b > 0$ such that the natural map

$$V_{a,b} \otimes_{\mathbb{C}} \mathcal{O}_{Y_0} \to S^{a\cdot b}(\mathcal{F}_0) \otimes j^*\mathcal{H}^b$$

is surjective where

$$V_{a,b} = H^0(Y, \delta_* S^{a\cdot b}(\mathcal{F}') \otimes \mathcal{H}^b) \cap H^0(Y_0, S^{a\cdot b}(\mathcal{F}_0) \otimes j^*\mathcal{H}^b).$$

Here the reader should keep in mind that $S^\eta(\mathcal{F}') := i_* S^\eta(i^*\mathcal{F}')$ and $det(\mathcal{F}') := i_* det(i^*\mathcal{F}')$ where $i : U' \to Y'$ is the largest open subscheme such that $i^*\mathcal{F}'$ is locally free.

In fact, (1.9) does not really coincide with the definition used in [14], II and III, since there we assumed Y to be compact. Nevertheless, the properties stated in [14], II, 2.4 and III, 2.4 which are similar to (1.7) and (1.8) remain true for "weakly positive with respect to". Again there is a close connection to ampleness:

Lemma 1.10 *Keeping the notations from (1.9) assume that both, $\mathcal{F}_0$ and $\mathcal{F}'$ are invertible. Then the following two conditions are equivalent:*
a) For some $\eta > 0$ the sheaf $\mathcal{F}_0^\eta \otimes j^*\mathcal{H}^{-1}$ is weakly positive with respect to

$$(Y', \mathcal{F}'^\eta \otimes \delta^*\mathcal{H}^{-1}).$$

b) The condition ii) in (1.9) holds true, and for $\mu >> 0$ the map

$$V_\mu \otimes_{\mathbb{C}} \mathcal{O}_{Y_0} \to \mathcal{F}_0^\mu$$

is surjective, where

$$V_\mu = H^0(Y, \delta_*\mathcal{F}'^\mu) \cap H^0(Y_0, \mathcal{F}_0^\mu),$$

and the induced morphism $\phi_0 : Y_0 \to \mathbb{P}(V_\mu)$ is an embedding.

Proof. Assume that a) holds true. Choosing $a = 2$ in (1.9) we find that

$$\mathcal{F}_0^{2\cdot\eta\cdot\beta} \otimes j^*\mathcal{H}^{-\beta}$$

is generated by sections which lie in $H^0(Y, \delta_*\mathcal{F}'^{2\cdot\eta\cdot\beta} \otimes \mathcal{H}^{-\beta})$ for some $\beta > 0$. We obtain a sujection

$$\oplus j^*\mathcal{H}^\beta \to \mathcal{F}_0^{2\cdot\eta\cdot\beta}.$$

For $\beta >> 0$, $j^*\mathcal{H}^\beta$ is globally generated by sections of $H^0(Y, \mathcal{H}^\beta)$ and those sections seperate points and tangent directions. Hence we obtain b).

Let us assume b). By [14], II, 2.4,a), we are allowed to blow up Y as long as the centers do not meet Y_0. Hence we can assume that ϕ_0 extends to a morphism $\phi : Y \to \mathbb{P}(V_\mu)$. Choosing μ big enough and replacing $\mathcal{F}'$ by some smaller invertible sheaf we can moreover assume that ϕ is an embedding and that $\mathcal{F}' = \delta^*\phi^*\mathcal{O}_{\mathbb{P}(V_\mu)}(1)$. Since a) is independent of the choice of $\mathcal{H}$ we can take $\mathcal{H} = \phi^*\mathcal{O}_{\mathbb{P}(V_\mu)}(1)$ and a) is trivial.

By (1.10) it makes sense to define

Definition 1.11 Let $\mathcal{F}, \mathcal{F}', Y, Y'$ and Y_0 be as in (1.9). Then we call $\mathcal{F}_0$ *ample with respect to* $(Y', \mathcal{F}')$ if condition (1.9, ii) holds true and if for some $\eta > 0$ and some ample invertible sheaf $\mathcal{H}$ on Y the sheaf $S^\eta(\mathcal{F}_0) \otimes j^*\mathcal{H}^{-1}$ is weakly positive with respect to $(Y', S^\eta(\mathcal{F}') \otimes \delta^*\mathcal{H}^{-1})$.

2 Singularities of divisors

In this section we recall and clarify some properties of the invariant e introduced in [3], §2, in order to measure singularities of divisors. At the end of this section we correct a mistake from [14], III.

Definition 2.1 Let V be a normal Gorenstein variety with at worst rational singularities, let $\mathcal{M}$ be an invertible sheaf on V and Γ be an effective divisor with $\mathcal{M} = \mathcal{O}_Y(\Gamma)$. We define:
a) $\mathcal{C}(\Gamma, e) = coker(\tau_*\omega_{V'}(-[\frac{\Gamma'}{e}]) \to \omega_V)$ where $\tau : V' \to V$ is a desingularization of V such that $\Gamma' = \tau^*\Gamma$ is a normal crossing divisor.
b) $e(\Gamma) = min\{e \in \mathbb{N} - \{0\}; \mathcal{C}(\Gamma, e) = 0\}$.
c) $e(\mathcal{M}) = sup\{e(\Gamma); \Gamma$ zero divisor of $s \in H^0(V, \mathcal{M})\}$.

By [13], 2.3, the cokernel $\mathcal{C}(\Gamma, e)$ does not depend on the desingularization $\tau : V' \to V$ choosen.

Assumptions 2.2 Throughout this section $f : X \to Y$ will be a flat Gorenstein morphism whose fibres $X_p := f^{-1}(p)$ are all reduced and normal with at most rational singularities. Γ denotes an effective Cartier divisor on X.

Proposition 2.3 *Assume in addition that Y is smooth, that X_p is not contained in Γ and let Δ be an effective normal crossing divisor on Y. Let $\tau : X' \to X$ be a desingularization such that $\Gamma' = \tau^*\Gamma$ as well as $\Delta' = \tau^*f^*\Delta$ and $\Gamma' + \Delta'$ are normal crossing divisors. Then X_p has a Zarisky open neighbourhood U such that*

$$\tau_*\omega_{X'}(-[\frac{\Gamma' + \Delta'}{e}]) \to \omega_X(-f^*[\frac{\Delta}{e}])$$

is surjective over U for $e \geq e(\Gamma|_{X_p})$.

Proof. We may assume that $[\frac{\Delta}{e}] = 0$. In fact, if $\Delta = \Delta_1 + e \cdot \Delta_2$ for effective divisors Δ_1 and Δ_2, then

$$f^*[\frac{\Delta}{e}] = f^*[\frac{\Delta_1}{e}] + f^*\Delta_2$$

and $\quad \frac{\Gamma'+\Delta'}{e}] = [\frac{\Gamma'+\tau^*f^*\Delta_1}{e} + \tau^*f^*\Delta_2] = [\frac{\Gamma'+\tau^*f^*\Delta_1}{e}] + \tau^*f^*\Delta_2.$

By projection formula we can replace Δ by Δ_1.

Let D be a smooth Cartier divisor containing p. If $p \in \Delta$, we choose D to be a component of Δ and we write α for the multiplicity of D in Δ. Then

$$f|_H : H := f^{-1}(D) \to D$$

fulfills again the assumptions made in (2.2). We can assume that the proper transform H' of H under τ is non singular and that H' intersects $\Gamma' + \Delta''$ transversally for

$$\Delta'' = \Delta' - \tau^* f^* \alpha \cdot D = \Delta' - \alpha \cdot \tau^* H.$$

By induction on $dim(Y)$ we may assume that

$$\tau_* \omega_{H'}(-[\frac{(\Gamma' + \Delta'')|_{H'}}{e}]) \to \omega_H(-f^*[\frac{(\Delta - \alpha \cdot D)|_D}{e}]) = \omega_H$$

is surjective over $W \cap H$ for some open neighbourhood W of X_p in X.

We have $0 \leq \alpha < e$ and

$$[\frac{\Gamma' + \Delta'}{e}] \leq [\frac{\Gamma' + \Delta''}{e}] + [\frac{\alpha \cdot \tau^* H}{e}] + (\tau^* H - H')_{red} \leq [\frac{\Gamma' + \Delta''}{e}] + (\tau^* H - H').$$

Therefore there is an inclusion

$$\omega_{X'}(-[\frac{\Gamma' + \Delta''}{e}] + H') \to \omega_{X'}(-[\frac{\Gamma' + \Delta'}{e}] + \tau^* H).$$

As in [3], 2.3, we consider the commutative diagram

$$\begin{array}{ccccc}
\tau_* \omega_{X'}(-[\frac{\Gamma'+\Delta''}{e}] + H') & \xrightarrow{\alpha} & \tau_* \omega_{H'}(-[\frac{(\Gamma'+\Delta'')|_{H'}}{e}]) & \xrightarrow{\beta} & \omega_H \\
\downarrow & & & & \| \\
\tau_* \omega_{X'}(-[\frac{\Gamma'+\Delta'}{e}]) \otimes \mathcal{O}_X(H) & \xrightarrow{\gamma} & \omega_X(H) & \longrightarrow & \omega_H
\end{array}$$

By [13], 2.3, α is surjective and hence $\beta \circ \alpha$ is surjective over $H \cap W$. Therefore we can find a neighbourhood U of X_p in W such that γ is surjective over U.

Corollary 2.4 *Keeping the notations from (2.2) we assume in addition that Y is a normal Gorenstein variety with at worst rational singularities and that X_p is not contained in Γ. Then X_p has a neighbourhood U with*

$$e(\Gamma|_U) \leq e(\Gamma|_{X_p}).$$

Proof. If Y is non singular this is nothing but (2.3) for $\Delta = 0$. In general let $\delta : Y' \to Y$ be a desingularization and

$$\begin{array}{ccc}
X' & \xrightarrow{\delta'} & X \\
{\scriptstyle f'}\downarrow & & \downarrow \\
Y' & \xrightarrow{\delta} & Y
\end{array}$$

be the fibre product. (2.3) applied to f' and all $p' \in \delta^{-1}(p)$ gives the existence of an open neighbourhood U' of $f'^{-1}\delta^{-1}(p) = \delta'^{-1}(X_p)$ with

$$e(\delta'^*\Gamma|_{U'}) \leq e(\delta'^*\Gamma|_{X'_{p'}}) = e(\Gamma|_{X_p}) = e.$$

Of course we can choose $U' = \delta'^{-1}(U)$ for an open neighbourhood U of X_p. If $\tau : X'' \to X'$ is a desingularization and $\Gamma'' = \tau^*\delta'^*\Gamma$ a normal crossing divisor then

$$\tau_*\omega_{X''}(-[\frac{\Gamma''}{e}]) \to \omega_{X'}$$

is an isomorphism over U' and

$$\delta'_*\tau_*\omega_{X''}(-[\frac{\Gamma''}{e}]) \to \delta'_*\omega_{X'}$$

is an isomorphism over U. By flat base change and projection formula we have

$$\delta'_*\omega_{X'} = \omega_{X/Y} \otimes \delta'_* f'^*\omega_{Y'} = \omega_{X/Y} \otimes f^*\delta_*\omega_{Y'} = \omega_X.$$

Proposition 2.5 *In addition to (2.2) assume that f is projective with connected fibres and that $e \geq e(\mathcal{O}_X(\Gamma)|_{X_p})$. If Y is a normal Gorenstein variety with at worst rational singularities and if X_p is not contained in the support of $\mathcal{C}(\Gamma, e)$, then there exists an open neighbourhood U of p in Y such that $\mathcal{C}(\Gamma|_{f^{-1}(U)}, e) = 0$.*

Proof. If Γ does not contain X_p, then (2.5) follows from (2.4). In general we have

Claim 2.6 There exist a desingularization $\delta : Y' \to Y$ and an effective normal crossing divisor Δ on Y' with: let

$$\begin{array}{ccc} X' & \xrightarrow{\delta'} & X \\ {\scriptstyle f'}\downarrow & & \downarrow{\scriptstyle f} \\ Y' & \xrightarrow{\delta} & Y \end{array}$$

be the fibre product and $\Gamma' = \delta'^*\Gamma - f'^*\Delta$. Then Γ' is an effective divisor which does not contain any fibre $X'_{p'} = f'^{-1}(p')$.

Proof. Of course this follows from the "flattening" of Hironaka. However in this simple situation one can as well argue in the following way:

In order to prove (2.6) we can replace Γ by $\Gamma + H$ for an ample divisor H. Hence we may assume that $\mathcal{M} = \mathcal{O}_X(\Gamma)$ has no higher cohomology on the fibres and hence that $f_*\mathcal{M}$ is locally free and compatible with base change. If $s : \mathcal{O}_Y \to f_*\mathcal{M}$ is the direct image of the section of $\mathcal{M}$ whose zero divisor is Γ, then we just have to choose $\delta : Y' \to Y$ such that the zero locus of $\delta^*(s)$ becomes a normal crossing divisor Δ. In particular

$$\mathcal{O}_{Y'}(\Delta) \to \delta^* f_*\mathcal{M} = f'_*\delta'^*\mathcal{M}$$

splits locally and we get (2.6).

Since $\mathcal{O}_{X'}(\Gamma')|_{X'_{p'}} \simeq \mathcal{O}_X(\Gamma)|_{X_p}$, for all $p' \in \delta^{-1}(p)$, we have $e \geq e(\Gamma'|_{X'_{p'}})$. Let us choose a desingularization $\tau : X'' \to X'$ such that $\Gamma'' + \Delta''$ is a normal crossing divisor for $\Gamma'' = \tau^*\Gamma'$ and $\Delta'' = \tau^* f'^*\Delta$. By (2.3) there is a neighbourhood W' of $f'^{-1}\delta^{-1}(p)$ such that

$$\tau_*\omega_{X''}(-[\frac{\Gamma'' + \Delta''}{e}]) \to \omega_{X'}(-f'^*[\frac{\Delta}{e}])$$

is an isomorphism over W'. Since τ is proper one can take $W' = \delta'^*(W)$ for some neighbourhood W of X_p. Hence the cokernel of

$$\delta'_*\tau_*\omega_{X''}(-[\frac{\Gamma'' + \Delta''}{e}]) \to \omega_X$$

is over W isomorphic to

$$\mathcal{C} = coker(\delta'_*\omega_{X'}(-f'^*[\frac{\Delta}{e}]) \to \omega_X).$$

By flat base change

$$\delta'^*\omega_{X'}(-f'^*[\frac{\Delta}{e}]) = \omega_{X/Y} \otimes f^*\delta_*\omega_{Y'}(-[\frac{\Delta}{e}])$$

and
$$\mathcal{C} = f^*(coker(\delta_*\omega_{Y'}(-[\tfrac{\Delta}{e}]) \to \omega_Y)).$$

Since we assumed that X_p does not lie in the support of $\mathcal{C}(\Gamma, e)$, for sufficiently small W around X_p one has

$$\mathcal{C}(\Gamma, e)|_W = \mathcal{C}|_W = 0.$$

Since f is proper, W contains $f^{-1}(U)$ for some open neighbourhood U of p.

Theorem 2.7 *Let $f : X \to Y$ be a proper morphism satisfying (2.2) and Γ be an effective divisor not containing any fibre of f. Then the function $e(\Gamma|_{X_y})$ is upper semicontinuous on Y.*

Proof. For $p \in Y$ given, let $e = e(\Gamma|_{X_p})$. Define

$$\Delta := \{y \in Y; e(\Gamma|_{X_y}) > e\}.$$

We have to show that p does not lie in the Zariski closure $\overline{\Delta}$ of Δ. Assume that $p \in \overline{\Delta}$ and let $\sigma : T \to \overline{\Delta}_0$ be the desingularization of some component $\overline{\Delta}_0$ of $\overline{\Delta}$ containing p. If $g : S \to T$ is the pullback of f and B the transform of Γ on S, then g and B satisfy the assumptions made in (2.3). Hence, if $\tau : S' \to S$ is a desingularization and $B' = \tau^* B$ a normal crossing divisor, then

$$\tau_*\omega_{S'}(-[\frac{B'}{e}]) \to \omega_S$$

will be an isomorphism over some open neighbourhood U of $g^{-1}(\sigma^{-1}(p))$. Since g is proper, U contains $g^{-1}(W)$ for some neighbourhood W of $\delta^{-1}(p)$, and for simplicity we may assume $W = T$. Let T_0 be the open subvariety of T over which $g' = g \circ \tau : S' \to T$ is smooth and B' is a relative normal crossing divisor. In contradiction to our assumption we have:

Claim 2.8 For $t \in T_0$ *one has* $e(B|_{g^{-1}(t)}) \leq e$.

Proof. If D is a smooth divisor passing through t and $H = g^{-1}(D)$, then $H' = \tau^{-1}(H)$ is irreducible and smooth and B' intersects H' transversally.

We have

$$\begin{array}{ccc} \tau_*\omega_{S'}(-[\frac{B'}{e}] + H') & \longrightarrow & \tau_*\omega_{H'}(-[\frac{B'|_{H'}}{e}]) \\ \| & & \downarrow \\ \tau_*\omega_{S'}(-[\frac{B'}{e}]) \otimes \mathcal{O}_S(H) & & \\ \| & & \downarrow \\ \omega_S(H) & \longrightarrow & \omega_H \end{array}$$

and therefore $\tau_*\omega_{H'}(-[\frac{B'|_{H'}}{e}]) = \omega_H$. Since $H' \to D$ is again smooth and $B'|_{H'}$ is a relative normal crossing divisor we can repeat this step until we obtain

$$\tau_*\omega_{g'^{-1}(t)}(-[\frac{B'|g'^{-1}(t)}{e}]) = \omega_{g'^{-1}(t)}.$$

Theorem 2.9 *Let Z be a projective normal Gorenstein variety with at most rational singularities and $X = Z \times \ldots \times Z$ the r-fold product. Let $\mathcal{L}$ be an invertible sheaf on Z and $\mathcal{M} = \otimes_{i=1}^{r} pr_i{}^*\mathcal{L}$. Then $e(\mathcal{M}) = e(\mathcal{L})$.*

Proof. Obviously $e(\mathcal{M}) \geq e(\mathcal{L}) = e$. Let Γ be any effective divisor with $\mathcal{M} = \mathcal{O}_X(\Gamma)$. By induction we may assume that (2.9) holds true for $(r-1)$-fold products. Hence (2.5) applied to $pr_i : X \to Z$ tells us that the support of $\mathcal{C}(\Gamma, e)$ is $pr_i^{-1}(T_i)$ for some subscheme T_i of Z. Since this holds true for all projections, $\mathcal{C}(\Gamma, e)$ must be zero.

In [3], 2.3, we obtained for smooth Z and very ample sheaves $\mathcal{L}$ that

$$e(\mathcal{M}^\nu) = e(\mathcal{L}^\nu) \leq \nu \cdot c_1(\mathcal{L})^{dim Z} + 1.$$

One has the slightly more general

Corollary 2.10 *If in (2.9) Z is non singular and $\mathcal{H}$ a very ample invertible sheaf on Z, then*

$$e(\mathcal{M}) = e(\mathcal{L}) \leq c_1(\mathcal{H})^{dim Z - 1} \cdot c_1(\mathcal{L}) + 1.$$

Proof. By (2.9) it is enough to verify the inequality. If Γ is the zero set of a section of $\mathcal{L}$, H a smooth divisor with $\mathcal{O}_Z(H) = \mathcal{H}$, then we choose $\tau : Z' \to Z$ such that Γ' is a normal crossing divisor and such that the proper transform H' of H in Z' intersects Γ' transversally. As in (2.3) we have the diagram

$$\begin{array}{ccccc} \tau_*\omega_{Z'}(-[\frac{\Gamma'}{e}] + H') & \xrightarrow{\alpha} & \tau_*\omega_{H'}(-[\frac{\Gamma'|_{H'}}{e}]) & \xrightarrow{\beta_H} & \omega_H \\ \downarrow & & & & \| \\ \tau_*\omega_{Z'}(-[\frac{\Gamma'}{e}]) \otimes_{\mathcal{O}_Z} \mathcal{H} & \xrightarrow{\beta} & \omega_Z(H) & \longrightarrow & \omega_H \end{array}$$

where again α is surjective. By induction we can assume that for

$$e \geq c_1(\mathcal{H}|_H)^{dim(H)-1} \cdot c_1(\mathcal{L}) + 1$$

β_H is surjective. Moving H we obtain 2.10.

Corollary 2.11 *Assume in (2.9) that $\mathcal{L}$ is an ample invertible sheaf on Z and that there exist a desingularization $\tau : Z' \to Z$ and an effective exceptional divisor E such that $\tau^*\mathcal{L} \otimes \mathcal{O}_{Z'}(-E)$ is very ample. Then*

$$e(\mathcal{M}) = e(\mathcal{L}) \leq c_1(\mathcal{L})^{dimZ} + 1.$$

Proof. Obviously $e(\mathcal{L}) \leq e(\tau^*\mathcal{L})$ and $E \cdot (c_1(\tau^*\mathcal{L}) - E)^j \cdot c_1(\tau^*\mathcal{L})^{dimZ-1-j} \geq 0$ for all $0 \leq j \leq dimZ - 1$. Hence

$$e(\mathcal{L}) \leq (c_1(\tau^*\mathcal{L}) - E)^{dimZ-1} \cdot c_1(\tau^*\mathcal{L}) + 1 \leq c_1(\mathcal{L})^{dimZ} + 1.$$

Remark 2.12 In [14], III, 2.2 the second author claimed that (2.10) holds true for all Z with rational Gorenstein singularities. He overlooked that a hyperplane section through a rational singularity might have non rational singularities. In fact, we doubt that (2.10) can be generalized in that way. However, the results of [14], III, are not really affected. It was only used that $e(\mathcal{H}^\nu)$ is bounded for all pairs $(X, \mathcal{H}) \in \mathcal{M}_h(\mathbb{C})$ where $\mathcal{M}_h$ is a bounded moduli functor of polarized varieties with rational Gorenstein singularities. This follows from (2.7) anyway. (2.11) gives a second way to correct this ambiguity: In [14], III, just before (1.4), one has to add (see also §5):

1.3': *Since $\mathcal{M}''_h$ is bounded we may even choose ν large enough such that for all $(F, \mathcal{H}) \in \mathcal{M}''_h(\mathbb{C})$ one has:*

> *(*) There exist a desingularization $\tau : F' \to F$ and an effective exceptional divisor E on F' such that $\tau^*\mathcal{H} \otimes \mathcal{O}_{F'}(-E)$ is very ample.*

The same property (*) with $\mathcal{H}$ replaced by $\mathcal{M}_0|_F$ has to be added to (2.7, a) in [14], III, (See Remark 3.10, c)).

3 Ampleness of certain direct image sheaves

Notations 3.1 Throughout this section $f : X \to Y$ will denote a flat projective Gorenstein morphism between reduced quasi-projective schemes whose fibres $X_p = f^{-1}(p)$ are irreducible normal reduced varieties with at worst rational singularities.

Definition 3.2 For $f : X \to Y$ as above let $\mathcal{L}$ be an invertible sheaf on X. We will call $\mathcal{L}$ *relatively semi-ample over* Y *(or relatively semi-ample for* f*) if for some* $N > 0$ *the map*

$$f^* f_* \mathcal{L}^N \to \mathcal{L}^N$$

is surjective.

Theorem 3.3 *Using the notations (3.1) let* $\mathcal{L}$ *be an invertible sheaf on* Y, Γ *be an effective Cartier divisor on* X *and* $e \in \mathbb{N} - \{0\}$. *Assume that:*
a) For $p \in Y$ *the divisor* Γ *does not contain* X_p *and* $e(\Gamma|_{X_p}) \leq e$.
b) $\mathcal{L}^e(-\Gamma)$ *is relatively semi-ample over* Y.
Then one has:
i) For $i \geq 0$, $R^i f_*(\mathcal{L} \otimes \omega_{X/Y})$ *is locally free and compatible with arbitary base change.*
ii) If for some $M_0 > 0$ *and all multiples* M *of* M_0 *the sheaf* $f_*(\mathcal{L}^e(-\Gamma))^M$ *is locally free and weakly positive over* Y, *then* $f_*(\mathcal{L} \otimes \omega_{X/Y})$ *is weakly positive over* Y.

Remark 3.4 In fact, (3.3, ii) is quite similar to the main technical result of [14], III. There however we assumed in Theorem 2.6, that beside of (3.3, a) we know that $\mathcal{L}^e(-\Gamma)$ is semi-ample. Moreover we assumed $f_*(\mathcal{L} \otimes \omega_{X/Y})$ to be locally free and compatible with arbitary base change, which by part i) is automatically true.

The proof of (3.3, i) is mainly due to J. Kollár, as explained in [14], II, 2.8, 4 (see also [7]).

Proof of 3.3, i. By "Cohomology and base change" (see [10]) it is enough to show that for $i \geq 0$ the sheaves $R^i f_*(\mathcal{L} \otimes \omega_{X/Y})$ are locally free. Moreover (loc. cit.) there is a finite complex $\mathcal{E}^\cdot$ of locally free sheaves such that for a coherent sheaf $\mathcal{C}$ on Y the i-th cohomology of $\mathcal{E}^\cdot \otimes \mathcal{C}$ is $R^i f_*(\mathcal{L} \otimes \omega_{X/Y} \otimes f^*\mathcal{C})$. Hence, to verify the local freeness we may assume that Y is a curve and, taking for $\mathcal{C}$ the integral closure of $\mathcal{O}_Y$, that Y is non singular. If $\tau : X' \to X$ is a desingularization and $\Gamma' = \tau^*\Gamma$ a normal crossing divisor then $R^i \tau_* \omega_{X'}(-[\frac{\Gamma'}{e}]) = 0$ for $i > 0$ (see [13], 2.3) and $\tau_* \omega_{X'}(-[\frac{\Gamma'}{e}]) = \omega_X$ by assumption a) and (2.3). Hence for $\mathcal{L}' = \tau^*\mathcal{L}$ and $f' = f \circ \tau$ we have

$$R^i f'_*(\mathcal{L}'(-[\frac{\Gamma'}{e}]) \otimes \omega_{X'/Y}) = R^i f_*(\mathcal{L} \otimes \omega_{X/Y}).$$

Now the proof ends, as usual, by using Kollár-Tankeev's vanishing theorem: Assume that for some $i \geq 0$, $R^i f'_*(\mathcal{L}'(-[\frac{\Gamma'}{e}]) \otimes \omega_{X'})$ is not locally free. Then for some ideal $I = \mathcal{O}_Y(-q)^a$ on Y,

$$\alpha_i : R^i f'_*(\mathcal{L}'(-[\frac{\Gamma'}{e}]) \otimes \omega_{X'}) \otimes I \to R^i f'_*(\mathcal{L}'(-[\frac{\Gamma'}{e}]) \otimes \omega_{X'})$$

will have a kernel. In order to get a contradiction we can replace e and Γ by some common multiple such that $f'^*f'_*\mathcal{L}^e(-\Gamma') \to \mathcal{L}^e(-\Gamma')$ is surjective. Moreover we can compactify Y and X' and assume that Γ' extends to a normal crossing divisor on the compactification. Replacing $\mathcal{L}$ by $\mathcal{L} \otimes f^*\mathcal{H}$ for some very ample sheaf $\mathcal{H}$ on Y, we can assume moreover that the sheaves $R^i f'_*(\mathcal{L}'(-[\frac{\Gamma'}{e}] \otimes \omega_{X'})$ have no higher cohomology on Y and that for some N_0 and all multiples N of N_0, $\mathcal{L}'^e(-\Gamma')^N \otimes f^*I$ is generated by its global sections.

Hence, if D is the zero divisor of a general section of this sheaf then

$$B = D + N \cdot \Gamma' + f^*a \cdot q$$

is a normal crossing divisor and $[\frac{B}{N \cdot e}] = [\frac{\Gamma'}{e}]$ for N sufficiently big. The non injectivity of α_i implies that

$$H^i(X', \mathcal{L}'(-[\frac{B}{N \cdot e}]) \otimes \omega_{X'}) \to H^i(X', \mathcal{L}'(-[\frac{B}{N \cdot e}]) \otimes \omega_{X'}(f^*a \cdot q))$$

is not injective. However, this contradicts [2], 3.3,1, where we have shown, that the dual of this map

$$H^{n-i}(X', \mathcal{L}'(-[\frac{B}{N \cdot e}])^{-1} \otimes \mathcal{O}_{X'}(-f^*a \cdot q)) \to H^{n-i}(X', \mathcal{L}'(-[\frac{B}{N \cdot e}])^{-1})$$

is surjective for all i.

Proof of 3.3, ii. The assumptions a) and b) are compatible with arbitary base change and the assumption added in ii) is compatible with flat base change. Using (3.3, i) and the equivalence of a) and c) in (1.7), it is enough to show that for an ample invertible sheaf $\mathcal{A}$ on Y the sheaf $f_*(\mathcal{L} \otimes \omega_{X/Y}) \otimes \mathcal{A}$ is weakly positive over Y. For some multiple M of M_0 the map

$$f^*f_*(\mathcal{L}^e(-\Gamma)^M) \otimes f^*\mathcal{A}^{M \cdot e} \to (\mathcal{L}^e(-\Gamma) \otimes f^*\mathcal{A}^e)^M$$

will be surjective by assumption b) and by the assumption made in ii)

$$S^b(f_*(\mathcal{L}^e(-\Gamma))^M) \otimes \mathcal{A}^{M \cdot b \cdot e}$$

is globally generated for $b >> 0$. Hence, $(\mathcal{L} \otimes f^*\mathcal{A})^e(-\Gamma)$ is semi-ample. (3.3, i) allows to apply [14], III, 2.6, to obtain the weak positivity over Y of

$$f_*((\mathcal{L} \otimes f^*\mathcal{A}) \otimes \omega_{X/Y}) = \mathcal{A} \otimes f_*(\mathcal{L} \otimes \omega_{X/Y}).$$

In some way (3.3, ii) can be seen as a generalization of [3], 1.7. As in [3], 1.9, we obtain:

Proposition 3.5 *Let $f : X \to Y$ be as in (3.1) and let $\mathcal{L}$ be an invertible sheaf on X. Assume that:*

a) $\mathcal{L}$ *is relatively semi-ample over* Y.
b) For some $M > 0$ *and all multiples* M *of* M_0 *the sheaf* $f_*(\mathcal{L}^M)$ *is locally free and weakly positive over* Y.
c) For some $N > 0$ *there is an ample invertible sheaf* $\mathcal{A}$ *on* Y *and an effective Cartier divisor* Γ *on* X, *not containing any fibre of* f, *with* $\mathcal{L}^N = f^*\mathcal{A} \otimes \mathcal{O}_X(\Gamma)$.
Then $f_*(\mathcal{L} \otimes \omega_{X/Y})$ *is ample.*

Addendum 3.6 *Under the assumptions of (3.5) let* $e \geq sup\{N, e(\Gamma|_{X_p})$ *for* $p \in Y\}$ *(which exists by (2.7)). Then the ampleness of* $f_*(\mathcal{L} \otimes \omega_{X/Y})$ *is measured by (see (1.4)) :*

$$f_*(\mathcal{L} \otimes \omega_{X/Y}) \succeq \frac{1}{e} \cdot \mathcal{A}.$$

Proof. We have to show that $S^e(f_*(\mathcal{L} \otimes \omega_{X/Y})) \otimes \mathcal{A}^{-1}$ is weakly positive over Y. Using the equivalence of a) and b) in (1.5) we are allowed to replace Y by a finite flat cover $\tau : Y' \to Y$ such that $\mathcal{O}_Y \to \tau_*\mathcal{O}_{Y'}$ splits. Hence, using (1.6, b) we may assume that $\mathcal{A}$ is the e-th power of some invertible sheaf $\mathcal{A}'$ on Y and we have to show that $f_*(\mathcal{L}' \otimes \omega_{X/Y})$ is weakly positive over Y for $\mathcal{L}' = \mathcal{L} \otimes f^*\mathcal{A}'^{-1}$. We have $\mathcal{L}'^e(-\Gamma) = \mathcal{L}^{e-N}$ and the assumptions a) and b) of (3.3) hold true for $\mathcal{L}'$ and Γ. The additional assumption made in (3.3, ii) is nothing but b) in (3.5) if $N < e$ and, since $f_*\mathcal{O}_X = \mathcal{O}_Y$, it is trivial for $N = e$. Therefore we can apply (3.3, ii) to end the proof of (3.3) and (3.4).

Just copying the arguments used in [3], 2.4, we can now prove theorem (0.1) as well as

Addendum 3.7 *Using the notations and assumptions from (0.1) we have:*
Let $e \geq sup\{\mu, e(\omega_{X_p}^{\mu\cdot(\eta-1)})$ *for* $p \in Y\}$. *Then*

$$f_*\omega_{X/Y}^{\eta} \succeq \frac{\eta - 1}{e \cdot rank(f_*\omega_{X/Y}^{\mu})} \cdot det(f_*\omega_{X/Y}^{\mu}).$$

Remark 3.8 In particular if f is smooth, if $\omega_{X/Y}$ is relatively ample over Y and if Y is connected, let $N > 0$ be an integer such that $\omega_{X_p}^N$ is very ample for all p. Then we can choose

$$e = \mu \cdot (\eta - 1) \cdot N^{dim X_p - 1} \cdot c_1(\omega_{X_p}^{dim X_p}) + 1$$

for any $p \in Y$. If f is not smooth but $\omega_{X/Y}$ relatively ample, then the same choice of e is possible, if we take N to be big enough, such that the condition (*) in (2.12) holds true for $\mathcal{H} = \omega_{X/Y}^N$.

Proof of 3.7 and 0.1 Set $r = rank(f_*\omega_{X/Y}^{\mu})$ and consider $f^r : X^r \to Y$, where X^r is the r-fold product of X over Y. Of course, f^r is again a flat projective Gorenstein morphism and the fibres of f^r still have at worst rational singularities (see [14], III, 2.9, for example). By flat base change one finds that

$$\omega_{X^r/Y} = \otimes_{i=1}^r pr_i^* \omega_{X/Y}$$

is again relatively semi-ample over Y. By [14], II, Theorem 2.7, $f^r_* \omega^\gamma_{X^r/Y}$ is weakly positive over Y for all $\gamma > 0$. In fact, there we added some assumptions on base change properties, which by (3.3,i) are no longer necessary. In order to apply (3.5) and (3.6) to $\mathcal{L} = \omega^{\eta-1}_{X/Y}$ we consider the natural inclusion

$$det(f_* \omega^\mu_{X/Y}) \to \otimes^r f_* \omega^\mu_{X/Y} = f^r_* \omega^\mu_{X^r/Y}.$$

Since this inclusion splits locally, the zero divisor Γ of

$$f^{r*} det(f_* \omega^\mu_{X/Y})^{\eta-1} \to \omega^{\mu\cdot(\eta-1)}_{X^r/Y}$$

does not contain any fibre and by (2.9)

$$e(\Gamma|_{X_p \times \ldots \times X_p}) \leq e(\omega^{\mu\cdot(\eta-1)}_{X_p \times \ldots \times X_p)}) = e(\omega^{\mu\cdot(\eta-1)}_{X_p}).$$

By (3.5) and (3.6) we obtain

$$f^r_*(\omega^\eta_{X^r/Y}) = \otimes^r f_* \omega^\eta_{X/Y} \succeq \frac{1}{e} \cdot det(f_* \omega^\mu_{X/Y})^{\eta-1}$$

and hence

$$S^r(f_* \omega^\eta_{X/Y}) \succeq \frac{1}{e} \cdot det(f_* \omega^\mu_{X/Y})^{\eta-1}.$$

By definition of $\succeq$ in (1.4) one obtains:

$$f_* \omega^\eta_{X/Y} \succeq \frac{1}{e \cdot r} \cdot det(f_* \omega^\mu_{X/Y})^{\eta-1}.$$

For application to moduli of polarized varieties we need a second application of (3.3) and (3.5) generalizing (0.1) (by taking $\mathcal{M} = \mathcal{O}_X$).

Theorem 3.9 *For $f : X \to Y$ as in (3.1) let $\mathcal{M}$ be an invertible sheaf on X. Assume that for some $e \in \mathbb{N} - \{0\}$ one has:*
a) $\mathcal{M}$ is relatively semi-ample over Y.
b) $f_* \mathcal{M}$ is locally free of rank r'.
c) $e \geq e(\mathcal{M}|_{X_p})$ for all $p \in Y$.
d) $\mathcal{M} \otimes \omega^e_{X/Y}$ is relatively semi-ample over Y.
e) $f_*(\mathcal{M} \otimes \omega^e_{X/Y})^N$ is locally free for all $N > 0$.
Then one has:
i) $(\otimes^{r'} f_*(\mathcal{M} \otimes \omega^e_{X/Y})) \otimes det(f_* \mathcal{M})^{-1}$ is weakly positive over Y.
ii) If for some $\mu > 0$ the sheaf $det(f_*(\mathcal{M} \otimes \omega^{e+1}_{X/Y})^\mu)^{r'} \otimes det(f_* \mathcal{M})^{-\mu\cdot r(\mu)}$ is ample for $r(\mu) = rank(f_*(\mathcal{M} \otimes \omega^{e+1}_{X/Y})^\mu)$, then $(\otimes^{r'} f_*(\mathcal{M} \otimes \omega^{e+1}_{X/Y})) \otimes det(f_* \mathcal{M})^{-1}$ is ample.

Remarks 3.10 a) As in (3.5) and (3.7) one can give effective bounds on the degree of ampleness in part ii) of (3.9).

b) Part i) of (3.9) is a straightforward generalization of [14], III, 2.7. However, since we weakend the assumptions we sketch the proof.

c) If as, in [14], III, 2.7, we assume that $\mathcal{M}$ is ample we can give bounds for e: Assume that for all $p \in Y$ there exists a desingularization $\tau : X'_p \to X_p$ and an effective exceptional divisor E_p on X'_p such that $\tau^*\mathcal{M}|_{X_p} \otimes \mathcal{O}_{X'_p}(-E_p)$ is very ample, then (3.9, c) can be replaced by the assumption

c')
$$e \geq c_1(\mathcal{M}|_{X_p})^{dim X_p} + 1.$$

Proof of 3.9 Obviously the assumptions are compatible with flat base change. Using (1.6, b) and (1.5) we can assume that $det(f_*\mathcal{M}) = \lambda'^{r'}$ for some invertible sheaf λ' on Y. Replacing $\mathcal{M}$ by $\mathcal{M} \otimes f^*\lambda'^{-1}$ we may as well assume that $det(f_*\mathcal{M}) = \mathcal{O}_Y$. We have to show in i) that $f_*(\mathcal{M} \otimes \omega^e_{X/Y})$ is weakly positive over Y and in ii) that the ampleness of $\lambda_\mu = det(f_*(\mathcal{M} \otimes \omega^{e+1}_{X/Y})^\mu)$ implies the ampleness of $f_*(\mathcal{M} \otimes \omega^{e+1}_{X/Y})$.

For $r \in \mathbb{N}$ let $f^r : X^r \to Y$ be the r-fold product of X over Y. We write

$$\mathcal{N} = \otimes^r_{i=1} pr_i^* \mathcal{M}.$$

In order to prove i) we choose $r = r'$. Hence $\mathcal{N}$ has a section induced by

$$det(f_*\mathcal{M}) = \mathcal{O}_Y \to f^r_*\mathcal{N} = \otimes^r f_*\mathcal{M}.$$

Let Γ be the zero divisor. Γ does not contain any fibre of f^r and for $N > 0$ and $\Gamma' = N \cdot \Gamma$ we have by (2.9) and by definition

$$e(\Gamma'|_{X_p \times \ldots \times X_p}) \leq N \cdot e(\Gamma|_{X_p \times \ldots \times X_p}) \leq N \cdot e(\mathcal{M}|_{X_p}) \leq N \cdot e.$$

Let $\mathcal{H}$ be an ample invertible sheaf and $m \geq 0$ be an integer. For $e' = N \cdot e$ let us consider the sheaf

$$\mathcal{L} = \mathcal{N}^N \otimes \omega^{e'-1}_{X^r/Y} \otimes f^{r*}\mathcal{H}^{m \cdot (e'-1) \cdot r}.$$

Then
$$\mathcal{L}^{e'}(-\Gamma') = (\mathcal{N}^N \otimes \omega^{e'}_{X^r/Y} \otimes f^{r*}\mathcal{H}^{m \cdot r \cdot e'})^{(e'-1)}$$

and the assumptions a) and b) of (3.3) hold true. By (1.8) weak positivity is compatible with tensor products and (3.3) implies:

Claim 3.11 If for some $M_0 > 0$ and all multiples M of M_0 the sheaf

$$f_*(\mathcal{M} \otimes \omega^e_{X/Y})^{N \cdot M} \otimes \mathcal{H}^{m \cdot e \cdot N \cdot M}$$

is weakly positive over Y, then

$$f_*(\mathcal{M} \otimes \omega^e_{X/Y})^N \otimes \mathcal{H}^{m \cdot (e \cdot N - 1)}$$

is weakly positive over Y.

Since $\mathcal{M} \otimes \omega^e_{X/Y}$ is relatively semi-ample over Y we can find some N_0 such that for all multiples N of N_0 and $M >> 0$ the multiplication maps

$$\alpha(N,M) : S^M(f_*(\mathcal{M} \otimes \omega^e_{X/Y})^N) \to f_*(\mathcal{M} \otimes \omega^e_{X/Y})^{N \cdot M}$$

are surjective. For those N and

$$m = Min\{\mu > 0; f_*(\mathcal{M} \otimes \omega^e_{X/Y})^N \otimes \mathcal{H}^{\mu \cdot e \cdot N} \text{ weakly positive over } Y\}$$

the surjectivity of $\alpha(N, M)$ implies the weak positivity over Y of

$$f_*(\mathcal{M} \otimes \omega^e_{X/Y})^{N \cdot M} \otimes \mathcal{H}^{m \cdot e \cdot N \cdot M}$$

for all $M >> 0$. Hence (3.11) gives that

$$f_*(\mathcal{M} \otimes \omega^e_{X/Y})^N \otimes \mathcal{H}^{m \cdot e \cdot N - m}$$

is weakly positive over Y. By the choice of m this implies that

$$(m-1) \cdot e \cdot N < m \cdot e \cdot N - m$$

or that $m < e \cdot N$. Hence $f_*(\mathcal{M} \otimes \omega^e_{X/Y})^N \otimes \mathcal{H}^{e^2 \cdot N^2}$ is weakly positive. Since this holds for all finite flat coverings of Y as well we obtain by (1.7) the weak positivity of $f_*(\mathcal{M} \otimes \omega^e_{X/Y})^N$. Applying (3.11) for $m = 0, N = 1$ and $M_0 = N_0$ defined above, we obtain part i) of (3.9).

To prove ii) we take $r = r' \cdot r(\mu)$ and $\mathcal{L} = \mathcal{N} \otimes \omega^e_{X^r/Y}$. We have natural inclusions, splitting locally

$$\mathcal{O}_Y = det(f_*\mathcal{M})^{r(\mu)} \to f^r_*\mathcal{N} = \otimes^{r' \cdot r(\mu)} f_*\mathcal{M}$$

and $$\lambda^{r'}_\mu \to f^r_*(\mathcal{N} \otimes \omega^{e+1}_{X^r/Y})^\mu = \otimes^{r' \cdot r(\mu)} f_*(\mathcal{M} \otimes \omega^{e+1}_{X/Y})^\mu.$$

If Δ_1 and Δ_2 denote the corresponding zero-divisors on X^r, then $\Delta_1 + \Delta_2$ does not contain any fibre of f^r and

$$\mathcal{L}^{(e+1) \cdot \mu} = (\mathcal{N} \otimes \omega^{e+1}_{X^r/Y})^{e \cdot \mu} \otimes \mathcal{N}^\mu = f^{r*} \lambda^{r' \cdot e}_\mu \otimes \mathcal{O}_X(e \cdot \Delta_2 + \mu \cdot \Delta_1).$$

Hence for $\mathcal{A} = \lambda^{r' \cdot e}_\mu$ and $N = (e+1) \cdot \mu$ the assumption (3.5, c) holds true. The assumption (3.5, b) is just part i) and e) of (3.9) and (3.5, a) is implied by (3.9, d). Hence

$$f^r_*(\mathcal{N} \otimes \omega^{e+1}_{X^r/Y}) = \otimes^r f_*(\mathcal{M} \otimes \omega^{e+1}_{X/Y})$$

is ample.

4 Ample sheaves on moduli schemes of canonically polarized manifolds

Let us consider the moduli functor $\mathcal{C}'_h$ of canonically polarized normal Gorenstein varieties with Hilbert polynomial h and with at worst rational singularities. Hence for a scheme S defined over $\mathbb{C}$,

$\mathcal{C}'_h(S) = \{f : X \to S; f$ flat, projective, Gorenstein; $\omega_{X/Y}$ relatively ample for f; all fibres F of f are irreducible normal varieties with at most rational singularities and $h(\nu) = \chi(\omega_F^\nu)\}/\simeq$.

Let $\mathcal{C}_h$ be a submoduli functor of $\mathcal{C}'_h$ such that $\mathcal{C}_h$ is bounded, separated and such that for $f : X \to S \in \mathcal{C}'_h(S)$ the subset $S_0 = \{s \in S; f^{-1}(s) \in \mathcal{C}_h(Spec\mathbb{C})\}$ is constructible in S.

By [6] we can choose $\mathcal{C}_h = \mathcal{C}'_h$, if $n = deg(h) \leq 2$, and in all dimensions

$$\mathcal{C}_h(S) = \{f \in \mathcal{C}'_h(S); f \text{ smooth }\}$$

will work by "Makusaka's big theorem". In any case we have (see [14], I, §1 and II, §6)

Assumption 4.1
i) There exists $\nu > 1$ such that for all $F \in \mathcal{C}_h(Spec(\mathbb{C}))$ the sheaf ω_F^ν is very ample.
ii) There exists a Hilbert scheme H and a universal family $g : \mathcal{X} \to H \in \mathcal{C}_h(H)$ together with an H-isomorphism

$$\mathbb{P}(g_*\omega^\nu_{\mathcal{X}/H}) \cong \mathbb{P}^{r-1} \times H.$$

iii) The action of $G = \mathbb{P}Gl(r, \mathbb{C})$ on H obtained by "change of coordinates in $\mathbb{P}^{r-1}$" is proper.
iv) Let us write $\lambda_\eta = det(g_*\omega^\eta_{\mathcal{X}/H})$ and $r(\eta) = rank(g_*\omega^\eta_{\mathcal{X}/H})$. Then for $\mu >> 0$ the sheaf $\mathcal{L}_0 = \lambda^{r(\nu)}_{\nu\cdot\mu} \otimes \lambda_\nu^{-\mu\cdot r(\nu\cdot\mu)}$ is ample on H.

Corollary 4.2 *λ_ν is ample on H.*

Proof. Of course we may assume H to be reduced. By [14], II, 2.7 the sheaf $g_*\omega^\nu_{\mathcal{X}/H}$ is weakly positive over H and hence λ_ν has the same property. By (1.2) $\lambda_{\nu\cdot\mu}$ is ample and as we have seen in (0.1), $g_*\omega^\eta_{\mathcal{X}/H}$ as well as λ_η will be ample for all $\eta > 1$ with $r(\eta) > 0$.

Let $\mathcal{L}$ be a G-linearized sheaf on H ([11], Def. 1.6). In [14], I, 5.2, we denoted by $H(\mathcal{L})^s$ the set of stable points of H with respect to $\mathcal{L}$ and under the G-action (see [11], Def. 1.7). Obviously the sheaves λ_η are G-linearized for all $\eta > 0$.

Corollary 4.3 *One has $H = H(\lambda_\nu)^s$.*

Proof. In [14], II, we proved that $H = H(\mathcal{L}_0 \otimes \lambda_\nu^\eta)^s$ for some $\eta >> 0$. However we only used "$\mathcal{L}_0$ ample" and not at all the specific shape of $\mathcal{L}_0$. Hence, if we start with λ_ν instead of $\mathcal{L}_0$ we obtain 4.3.

The stability criterion used in [14], II, was formulated in a more general set up in [15], 3.2.

By [11], 4.3 implies the existence of a geometric quotient H/G and H/G is embedded into some projective space by G-invariant sections of λ_ν^p for $p >> 0$. Therefore λ_ν^p descends to some ample sheaf on H/G, which we denote by $\lambda_\nu^{(p)}$. As it is shown in [11] $C_h = H/G$ is a coarse moduli scheme for $\mathcal{C}_h$. Hence we obtained:

Theorem 4.4 *Let $\mathcal{C}_h$ be the moduli functor of canonically polarized manifolds with Hilbert polynomial h (or any moduli functor satisfying (4.1)). Then there exists a coarse moduli scheme C_h and an ample invertible sheaf $\lambda_\nu^{(p)}$ for ν as in (4.1, i) and $p >> 0$, such that: For $f : X \to S \in \mathcal{C}_h(S)$ let $\varphi : S \to C_h$ be the induced morphism. Then*

$$\varphi^*\lambda_\nu^{(p)} = det(f_*\omega_{X/S}^\nu)^p.$$

Corollary 4.5 *For $f : X \to S \in \mathcal{C}_h(S)$ assume that $\varphi : S \to C_h$ is affine over its image. Then for all $\eta > 1$ with $h(\eta) > 0$ the sheaf $f_*\omega_{X/S}^\eta$ is ample.*

Proof. By (4.4) the sheaf $det(f_*\omega_{X/S}^\nu)$ is ample and (4.5) follows from (0.1).

It is easy to show, that for all $\eta > 0$ with $h(\eta) > 0$ the sheaves λ_η on H descend to invertible sheaves $\lambda_\eta^{(p(\eta))}$ on C_h for some $p(\eta) > 0$ (see for example [8]).

Corollary 4.6 *Let $\tau : C_h' \to C_h$ be the normalization. Then for all $\eta > 0$ with $h(\eta) > 0$ the sheaf $\tau^*(\lambda_\eta^{p(\eta))})$ is ample on C_h'.*

Proof. By [8], §2, there exist a finite cover $\varphi : Y \to C_h$ and $f : X \to Y \in \mathcal{C}_h(Y)$ such that φ is induced by f. We may assume that φ factors through $\varphi' : Y \to C_h'$. Hence

$$\varphi'^*\tau^*(\lambda_\eta^{(p(\eta))}) = (det f_*\omega_{X/Y}^\eta)^{p(\eta)}$$

is ample by (4.5) and by (1.5) ampleness descends to C_h'.

(4.6) suggests the following question, which, in fact, would have an affirmative answer if we could choose in the proof of (4.6) $\varphi : Y \to C_h$ such that $\mathcal{O}_{C_h} \to \varphi_*\mathcal{O}_Y$ splits.

Question 4.7 Are the sheaves $\lambda_\eta^{(p(\eta))}$ ample on C_h for all $\eta > 1$ with $h(\eta) > 0$?

5 Ample sheaves on moduli schemes of polarized manifolds

As in [14], III, §1 let us consider the moduli functor $\mathcal{M}'_h$ with

> $\mathcal{M}'_h(S) = \{(f : X \to S, \mathcal{H}); f$ flat, projective and Gorenstein and $\mathcal{H}$ invertible, relatively ample over S, such that: for all $p \in S$ $X_p = f^{-1}(p)$ is a normal variety with at worst rational singularities, $\chi(\mathcal{H}|_{X_p}) = h(p)$ and X_p is not uniruled $\}/\simeq$.

Differently from the definition of moduli of polarized varieties in [11] or [16], we define

$$(f : X \to S, \mathcal{H}) \simeq (f' : X' \to S, \mathcal{H}')$$

if there is an S-isomorphism $\tau : X \to X'$ and an invertible sheaf $\mathcal{B}$ on S such that

$$\tau^*\mathcal{H}' \simeq \mathcal{H} \otimes f^*\mathcal{B}.$$

We take $\mathcal{M}''_h$ to be a bounded and separated submoduli functor of $\mathcal{M}'_h$ such that for all S and all $(f : X \to S, \mathcal{H}) \in \mathcal{M}'_h(S)$ the subset

$$S_0 = \{s \in S; (f^{-1}(s), \mathcal{H}|_{f^{-1}(s)}) \in \mathcal{M}'_h(Spec(\mathbb{C}))\}$$

is constructible in S.

By [6] again, we can choose $\mathcal{M}''_h = \mathcal{M}'_h$, if $deg(h) = 2$ and by "Makusaka's big theorem"

$$\mathcal{M}''_h(S) = \{(f : X \to S, \mathcal{H}) \in \mathcal{M}'_h(S), f \text{ smooth }\}$$

will always work. By boundedness we can find some $\nu > 0$ such that for all $(F, \mathcal{H}) \in \mathcal{M}''_h(Spec(\mathbb{C})))$ the sheaf $\mathcal{H}^\nu$ is very ample. We can even choose ν big enough to have:

Assumption 5.1 There is some $\nu > 0$ such that for all $(F, \mathcal{H}) \in \mathcal{M}''_h(Spec(\mathbb{C}))$ one has:
a) There is a desingularization $\tau : F' \to F$ and an effective exceptional divisor E on F' such that $\tau^*\mathcal{H}^\nu \otimes \mathcal{O}_{F'}(-E)$ is very ample.
b) $H^i(F, \mathcal{H}^\nu) = 0$ for $i > 0$.
c) For all numerically effective sheaves $\mathcal{L}$ on F, the sheaf $\mathcal{H}^\nu \otimes \omega_F \otimes \mathcal{L}$ is very ample and without higher cohomology.

Proof. By boundedness, for some smooth Y, there is a family $(g : X \to Y, \mathcal{H}') \in \mathcal{M}''_h(Y)$ such that all $(F, \mathcal{H}) \in \mathcal{M}''_h(Spec(\mathbb{C}))$ occur as fibres. For a) we consider a desingularization of X. We find a) to be true for all $(F, \mathcal{H})$ over some dense open subscheme of Y. Repeating this for the complement we obtain a).

In a) we are allowed to replace ν by any multiple. Then b) is obvious and c) follows from [14], III, 1.3, if we replace ν by $\nu(n+1)$ for $n = deg(h)$.

Notations 5.2 For ν as in 5.1 let c be the highest coefficient of h and

$$e \geq (n!) \cdot c \cdot \nu^n + 2$$

for $n = deg(h)$. Especially $e \geq e(\mathcal{H}^\nu) + 1$ for all $(F, \mathcal{H}) \in \mathcal{M}''_h(Spec(\mathbb{C}))$ by (2.11). Let us choose

$$\mathcal{M}_h(S) = \{(f : X \to S, \mathcal{H}) \in \mathcal{M}''_h(S); \mathcal{H}^\nu \otimes \omega^e_{X/S} \text{ relatively very ample over } S\}.$$

As in [14], III, 1.4, we may assume, that $\chi(F, (\mathcal{H}^\nu \otimes \omega^e_F)^\eta)$ and $\chi(F, \mathcal{H}^{\nu \cdot \eta + 1} \otimes \omega^{e \cdot \eta}_F)$ are the same for all $(F, \mathcal{H}) \in \mathcal{M}_h(Spec(\mathbb{C}))$ regarded as polynomials in η.

By (5.1, c) $\mathcal{M}_h(Spec(\mathbb{C}))$ will contain all $(F, \mathcal{H}) \in \mathcal{M}''_h(Spec(\mathbb{C}))$ with ω_F numerically effective. But, since we do not know wether "ω_F nef" is a constructible condition we are not able to consider just

$$\mathcal{M}_h^{nef}(S) = \{(f : X \to S, \mathcal{H}); \omega_{X/Y} \text{ numerically effective on each fibre }\}.$$

However, the functor $\mathcal{M}_h^{sa}$ can replace $\mathcal{M}_h$ in the results following where

$$\mathcal{M}_h^{sa} = \{(f : X \to X, \mathcal{H}); \omega_{X/Y} \text{ relatively semi-ample over } S\}.$$

Lemma 5.3 *Using the notations and assumption introduced above one has for all*

$$(f : X \to S, \mathcal{H}) \in \mathcal{M}_h(S) :$$

a) $\mathcal{H}^\nu$ is relatively ample over S and $f_*\mathcal{H}^\nu$ is locally free of rank r'.
b) $\mathcal{H}^\nu \otimes \omega_{X/S}^e$ is relatively ample over S and $f_*(\mathcal{H}^\nu \otimes \omega_{X/S}^e)^\eta$ is locally free of rank $r(\eta)$ for $\eta > 0$.
c) $f_*(\mathcal{H}^\nu \otimes \omega_{X/S}^e)$ is weakly positive over Y.
d) If for some $\mu > 0$ the sheaf $\lambda_\mu = det(f_*(\mathcal{H}^\nu \otimes \omega_{X/S}^e)^\mu)^{r'} \otimes det(f_*\mathcal{H}^\nu)^{-\mu \cdot r(\mu)}$ is ample, then $(\otimes^{r'} f_*(\mathcal{H}^\nu \otimes \omega_{X/S}^e)) \otimes det(f_*\mathcal{H}^\nu)^{-1}$ is ample.

Proof. a) holds true since we have no higher cohomology along the fibres and everything is compatible with base change. Moreover $\mathcal{H}^\nu \otimes \omega_{X/S}^e$ is relatively ample over S by definition of $\mathcal{M}_h$. Hence $\mathcal{H}^{\nu \cdot \eta} \otimes \omega_{X/S}^{e'}$ is relatively ample over S for $o \le e' \le \eta \cdot e$. By (3.3, i) all the sheaves

$$f_*(\mathcal{H}^\nu \otimes \omega_{X/S}^{e'})^\eta$$

are locally free for $e' = e$ or $e' = e - 1$. Since $e \ge e(\mathcal{H}|_F) + 1$ for all fibres F of f, c) and d) are implied by (3.9, i) and (3.9, ii).

Let H be the Hilbert scheme considered in [14], III, 1.5, d. Especially we have a "universal family"

$$(g : \mathcal{X} \to H, \mathcal{H}) \in \mathcal{M}_h(H)$$

and an isomorphism

$$\varphi : \mathbb{P}(g_*(\mathcal{H}^\nu \otimes \omega_{\mathcal{X}/H}^e)) \to \mathbb{P}^{r-1} \times H.$$

Again, $G = \mathbb{P}Gl(r, \mathbb{C})$ acts on H properly and G/H will be a candidate for M_h (see [11]).

Lemma 5.4 *For ν as in (5.1) the sheaf λ_1 is ample on H (see (5.3) for the notations).*

Proof. For $\mu >> 0$ the Plücker coordinates give an embedding of H into some projective space and the induced ample sheaf is

$$\mathcal{L}_0 = det(g_*(\mathcal{H}^\nu \otimes \omega_{\mathcal{X}/H}^e))^{-\mu \cdot r(\mu)} \otimes det(g_*(\mathcal{H}^{\mu \cdot \nu} \otimes \omega_{\mathcal{X}/H}^{e \cdot \mu}))^{r(1)}.$$

Hence $\mathcal{L}_0^{r'} = \lambda_1^{-\mu \cdot r(\mu)} \otimes \lambda_\mu^{r(1)}$ is ample. By (3.9, i) λ_1 as the determinant of a weakly positive sheaf is weakly positive over Y and hence λ_μ is ample. Since we have choosen

$e \geq e(\mathcal{H}) + 1$ for $(\mathcal{H}, F) \in \mathcal{M}_h(Spec(\mathbb{C}))$ we can apply (3.9, ii) for $(e-1)$ and we obtain the ampleness of

$$(\otimes^{r'} g_*(\mathcal{H}^\nu \otimes \omega^e_{X/H})) \otimes det(f_*\mathcal{H}^\nu)^{-1}$$

and hence of λ_1.

As in (4.3) and (4.4) we obtain

Corollary 5.5 *One has $H = H(\lambda_1)^s$.*

Theorem 5.6 *Let $\mathcal{M}_h$ be the moduli functor of (5.2), ν as in (5.1) and e as in (5.2). Then there exists a coarse moduli scheme M_h and an ample invertible sheaf $\lambda^{(p)}$ for some $p > 0$, such that:*
For $(f : X \to S, \mathcal{H}) \in \mathcal{M}_h(S)$ let $\varphi : S \to M_h$ be the induced morphism. Then

$$\varphi^*\lambda^{(p)} = (det(f_*(\mathcal{H}^\nu \otimes \omega^e_{X/S}))^{r'} \otimes det(f_*\mathcal{H}^\nu)^{-r})^p$$

for $r' = rank(f_*\mathcal{H}^\nu)$ and $r = rank\ f_*(\mathcal{H}^\nu \otimes \omega^e_{X/S})$.

Remarks 5.7 a) In fact, if one compares (5.4) with [14], III, 1.11, then the choice of e is slightly different. However in [14], III, we only used that $e \geq (n!) \cdot c \cdot \nu^n + 1$ and not, as stated there, that one has equality. This is obvious if one takes the stability criterion [15], 3.2.
b) Kollár [8] and Fujiki-Schumacher [4] developed independently methods to study sheaves on analytic moduli spaces, the first one by estimating the degree on complete curves of certain natural sheaves on moduli spaces, the two others by curvature estimates. Both methods give ampleness criteria for sheaves on compact subspaces of moduli spaces. The comparison of the results of [4] with those of this note should give some candidates beside of $\lambda^{(p)}$ for ample sheaves on M_h and some hope that question (4.7) has an affirmative answer.

Corollary 5.8 *For $(f : X \to S, \mathcal{H}) \in \mathcal{M}_h(S)$ assume that $\varphi : S \to M_h$ is affine over its image. Then for ν, e, r' as above the sheaf $\otimes^{r'} f_*(\mathcal{H}^\nu \otimes \omega_{X/S}) \otimes det(f_*\mathcal{H}^\nu)^{-1}$ is ample.*

Proof. Use (5.6) and (3.9, ii)

Notations 5.9 It is quite easy to see, that invertible G-linear sheaves on H have some power which descends to M_h (see [8] for example). Since $det(g_*\omega^\delta_{X/H})^q$ is G-invariant, for $q >> 0$ we can descend this sheaf to some sheaf $\gamma^{(q)}_\delta$ on M_h. Especially, if we have choosen $\mathcal{M}_h$ such that for some $\delta > 0$ and all $(F, \mathcal{H}) \in \mathcal{M}_h(Spec(\mathbb{C}))$ one has $\omega^\delta_F = \mathcal{O}_F$, then for $(f : X \to S, \mathcal{H}) \in \mathcal{M}_h(S)$ and the induced morphism $\varphi : S \to M_h$ one has

$$f^*\varphi^*\gamma^{(q)}_\delta = \omega^{q\cdot\delta}_{X/S}.$$

Corollary 5.10 *Assume that for some $\delta > 0$ and all $(F, \mathcal{H}) \in \mathcal{M}_h(Spec(\mathbb{C}))$ one has $\omega_F^\delta = \mathcal{O}_F$. Assume moreover that the integer e used in the definition of $\mathcal{M}_h$ is divisible by δ. Then $\gamma_\delta^{(q)}$ is ample on M_h.*

Proof. By (5.6) $\lambda^{(p)}$ is ample on M_h. For $e = \delta \cdot e'$ we have for

$$(f : X \to S, \mathcal{H}) \in \mathcal{M}_h(S)$$

that

$$f_*(\mathcal{H}^\nu \otimes \omega_{X/S}^e) = f_*(\mathcal{H}^\nu) \otimes (f_*\omega_{X/S}^\delta)^{e'}$$

and $\varphi^*\lambda^{(p)} = (f_*\omega_{X/S}^\delta)^{e' \cdot r' \cdot p}$. Hence $\lambda^{(p)} = \gamma_\delta^{(e' \cdot r' \cdot p)}$.

(5.10) generalizes the result of Pjatetskij-Šapiro and Šafarevich on moduli of $K3$ surfaces [12]. In fact, the quite simple ample sheaf obtained in [12] was one of the motivations to reconsider the proof of the existence of quasi-projective moduli schemes and to try to improve the ample sheaves obtained in [14], III.

Remark 5.11 a) Replacing "isomorphisms of polarizations" by "numerical equivalence of polarizations" (see for example [16], §4) one obtains the "right" functor $\mathcal{P}_h$ of polarized varieties. In [16] we constructed a coarse quasi-projective moduli space P_h for $\mathcal{P}_h$. The natural morphism

$$\Sigma : M_h \to P_h$$

is proper, and, replacing $\mathcal{H}$ by some high power and $h(t)$ by $h(\eta \cdot t)$, we may assume moreover that the fibres of Σ are connected. If for all $(F, \mathcal{H})$ one has

$$q(F) = h^0(F, \Omega_F^1) = 0,$$

then Σ will be an isomorphism and (5.6) gives an ample sheaf on P_h as well.
b) On the other hand, if $\mathcal{M}_h$ satisfies the assumptions of (5.10), then the ampleness of $\gamma_\delta^{(q)}$ just implies, that M_h can not contain projective curves C such that the induced family $(f : X \to C', \mathcal{H})$ over some finite cover C' of C satisfies $X \simeq F \times C'$. Hence Σ is finite and (5.10) gives an ample sheaf on P_h again.
c) Regarding the construction of P_h in §3 of [16], the fact that Σ is finite just means that one has:

> *If F is a manifold with $\omega_F^\delta = \mathcal{O}_F$, $\mathcal{H}$ ample invertible on F and ν sufficiently large (as in (5.1) for example), then for each $\mathcal{L} \in Pic^0(F)$ one can find an automorphism $\tau : F \to F$ such that $\tau^*(\mathcal{H}^\nu) = \mathcal{H}^\nu \otimes \mathcal{L}$ or $\mathcal{L} = \tau^*(\mathcal{H}^\nu) \otimes \mathcal{H}^{-\nu}$.*

6 Degenerate fibres

The ampleness criteria (0.1) and (3.9) are strong enough to be applied to moduli schemes. However, from the point of view of fibrespaces it seems natural to keep track of the behaviour of sections with respect to compactifications, as we also did in [3].

For simplicity we stay in the category of quasi-projective complex schemes, even if the result remain true for analytic spaces, if one modifies the definition of weak positivity, as it was done in [14], II, §5.

Since it is not at all understood how to get the natural compactification of the total space of a smooth morphism to a singular scheme, we have to return to the notations used in [14], II and III:

Assumptions 6.1 Let

$$\begin{array}{ccccc} X_0 & \xrightarrow{i} & X & \xleftarrow[\sigma]{} & X' \\ {\scriptstyle f_0}\downarrow & & {\scriptstyle f}\downarrow & & \downarrow{\scriptstyle f'} \\ Y_0 & \xrightarrow{j} & Y & \xleftarrow[\delta]{} & Y' \end{array}$$

be a commutative diagram of morphisms of reduced quasi-projective schemes such that
a) i and j are open dense embeddings and σ and δ are desingularizations
b) f and f' are surjective and $X_0 = f^{-1}(Y_0)$.
c) f_0 is flat and Gorenstein. All fibres $X_p = f_0^{-1}(p)$ are reduced irreducible normal varieties of dimension n with at worst rational singularities.
d) If (...) is a sheaf or a divisor on X (or Y) then $(...)_0$ will always denote the restriction to X_0 (or Y_0) and $(...)'$ will be the pullback to X' (or Y').
e) ω_{X_0/Y_0} is the dualizing sheaf of f_0 and $\omega_{X'/Y'} = \omega_{X'} \otimes f'^*\omega_{Y'}^{-1}$.

The main purpose of this chapter is to sketch the changes of the arguments employed in §3 and §4 to obtain:

Claim 6.2 *Under the assumption (6.1) assume that $f' : X' \to Y'$ is semi stable in codimension one (see [9], 4.6), that ω_{X_0/Y_0} is relatively semi-ample over Y_0 and that for some $\mu > 0$ the sheaf $det(f_{0*}\omega^{\mu}_{X_0/Y_0})$ is ample with respect to $(Y', det(f'_*\omega^{\mu}_{X'/Y'}))$. Then for all $\eta \geq 2$ with $f_{0*}\omega^{\eta}_{X_0/Y_0} \neq 0$, the sheaf $f_{0*}\omega^{\eta}_{X_0/Y_0}$ is ample with respect to $(Y', f'_*\omega^{\eta}_{X'/Y'})$.*

Claim 6.3 *Under the assumptions (6.1) assume that $f_0 : X_0 \to Y_0 \in \mathcal{C}_h(Y_0)$ and that the induced morphism $\varphi : Y_0 \to C_h$ is quasi-finite over its image. Then for all $\eta > 1$ with $k(\eta) > 0$ the sheaf $f_{0*}\omega^{\eta}_{X_0/Y_0}$ is ample with respect to $(Y', f'_*\omega^{\eta}_{X'/Y'})$.*

Contrary to [14], II and III, we do not assume here in (6.1) that Y, Y', X and X' are compact. The main reason is that (6.2) will be needed for partial compactifications in order to prove (6.3). Hence to give complete proofs of (6.2) and (6.3), one has to verify first, that neither in the definition of "weak positivity with respect to" nor in [14], II, 2.7, or [14], III, 2.6, the compactness of Y (or X) was really necessary.

Since this can not be done in all details here, and since we just give a coarse outline of the proofs, the reader should regard (6.2) and (6.3) with certain doubts.

Remark 6.4 In (6.2) and (6.3) the conditions (1.9, i and ii) hold true. In fact, using (3.3, i) one finds that $f_{0*}\omega^{\eta}_{X_0/Y_0}$ is locally free and compatible with base change and hence the inclusion of sheaves asked for is just given by

$$S^{\nu}(f_{0*}\omega^{\eta}_{X_o/Y_o}) \to S^{\nu}(f_{0*}\omega^{\eta}_{X_0/Y_0}) \otimes j^*\delta_*\mathcal{O}_{Y'} \simeq j^*\delta_*S^{\nu}(f_*\omega^{\eta}_{X'/Y'}).$$

Sketch of the proof of 6.2 First of all, the equivalence of a) and c) in (1.7) extends to "weakly positive with respect to" by [14], II, 2.4, b, and one can even assume there that $\tau : Z \to Y$ is flat. In [14], III, 2.6 the assumption that:

$(\mathcal{L}_0^e(-\Gamma_0))^N$ *is globally generated by* $H^0(X_0, (\mathcal{L}_0^e(-\Gamma_0))^N) \cap H^0(X', (\mathcal{L}'^e(-\Gamma'))^N)$ *over* X_0

can be replaced by:

a) $\mathcal{L}_0^e(-\Gamma_0)$ *is relatively semi-ample over* Y_0
b) *For some* $M_0 > 0$ *and all multiples* M *of* M_0 *the sheaf* $f_{0*}(\mathcal{L}_0^e(-\Gamma_0))^M$ *is locally free and weakly positive over* Y_0 *with respect to* $(Y', f'_*(\mathcal{L}'^e(-\Gamma'))^M)$.

The proof remains the same. In order to be able to apply [14], II, 2.4, b one has to use [14], II, 2.5. In the same way (3.5) can be modified to

Claim 6.5 *For* $f : X \to Y$ *as in (6.1) let* $\mathcal{L}$ *be an invertible sheaf on* X. *Assume that:*
a) $\mathcal{L}_0$ *is relatively semi-ample over* Y_0 .
b) For some $M > 0$ *and all multiples* M *of* M_0 *the sheaf* $f_{0*}(\mathcal{L}_0^M)$ *is locally free and weakly positive over* Y_0 *with respect to* $(Y', f'_*(\mathcal{L}'^M))$.
c) For some $N > 0$ *there is an ample invertible sheaf* $\mathcal{A}$ *on* Y *and an effective Cartier divisor* Γ *on* X, *not containing* X_p *for* $p \in Y_0$, *with* $\mathcal{L}^N = f^*\mathcal{A} \otimes \mathcal{O}_X(\Gamma)$.
Then $f_{0*}(\mathcal{L}_0 \otimes \omega_{X_o/Y_0})$ *is ample with respect to* $(Y', f'_*(\mathcal{L}' \otimes \omega_{X'/Y'}))$.

Now, using (2.10, c) of [14], III, the proof of (0.1) carries over to prove (6.2).

Sketch of the proof of 6.3 Using [9], 4.6 and [14], II, 1.10 and 2.4, a, we can assume that $f' : X' \to Y'$ is semi-stable in codimension one. Hence, by (6.2) it is enough to show that $det(f_{0*}\omega^{\mu}_{X_0/Y_0})$ is ample with respect to $(Y', det(f'_*\omega^{\mu}_{X'/Y'}))$.

If one forgets about the compactification, i.e. if $Y = Y_0$, (6.3) is obtained in (4.5) using the ampleness of λ_{ν} on C_h. However [14], I, §2 and §4 contain a direct proof of the ampleness of $det(f_{0*}\omega^{\nu}_{X_0/Y_0})$ in that case, parallel to methods from "Geometric Invariant Theory". The only necessary modification is that in §4 of [14], I, one uses the ampleness of $\mathcal{L}_0$ and (6.2) to get the ampleness of $\pi^*det(\mathcal{E})$.

However, since in (6.3) we want to allow degenerate fibres, one has to modify the arguments used to prove [14], II, 5.2. The necessary changes are more difficult to explain:

The "Ampleness Criterion" [14], II, 5.7, should be applied to the multiplication maps

$$S^{\mu}(f_{0*}\omega^{\nu}_{X_0/Y_0}) \to f_{0*}\omega^{\nu\cdot\mu}_{X_0/Y_0}$$

and

$$S^{\mu}(f'_{*}\omega^{\nu}_{X'/Y'}) \to f'_{*}\omega^{\nu\cdot\mu}_{X'/Y'},$$

hence for $s = 1$, $T_1 = S^{\mu}$ and $\mathcal{F}^{(1)}_0 = f_*\omega^{\nu}_{X_0/Y_0}$ in the notation of [14], II, 5.6. Using the notations from the proof of [14], II, 5.7 we have to consider

$$\mathcal{F} = \delta_*(f'_*\omega^{\nu}_{X'/Y'}) \cap j_*(f_{0*}\omega^{\nu}_{X_0/Y_0})$$

and

$$Q = \delta_*(f'_*\omega^{\nu\cdot\mu}_{X'/Y'}) \cap j_*(f_{0*}\omega^{\nu\cdot\mu}_{X'_0/Y'_0})$$

and we may assume that both sheaves are locally free. On some blowing up $\mathbb{P}'$ of $\mathbb{P} = \mathbb{P}(\mathcal{F})$ we found an effective divisor E, not meeting $\mathbb{P}_0$, such that

$$\pi'^*(det(Q)^a \otimes det(\mathcal{F})^{-a}) \otimes \mathcal{O}_{\mathbb{P}'}(E)|_{\mathbb{P}'-\tau^*(D)}$$

is ample on $\mathbb{P}' - \tau^*(D)$.

Since $det(\mathcal{F})$ is weakly positive over Y_0, $\pi'^*det(Q)^a$ will be ample over $\mathbb{P}_0 - D$. From (6.2) applied to be the pullback families over $\mathbb{P}' - \tau^*D$ we find that

$$\pi'^*(det(\mathcal{F}))^{a'} \otimes \mathcal{O}_{\mathbb{P}}(E')|_{\mathbb{P}'-\tau^*(D)}$$

will again be ample for some E' and $a' > 0$. As in [14], II, p 220, we will get for $\alpha >> 0$ and for an effective divisor D' supported in τ^*D that

$$\pi'^*(det(\mathcal{F}))^{\alpha} \otimes \mathcal{O}_{\mathbb{P}'}(E' + D')$$

is ample. As in [14], I, 4.7 one can descend this to obtain the ampleness of $det(\mathcal{F})$.

References

1. Esnault, H.: Classification des variétés de dimension 3 et plus. Sém. Bourbaki, Exp. 568, Février 1981. (Lecture Notes Math., Vol 901). Berlin-Heidelberg-New York: Springer 1981

2. Esnault, H., Viehweg, E.: Logarithmic De Rham complexes and vanishing theorems. Invent. math. 86, 161-194 (1986)

3. Esnault, H., Viehweg, E.: Effective bounds for semi positive sheaves and for the height of points on curves over complex function fields. Compos. Math. 76, 69-85 (1990)

4. Fujiki, A., Schumacher, G.: The moduli space of extremal compact Kähler manifolds and generalized Weil-Petersson metrics. Publ. RIMS. 26, 101-183 (1990)

5. Hartshorne, R.: Ample vector bundles. Publ. Math., Inst. Hautes Etud. Sci. 29, 63-94 (1966)

6. Kollár, J.: Toward moduli of singular varieties. Compos. Math. 56, 369-398 (1985)

7. Kollár, J.: Higher direct images of dualizing sheaves. Ann. Math. 123, 11-42 (1986)

8. Kollár, J.: Projectivity of complete moduli. J. Differ. Geom. 32, 235-268 (1990)

9. Mori, S.: Classification of higher-dimensional varieties. Algebraic Geometry. Bowdoin 1985. Proc. Symp. Pure Math. 46, 269-331 (1987)

10. Mumford, D.: Abelian Varieties. Tata Inst. Fund. Res., Bombay, and Oxford Univ. Press, 1970

11. Mumford, D., Fogarty, J.: Geometric Invariant Theory, Second Edition. (Ergebnisse der Math., Vol. 34). Berlin-Heidelberg-New York: Springer 1982

12. Pjatetskij-Šapiro, I.I., Šafarevich, I.R.: A Torelli theorem for algebraic surfaces of type K3. Math. USSR Izv. 5, 547-588 (1971)

13. Viehweg, E.: Vanishing theorems. J. Reine Angew. Math. 335, 1-8 (1982)

14. Viehweg, E.: Weak positivity and the stability of certain Hilbert points, I. Invent. Math. 96, 639-667 (1989), II. Invent. Math. 101, 191-223 (1990), III. Invent. Math. 101, 521-543 (1990)

15. Viehweg, E.: Positivity of sheaves and geometric invariant theory. Proc. of the "A.I. Maltsev Conf.", Novosibirsk, 1989, to appear

16. Viehweg, E.: Quasi-projective quotients by compact equivalence relations. Math. Annalen. to appear

SIMULTANEOUS CANONICAL MODELS OF DEFORMATIONS OF ISOLATED SINGULARITIES

SHIHOKO ISHII

Dedicated to Prof. Heisuke Hironaka on his sixtieth birthday

In this paper we continue our study of deformation of normal isolated singularities of dimension $n \geq 2$. In the previous article [I2], we obtain the upper semi-continuity of the m-genus δ_m ($m \in \mathbb{N}$) of an isolated singularity under a deformation. It is natural to consider whether another m-genus γ_m (m $\in$ N) has the same property or not. We show here that γ_m is also upper semi-continuous under a deformation $\pi : X \to D$ with the property:

(FG)a graded ring $\bigoplus_{m \geq 0} f_* \omega_{\tilde{X}}^m$ is finitely generated $\mathcal{O}_X$- algebra, where $f : \tilde{X} \to X$ is a resolution of the singularities of X.

From now on, we call a deformation with the property (FG) an (FG)-deformation. Next we show that an (FG)-deformation $\pi: X \to D$ admits the simultaneous canonical model, if $\gamma_m(X_\tau)$ is represented as $e_\tau m^n + O(m^{n-1})$ and e_τ is constant for every $\tau \in D$, where $O(m^{n-1})$ means the term which grows in order at most $n-1$.

Then we study a deformation of n-dimensional Gorenstein purely elliptic singularities of type (0, n-1), where the singularity of this type is considered as an n-dimensional analogue of a simple elliptic singularity [S]. Such a singularity deforms to either singularities of the same type or rational singularities . We prove that if $\pi : X \to D$ is an (FG)-deformation of the former type (i.e. every fiber X_τ has a Gorenstein purely elliptic singularity of type (0, n-1)), then it admits the simultaneous canonical model. Applying it to the case $n = 3$, we have that 95-classes of quasi-homogeneous hypersurface simple K3-singularities (i.e. 3-dimensional Gorenstein purely elliptic singularities of type (0,2)) classified by Yonemura [Y] cannot connect to each other under any (FG)-deformation.

As is well known, the condition (FG) is satisfied, if the Minimal Model Conjecture holds for a resolution $\tilde{X}$ of X [KMM,0.4.4]. So every deformation $\pi: X \to D$ of surface singularities over the unit disk $D \subset \mathbf{C}^1$ satisfies (FG) (cf.[M]). Therefore, for every deformation of normal surface singularities, we get:

(i) γ_m is upper semi-continuous, (ii)π admits the simultaneous Du Val resolution, if the self-intersection K_τ^2 of the numerical canonical divisor on the minimal resolution is constant and (iii)π admits the simultaneous Du Val resolution, if every fiber X_τ has a simple elliptic singularity. Here (ii) is proved by Laufer [L] in the case X_τ are Gorenstein singularities and its global version is also proved by Kollár and Shepherd-Barron [K-SR] in the case $\pi : X \to D$ is a projective morphism. They also proved the opposite implications under their situations, while the converse is not true in general.

This work has been stimulated by various discussions with Professors M.Tomari, K-i. Watanabe, K. Watanabe and other members of the seminar.

§1.The behavior of the invariants γ_m.

For a germ (V, v) of an analytic space at a point v we denote a sufficiently small Stein neighbourhood of v by V again, in order to save symbols.

Definition 1.1. Let (V, v) be a germ of normal singularity. We call a morphism $f : Y \to V$ a partial resolution, if it is a projective morphism isomorphic away from the singular locus on V and the space Y is normal.

Definition 1.2. Let (V, v) be a germ of a normal singularitiy. We call a morpohism $f : Y \to V$ the canonical model, if it is a partial resolution with at worst canonical singularities on Y and K_Y is relatively ample with respect to f.

Definition 1.3. Let $\pi : X \to D$ be a flat family of normal singularities over the unit disk D. We call a projective morphism $F : Y \to X$ the simultaneous canonical model, if the restriction $F_\tau : Y_\tau \to X_\tau$ is the canonical model for every $\tau \in D$, where $Y_\tau = F^{-1}(X_\tau)$ and $X_\tau = \pi^{-1}(\tau)$.

Definition 1.4. Let $\pi : X \to D$ be a flat family of normal singularities. We call π an (FG)-familay or an (FG)-deformation, if $\bigoplus_{m \geq 0} F_* \omega_Y^{[m]}$ is a finitely generated $\mathcal{O}_X$ -algebra for a resolution $F : Y \to X$ with at worst canonical singularities on Y.

Remark 1.5. (i) Here, the graded $\mathcal{O}_X$-algebra $\bigoplus_{m \geq 0} F_* \omega_Y^{[m]}$ is independent of the choice of a partial resolution $F : Y \to X$ where Y has at worst canonical singularities ([R]).

(ii) A singularity (V, v) has a canonical model if and only if $\bigoplus_{m \geq 0} F_* \omega_Y^{[m]}$ is a finitely generated $\mathcal{O}_V$-algebra, for F as above. In this case, the canonical model is isomorphic to the canonical projection $Proj \bigoplus_{m \geq 0} F_* \omega_Y^{[m]} \to V$ over V. So an (FG)-deformation is a deformation $\pi : X \to D$ whose total space X has the canonical model ([R], [KMM]).

(iii) The existence of the canonical models for every singularity is not yet proved. It is known that if the minimal model conjecture holds for a resolution of a singularity then the singularity has the canonical model (cf. [KMM]). Therefore 2 and 3-dimensional singularities have canonical models by Mori [M].

(iv) If an (FG)-deformation $\pi : X \to D$ has the simultaneous canonical model $F : Y \to X$, then F turns out to be the canonical model of the total space. In fact, by a result of Kawamata [KMM, 7.2.4], Y has at worst canonical singularities, while the relative ampleness of K_Y follows immediately from that of K_{Y_τ}'s $(\tau \in D)$.

Definition 1.6 (Knöller [K] and Watanabe [W1]). Let (V, v) be a germ of a normal isolated singularity of an analytic space of dimension $n \geq 2$.

(i) Let $f: Y \to V$ be a partial resolution where Y has at worst canonical singularities. For every $m \in \mathbf{N}$, we define an m-genus $\gamma_m(V, v)$ as follows:

$$\gamma_m(V,v) = dim_{\mathbf{C}}\ \omega_V^{[m]} \Big/ f_*\omega_Y^{[m]}.$$

(ii) Let $f: \tilde{V} \to V$ be a good resolution which means a resolution with the fiber $E = f^{-1}(x)_{red}$ a divisor of normal crossings. For every $m \in \mathbf{N}$, we define another kind of m-genus $\delta_m(V, v)$ as follows:

$$\delta_m(V,v) = dim_{\mathbf{C}}\ \omega_V^m / f_*\omega_{\tilde{V}}^m((m-1)E).$$

(iii) If an analytic space W has the singular locus S of dimension 0, we denote $\sum_{w \in S} \gamma_m(W, w)$ and $\sum_{w \in S} \delta_m(W, w)$ by $\gamma_m(W)$ and $\delta_m(W)$ respectively.

Remark 1.7. (i) The above definitions are independent of the choice of $f's$.

(ii) The asymptotic behaviors of $\gamma_m(V, v)$ and $\delta_m(V, v)$ are studied in [I3]. If $\gamma_m(V, v) \neq 0$ for some m, then $\gamma_m(V, v)$ grows in order n as a function in m. On the other hand, if $\delta_m(V, v) \neq 0$ for some m, then $\delta_m(V, v)$ grows in order either 0,1,..,$n-2$, or n (it skips the value $n-1$).

Definition 1.8. Let $\pi: X \to D$ be a flat family of normal isolated singularities and ϕ_m be the map $\omega_X^{[m]} \to \omega_{X_0}^{[m]}$, induced from the residue map $Res_{X_0} \frac{\cdot}{t}: \omega_X \to \omega_{X_0}$, where t is a local parameter at 0 on D. For every $m \in \mathbf{N}$, we define the m-th difference $\iota_m(X; X_0)$ as follows:

$$\iota_m(X;X_0) := dim_{\mathbf{C}}\ \omega_{X_0}^{[m]} \Big/ \phi_m\omega_X^{[m]} = dim_{\mathbf{C}}\ \omega_{X_0}^{[m]} \Big/ \omega_X^{[m]} \otimes \mathbf{C}(0).$$

Theorem 1. *Let $\pi: X \to D$ be an (FG)-deformation of normal isolated singularity (X_0, x) over the unit disk D. Then:*

(i) for every m, there exists a closed analytic set $0 \in S_m \subset D$ such that

$$\gamma_m(X_\tau) \leq \gamma_m(X_0, x) - \iota_m(X; X_0)$$

for all $\tau \in D_m = D - S_m$, which implies that the map $D \to \mathbf{N} \cup \{0\}$ $(\tau \mapsto \gamma_m(X_\tau))$ is upper semi-continuous;

(ii) the following are equivalent:

(iia) π admits the simultaneous canonical model on a neighbourhood of $0 \in D$;

(iib) the equality in (i) holds for every $m \in \mathbf{N}$;

(iic) there exists a positive integer r such that the equality in (i) holds for every mr $m \in \mathbf{N}$;

(iii) if there exists a positive integer r such that $\gamma_m(X_\tau)$ $(\tau \in D)$ is constant for every m divisible by r, then π admits the simultaneous canonical model;

(iv) if $\gamma_m(X_\tau)$ is represented as $e_\tau m^n + O(m^{n-1})$ and e_τ $(\tau \in D)$ is constant, then π admits the simultaneous canonical model.

Lemma 1.9. *Let $\pi: X \to D$ be a flat deformation of normal isolated singularity (X_0, x) over the unit disk D. Let $F: Y \to X$ and $f: W \to X_0$ be resolutions of X and $X_0 = \pi^{-1}(0)$, respectively. Let $\phi_m: \omega_X^{[m]} \to \omega_{X_0}^{[m]}$ be as in (1.8). Assume $f_* \omega_W^{[m]} \subset \phi_m(F_* \omega_Y^{[m]})$ for every $m \in \mathbf{N}$. Then the assertion (i) in Theorem 1 holds .*

Proof of Lemma 1.9. Denote the sheaf $\omega_X^{[m]} \big/ F_* \omega_Y^{[m]}$ by $\mathcal{F}_m$. Then $\pi_* \mathcal{F}_m$ is a coherent sheaf on D, since $\mathcal{F}_m$ is $\mathcal{O}_X / Ann(\mathcal{F}_m)$-coherent and the subspace defined by the ideal $Ann(\mathcal{F}_m)$ is a finite covering over D. Therefore, by Nakayama's Lemma, there exists a closed analytic set $S'_m \subset D$ such that $dim\ \pi_* \mathcal{F}_m \otimes_{\mathcal{O}_D} \mathbf{C}(\tau) \leq dim\ \pi_* \mathcal{F}_m \otimes_{\mathcal{O}_D} \mathbf{C}(0)$ for evey $\tau \in D - S'_m$. So, for the assertion (i), it is sufficient to prove that:

(1.9.1) $$dim\ \pi_* \mathcal{F}_m \otimes_{\mathcal{O}_D} \mathbf{C}(0) \leq \gamma_m(X_0, x) - \iota_m(X; X_0)$$

and there exists a closed analytic set $S''_m \subset D$ such that

$$(1.9.2) \qquad dim\ \pi_*\mathcal{F}_m \otimes_{\mathcal{O}_D} \mathbf{C}(\tau) = \gamma_m(X_\tau),\ for\ every\ \tau \in D - S''_m.$$

To prove (1.9.1), we remark the relations:

$$dim\ \pi_*\mathcal{F}_m \otimes_{\mathcal{O}_D} \mathbf{C}(0) = dim_{\mathbf{C}}\ \omega_X^{[m]} \otimes \mathbf{C}(0) \Big/ \phi_m F_* \omega_Y^{[m]}$$

$$= dim_{\mathbf{C}}\ \omega_{X_0}^{[m]} \Big/ \phi_m F_* \omega_Y^{[m]} - dim_{\mathbf{C}}\ \omega_{X_0}^{[m]} \Big/ \omega_X^{[m]} \otimes \mathbf{C}(0).$$

By the assumption $f_*\omega_W^{[m]} \subset \phi_m(F_*\omega_Y^{[m]})$ of the lemma we have

$$dim_{\mathbf{C}}\ \omega_{X_0}^{[m]} \Big/ \phi_m F_* \omega_Y^{[m]} \leq dim_{\mathbf{C}}\ \omega_{X_0}^{[m]} \Big/ f_*\omega_W^{[m]} = \gamma_m(X_0, x),$$

which completes the proof of (1.9.1). To show (1.9.2), it is sufficient to prove that there exists $S''_m \subset D$ such that, for $\tau \in D - S''_m$,

$$(1.9.3) \qquad \omega_X^{[m]} \otimes_{\mathcal{O}_D} \mathbf{C}(\tau) \simeq \omega_{X_\tau}^{[m]}$$

$$(1.9.4) \qquad F_*\omega_Y^{[m]} \otimes \mathbf{C}(\tau) \simeq f_{\tau *}\omega_{\tilde{X}_\tau}^{[m]}$$

where $f_\tau: \tilde{X}_\tau \to X_\tau$ is a resolution. Consider the exact sequence on $X - \Sigma$:

$$0 \to \omega_{X-\Sigma}^m \xrightarrow{\times t_\tau} \omega_{X-\Sigma}^m \xrightarrow{\varphi_\tau} \omega_{X_\tau - \Sigma}^m \to 0,$$

where t_τ is a local parameter at $\tau \in D$ and φ_τ is the map induced from the residue map $Res_{X_\tau}\frac{\cdot}{t_\tau}$: $\omega_{X-\Sigma} \to \omega_{X_\tau - \Sigma}$.

Denote the canonical inclusion $X - \Sigma \hookrightarrow X$ by i. By taking i_* of the previous exact sequence, we have:

$$0 \to \omega_X^{[m]} \xrightarrow{\times t_\tau} \omega_X^{[m]} \xrightarrow{\varphi_\tau} \omega_{X_\tau}^{[m]} \to R^1 i_* \omega_{X-\Sigma}^m \xrightarrow{\times t_\tau} R^1 i_* \omega_{X-\Sigma}^m.$$

If $R^1 i_* \omega_{X-\Sigma}^m$ is $\mathcal{O}_D$-torsion free on $D - S''_m$ for a suitable closed analytic subset S''_m, then (1.9.3) follows. In fact, the torsion-freeness implies that the last arrow is injective for $\tau \in D - S''_m$, which yields the surjectivity of φ_τ and therefore the assertion (1.9.3). Now we are going to prove that $R^1 i_* \omega_{X-\Sigma}^m$ is $\mathcal{O}_D$-torsion free away from a closed analytic subset of D. Let $F: Y \to X$ be a resolution of X obtained by blowing up of an ideal $\mathcal{I}$ such that $Supp\, \mathcal{O}_X/\mathcal{I} = \Sigma$. Let E be the divisor on Y such that $\mathcal{I}\mathcal{O}_Y = \mathcal{O}_Y(-E)$. Then $R^1 i_*\omega^m_{X-\Sigma} = \varinjlim_r R^1 F_*\omega^m_Y(rE)$. By the relative ampleness of $\mathcal{O}_Y(-E)$ with respect to F, we can prove that there exists $r_0 > 0$ such that $R^1 F_*(\omega^m_Y(rE) \otimes \mathcal{O}_{(r-r_0+1)E})$ is torsion free (is zero, if $n \geq 3$) for every $r > r_0$. Therefore every torsion element of $R^1 F_*\omega^m_Y(rE)$ is in the image of $R^1 F_*\omega^m_Y((r_0-1)E)$ for $r > r_0$. Thus every torsion element of $R^1 i_*\omega^m_{X-\Sigma}$ is in the image of $R^1 F_*\omega^m_Y((r_0-1)E)$ which is $\mathcal{O}_D$-coherent. Since a coherent sheaf is torsion free outside a closed analytic subset, we have the desired statement. To prove (1.9.4), we have to consider the exact sequences:

$$0 \to \omega^m_Y \xrightarrow{\times t_\tau} \omega^m_Y \xrightarrow{\varphi_\tau} \omega^m_{Y_\tau} \to 0$$

for $\tau \in D$. By taking F_* we obtain the following exact sequences:

$$0 \to F_*\omega^m_Y \xrightarrow{\times t_\tau} F_*\omega^m_Y \xrightarrow{\varphi_\tau} F_*\omega^m_{Y_\tau} \to R^1 F_*\omega^m_Y \xrightarrow{\times t_\tau} R^1 F_*\omega^m_Y.$$

Since $R^1 F_*\omega^m_Y$ is $\mathcal{O}_D$-coherent, it is torsion free on $D - S$ for a closed analytic subset S of D, which implies that the last arrow is injective for $\tau \in D - S$. Therefore, $F_*\omega^m_Y \otimes_{\mathcal{O}_D} \mathbf{C}(\tau) \simeq F_*\omega^m_{Y_\tau}$ for $\tau \in D - S$. Here we may assume that the restriction morpohism $F_\tau : Y_\tau \to X_\tau$ is a resolution for $\tau \in D - S$ by replacing S with a suitably big one. Remarking that $f_{\tau *}\omega^{[m]}_{\tilde{X}_\tau}$ is independent of the choice of a resolution $f_\tau : \tilde{X}_\tau \to X_\tau$, we have the isomorphism in (1.9.4). This completes the proof of Lemma 1.9.

Proof of Theorem 1. Since $\pi : X \to D$ is an (FG)-deformation, X has the canonical model $F : Y \to X$. Let $f : W \to X_0$ be a resolution of X_0 which factors through the normalization $\tilde{Y}_{00}$ of the proper transform Y_{00} of X_0 in Y. For the assertion (i), it is sufficient to prove that $f_*\omega^{[m]}_W \subset \phi_m(F_*\omega^{[m]}_Y)$ for every $m \in \mathbf{N}$ by Lemma 1.9. First claim that $F_*\omega^{[m]}_Y \otimes_{\mathcal{O}_D} \mathbf{C}(0) = F_*(\omega^{[m]}_Y \otimes_{\mathcal{O}_Y} \mathcal{O}_{Y_0})$, where Y_0 is the total inverse image $F^{-1}(X_0)$ of X_0. from the exact sequence:

$$0 \to \omega^{[m]}_Y \xrightarrow{\times t} \omega^{[m]}_Y \to \omega^{[m]}_Y \otimes \mathcal{O}_{Y_0} \to 0,$$

we have the following exact sequence:

$$0 \to F_*\omega^{[m]}_Y \xrightarrow{\times t} F_*\omega^{[m]}_Y \to F_*(\omega^{[m]}_Y \otimes \mathcal{O}_{Y_0}) \to R^1 F_*\omega^{[m]}_Y,$$

where t is a local parameter at $0 \in D$. Here $R^1 F_*\omega_Y^{[m]} = 0$ by [KMM, Theorem1.2.5]. Thus we have the equality as claimed.

Consider the following diagram:

$$\begin{array}{ccccc}
\omega_{X_0}^{[m]} & \xleftarrow{\;a\;} & f_*\omega_W^{[m]} & \xrightarrow{\;b\;} & F_*\sigma_*\omega_{\tilde{Y}_{00}} \\
\uparrow c & & d \,\downarrow & & g \,\downarrow \\
\phi_m\omega_X^{[m]} & \xleftarrow[e]{} & \phi_m F_*\omega_Y^{[m]} & \longleftarrow & (F_*\omega_Y^{[m]}) \otimes \mathbf{C}(0),
\end{array}$$

where a, c and e are inclusions. In order to show the existence of the inclusion d, it is sufficient to prove that there exists a homomorphism g which makes the diagram commutative. For the finite morphism $\sigma\colon \tilde{Y}_{00} \to Y_0$, we have the natural injective map $\sigma_*\omega_{\tilde{Y}_{00}}^{[m]} \to \omega_Y^{[m]} \otimes \mathcal{O}_{Y_0}$ by [N, Lemma 1]. Taking F_*, we get the homomorphism $g\colon F_*\sigma_*\omega_{\tilde{Y}_{00}}^{[m]} \to F_*(\omega_Y^{[m]} \otimes \mathcal{O}_{Y_0}) = (F_*\omega_Y^{[m]}) \otimes \mathbf{C}(0)$ as desired.

Now we are going to prove (ii). Let $F : Y \to X$ and $f : \tilde{X}_0 \to X_0$ be resolutions of X and X_0 respectively. First we prove that the equality $\gamma_m(X_\tau) = \gamma_m(X_0, x) - \iota_m(X; X_0)$ in (i) is equivalent to the equality

$$F_*\omega_Y^{[m]} \otimes \mathbf{C}(0) = f_*\omega_{\tilde{X}_0}^{[m]}. \tag{1.1}$$

Denote the sheaf $\omega_X^{[m]} \big/ F_*\omega_Y^{[m]}$ by $\mathcal{F}_m$ as in Lemma 1.9. Since in the proof of Lemma 1.9 we have inequalities:

$$dim\pi_*\mathcal{F}_m \otimes \mathbf{C}(\tau) \leq dim\pi_*\mathcal{F}_m \otimes \mathbf{C}(0) \leq dim\ \omega_{X_0}^{[m]} \big/ f_*\omega_{\tilde{X}_0}^{[m]} - dim\ \omega_{X_0}^{[m]} \big/ \phi_m\omega_X^{[m]}, \tag{1.2}$$

the equality in (i) to hold is equivalent to the two equalities in (1.2) to hold. The first one means that $\pi_*\mathcal{F}_m$ is a locally free $\mathcal{O}_D$-coherent module and we obtain

$$F_*\omega_Y^{[m]} \otimes_{\mathcal{O}_D} \mathbf{C}(0) = \phi_m F_*\omega_Y^{[m]}, \tag{1.3}$$

by the definition of $\mathcal{F}_m$. The second one means:

$$\phi_m F_*\omega_Y^{[m]} = f_*\omega_{\tilde{X}_0}^{[m]}. \tag{1.4}$$

Combining (1.3) and (1.4), we get the desired equality. In order to show the equivalences of (ii), it is sufficient to prove (iic)→ (iia), and (iia)→ (iib). Assume (iic), then by the above argument we have $F_*\omega_Y^{[mr]} \otimes \mathbf{C}(0) = f_*\omega_{\tilde{X}_0}^{[mr]}$ for every $m \in \mathbf{N}$. Therefore $Proj \oplus_m F_*\omega_Y^{[mr]} \otimes \mathbf{C}(0) = Proj \oplus_m f_*\omega_{\tilde{X}_0}^{[mr]}$, which means that the inverse image of X_0 in the canonical model $F' : Y' = Proj \oplus_m F_*\omega_Y^{[mr]} \to X$ of X is the canonical model of X_0. On the other hand, the restriction $F'_\tau : Y'_\tau \to X_\tau$ is the canonical model of X_τ for general $\tau \in D - \{0\}$. In fact, the F'_τ-ampleness of $K_{Y'_\tau}$ follows from F'-ampleness of K_Y and the canonicalness of the singularities of Y'_τ is proved as follows: we can shrink D so that $Y'^* \to D^* = D - \{0\}$ admits a simultaneous resolution $h : \tilde{Y} \to Y'^*$; since Y'^* has at worst canonical singularities, we have $K_{\tilde{Y}} = h^*K_{Y'^*} + \sum a_i E_i$ with $a_i \geq 0$ for every exceptional component E_i; as E_i's are all horizontal, we obtain $K_{\tilde{Y}_\tau} = h^*K_{Y'^*_\tau} + \sum a_i E_{i\tau}$ for $\tau \in D^*$, which implies that the singularities of Y'^*_τ are canonical. So F' turns out to be the simultaneous canonical model on a neighbourhood of $0 \in D$. Next assume (iia) and let $F : Y \to X$ be the simultaneous canonical model. then it is the canonical model of X (cf. Remark 1.5 (iv)). By the vanishing property $R^1 F_*\omega_Y^{[m]} = 0$ of the canonical model ([KMM, Theorem 1.2.5]), the equality $F_*\omega_Y^{[m]} \otimes \mathbf{C}(0) = F_*(\omega_Y^{[m]} \otimes \mathbf{C}(0))$ holds. Here the right hand side coincides with $F_*\omega_{Y_0}^{[m]}$. In fact, as $\omega_Y^{[m]}$ is a Cohen-Macaulay $\mathcal{O}_Y$-module (cf.[T,4.6.1]), we have $\omega_{Y_0}^{[m]} = \omega_Y^{[m]} \otimes_{\mathcal{O}_D} \mathbf{C}(0)$ for every $m \in \mathbf{N}$. Since $Y_0 \to X_0$ is the canonical model, we obtain the equality (1.1) for every m. Thus we get (iib). This completes the proof of (ii).

The proof of (iii) is now obvious by (ii). Next we are going to prove (iv). Let $F: Y \to X$ be a resolution of X such that the restriction $F_0: Y_{00} \to X_0$ of F to the proper transform Y_{00} of X_0 is a resolution of X_0 and F_0 factors through the blowing-up by the maximal ideal of x in order to apply an argument of [T-W]. Decompose the exceptional divisor E on Y_{00} into irreducible components $E_1, E_2, .., E_s$. Assume π does not admit the simultaneous canonical model. Then by (ii), there is a number $r > 0$ with $\gamma_r(X_\tau) < \gamma_r(X_0, x) - \iota_r(X; X_0)$. By the first argument in the proof of (ii), at least one of the equalities (1.3), (1.4) fails for $m = r$. That is to say, $\pi_*\mathcal{F}_r$ is not locally free or the injection $F_{0*}\omega_{Y_{00}}^{[r]} \hookrightarrow \phi_m F_*\omega_Y^{[r]}$ is not bijective. If $F_{0*}\omega_{Y_{00}}^{[r]} \hookrightarrow \phi_m F_*\omega_Y^{[r]}$ is not bijective, we can take an element $\theta \in \phi_m F_*\omega_Y^{[r]} \subset \omega_{X_0}^{[r]}$ such that $\theta \notin F_{0*}\omega_{Y_{00}}^{[r]}$. Define an $\mathcal{O}_X$-homomorphism $\psi_m: \mathcal{O}_{X_0} \to \phi_m F_*\omega_Y^{[mr]}$ by $g \mapsto g \cdot \theta^m$ for every $m \in \mathbf{N}$. Denote the inverse image $\psi_m^{-1}(F_{0*}\omega_{Y_{00}}^{[mr]})$ by $\mathcal{I}^{(m)}$. Then $\mathcal{I}^{(m)} = F_{0*}\mathcal{O}_{Y_{00}}(\sum_i min\{m\nu_{E_i}(\theta), 0\}E_i)$, where ν_{E_i} is the valuation at E_i. Here by the definition of θ there exists a component E_i such that $\nu_{E_i}(\theta) < 0$. By Tomari and Watanabe's Lemma ([T-W] or [I3, Lemma 1.5]), $dim\ \mathcal{O}_{X_0}/\mathcal{I}^{(m)} \geq \alpha m^n$ for a suitable $\alpha > 0$. Since ψ_m introduce the inclusion $\mathcal{O}_{X_0}/\mathcal{I}^{(m)} \subset \phi_m F_* \omega_Y^{[mr]} \big/ F_{0*}\omega_{Y_{00}}^{[mr]}$, we have the inequality $dim\ \phi_m F_* \omega_Y^{[mr]} \big/ F_{0*}\omega_{Y_{00}}^{[mr]} \geq \alpha m^n$. This proves the second inequality of the following:

$$\gamma_{mr}(X_\tau) \leq dim \pi_* \mathcal{F}_{mr} \otimes \mathbf{C}(0) \leq \gamma_{mr}(X_0, x) - \iota_{mr}(X; X_0) - \alpha m^n.$$

Therefore we obtain that $e_\tau \leq e_0 - \alpha/r^n$, which is a contradiction to the constantness of e_τ. Now we get that $F_{0*}\omega_{Y_{00}}^{[m]} = \phi_m F_* \omega_Y^{[m]}$, for every $m \in \mathbf{N}$. Next suppose $\pi_* \mathcal{F}_r$ is not locally free in a neighbourhood of $0 \in D$, then there exists a torsion element η in $\pi_* \mathcal{F}_r$ with the image $\eta_0 \neq 0$ in $\pi_* \mathcal{F} \otimes_{\mathcal{O}_D} \mathbf{C}(0) = \pi_*\ \omega_X^{[r]} \otimes \mathbf{C}(0) \big/ \phi_r F_* \omega_Y^{[r]}$. Denote an element of $\omega_X^{[r]} \otimes \mathbf{C}(0)$ corresponding to η_0 by η_0 again. Then $\eta_0 \notin \phi_r F_* \omega_Y^{[r]} = F_{0*}\omega_{Y_{00}}^{[r]}$. Define $\lambda_m : \mathcal{O}_{X_0} \to \omega_X^{[mr]} \otimes \mathbf{C}(0)$ by $g \mapsto g \cdot \eta_0^m$. Let $\mathcal{J}^{(m)}$ be the inverse image $\lambda_m^{-1}(F_{0*}\omega_{Y_{00}}^{[mr]})$, then $dim\ \mathcal{O}_{X_0}/\mathcal{J}^{(m)} \geq \beta m^n$ for a suitable $\beta > 0$, by the same argument as above. Every element of the image of $\mathcal{O}_{X_0}/\mathcal{J}^{(m)} \hookrightarrow \pi_*(\omega_X^{[mr]} \otimes \mathbf{C}(0) \big/ F_{0*}\omega_{Y_{00}}^{[mr]}) = \pi_* \mathcal{F}_{mr} \otimes \mathbf{C}(0)$ gives a torsion element of $\pi_* \mathcal{F}_{mr}$. Therefore

$$\begin{aligned}\gamma_{mr}(X_\tau) &\leq dim \pi_* \mathcal{F}_{mr} \otimes \mathbf{C}(0) - \beta m^n \\ &\leq \gamma_{mr}(X_0, x) - \iota_{mr}(X; X_0) - \beta m^n.\end{aligned}$$

This leads us to a contradiction $e_\tau \leq e_0 - \beta/r^n$ too. [Q.E.D. of Theorem 1]

Corollary 1.10. *Let $\pi : X \to D$ be a deformation of surface singularities over the unit disk D with constant K_τ^2 $(\tau \in D)$, where K_τ is the $\mathbf{Q}$-divisor on the minimal resolution $\tilde{X}_\tau$ of X_τ which has the support on the exceptional set and is numerical equivalent to $K_{\tilde{X}_\tau}$. Then π admits the simultaneous canonical model.*

Proof. By [Mo],

$$\gamma_m(X_\tau) = -\frac{1}{2}m(m-1)K_\tau^2 + \varepsilon(m),$$

where $\varepsilon(m)$ is bounded for $m \in \mathbf{N}$. Since $\pi : X \to D$ is an (FG)-deformation by [M], we can apply Theorem 1, (iv).

Corollary 1.11. *Let $\pi: X \to D$ be an (FG)-deformation of normal isolated Gorenstein singularities of dimension $n \geq 2$ over the unit disk D. Assume each fiber X_τ $(\tau \in D)$ has the canonical model $\tilde{X}_\tau \to X_\tau$ and denote the canonical divisor on $\tilde{X}_\tau$ by K_τ. Then the following are equivalent:*

(i) π admits the simultaneous canonical model,

(ii) K_τ^n $(\tau \in D)$ is constant, and

(iii) for each $m \in \mathbf{N}$, $\gamma_m(X_\tau)$ $(\tau \in D)$ is constant.

Proof. First of all, one sees that $\iota_m(X; X_0) = 0$, since X is also a Gorenstein space. Therefore the equivalence between (i) and (iii) comes from the equivalence between (iia) and (iib) in Theorem 1. Since $\gamma_m(X_\tau)$ is represented as

$$-\frac{1}{n!}K_\tau^n m^n + O(m^{n-1}),$$

the implication (iii)→ (ii) is obvious, while (ii) → (i) follows from Theorem 1, (iv).

Remark 1.12. Without Gorenstein condition, (i) does not necessarily imply (ii) in Corollary 1.11. In [P, § 8], we can see an example of deformation $\pi: X \to D$ with $K_0^2 = -1$ and $K_\tau^2 = 0$ for $\tau \in D - \{0\}$, which admits the simultaneous canonical model.

§2. (FG)-deformations of purely elliptic singularities.

A normal isolated singularity (V, v) is called a purely elliptic singularity if $\delta_m(V, v) = 1$ for every $m \in \mathbf{N}$ (cf. Definitioin 1.6, (ii)). In this section we study an (FG)-deformation of Gorenstein purely elliptic singularities. First we summarize basic notion developed in [I1] on Gorenstein purely elliptic singularities. Let $f : \tilde{V} \to V$ be a good resolution of an isolated singularity (V, v) and $E = \sum_{i=1}^{r} E_i$ be the irreducible decomposition of the reduced exceptional divisor.

Proposition 2.1([I1]). *A Gorenstein isolated singularity (V, v) is purely elliptic if and only if $K_{\tilde{V}} = f^*K_V + \sum_{i=1}^{r} m_i E_i$, $m_i \geq -1$ for every i and $m_i = -1$ for some i.*

We call the reduced effective divisor E_J consisting of the components E_i with $m_i = -1$ the essential divisor on $\tilde{V}$. Since $Gr_F^0 H^{n-1}(E_J) \simeq H^{n-1}(E_J, \mathcal{O}_{E_J}) = H^{n-1}(E, \mathcal{O}_E) \simeq \mathbf{C}$, where F is the Hodge filtration on $H^{n-1}(E_J, \mathbf{C})$, there is only one i $(0 \leq i \leq n-1)$ such that the $(0, i)$-Hodge component of $Gr_F^0 H^{n-1}(E_J)$ is not zero. This number i is independent of the choice of a good resolution f. So, we say that the singularity is of type (0, i).

For example, a 2-dimensional purely elliptic singularity of type (0, 1) (resp. type (0,0)) is a simple elliptic singularity (resp. a cusp singularity). In general, n-dimensional Gorenstein purely elliptic singularity of type (0, n-1) admits only one essential component on any good resolution. For $n = 2$, the essential component of a simple elliptic singularity is an elliptic curve. For $n = 3$, the essential component of a Gorenstein purely elliptic singularity of type (0,2) is birational to a K3-surface. A singularity of this type is called a simple K3-singularity and studied in [I-W],[W2], [T] and [Y]. It is characterized by the fact that its $\mathbf{Q}$-factorial terminal modification has a normal K3-surface as the exceptional divisor([I-W]). Yonemura classifies in [Y] quasi-homogeneous hypersurface simple K3-singularities into 95-classes by the weights independently by Fletcher [F]. His list is bijective to that of weighted K3-surfaces announced by Reid [R] and also bijective to a list of weighted $\mathbf{Q}$-Fano 3-folds made by Fletcher [F].

Proposition 2.2. *Let $\pi : X \to D$ be a flat family of normal isolated singularities of dimension $n \geq 2$ on the unit disk D. Assume (X_0, x) is as Gorenstein purely elliptic singularity. Then, by shrinking D and X sufficiently,* *(i)* X_τ $(\tau \in D)$ *have Gorenstein singularities, more precisely, the singularities on* X_τ *are either all rational or one purely elliptic and others rational and*

(ii) if π *has a relative projective compactification* $\bar{\pi} : \bar{X} \to D$ *(i.e.* $\bar{X}_\tau$ *is a projective variety and contains* X_τ *as an open subset) such that every fiber* $\bar{X}_\tau$ *contains only one singular point* x_τ *which is purely elliptic, then the type (0, k) of* (X_0, x) *and the type (0, j) of general* (X_τ, x_τ) *have the relation* $j \geq k$.

Proof. The statement (i) is proved in [I2, Proposition 4.4]. We will prove (ii). In the following, we denote a projective morphism which is a compactification π by the same symbols $\pi : X \to D$. Let $F : \tilde{X} \to X$ be a good resolution of singularities and E be the reduced exceptional divisor on $\tilde{X}$. By shrinking D sufficiently, we may assume that every component of $E|_{D^*}$ is horizontal (i.e. dominating D^*), and $\{H^{n-1}(E_\tau)\}_{\tau \in D^*}$ admits the variation of mixed Hodge structures on $D^* = D - \{0\}$. Therefore there is an integer j such that the purely elliptic singularity of X_τ is of type (0, j) for every $\tau \in D^*$, by the definition of the type of a singularity. If the singularity on X_τ $(\tau \in D^*)$ is of type (0, n-1), the assertion obviously holds. Now we assume that the singularity on X_τ $(\tau \in D^*)$ is of type (0, j) with $0 \leq j \leq n-2$. By the exact sequence:

$$H^{n-1}(\tilde{X}_\tau, \mathbf{C}) \to H^{n-1}(E_\tau, \mathbf{C}) \to H^n(X_\tau, \mathbf{C}) \to H^n(\tilde{X}_\tau, \mathbf{C}),$$

for $\tau \in D^*$, we have

$$\mathbf{C} = Gr_F^0 Gr_j^W H^{n-1}(E_\tau) \simeq Gr_F^0 Gr_j^W H^n(X_\tau), \tag{2.2.1}$$

where W and F are the weight filtration and the Hodge filtration, respectively. In fact, $H^i(\tilde{X}_\tau)$ admits the Hodge structure of pure weight i, and therefore the both ends of the exact sequence do not make any contribution to the (0, j)-components of the two middle terms. Let $H^i(X_\infty)$ be the cohomology group with the limit Hodge structure defined by $\pi|_{D^*} : X|_{D^*} \to D^*$ (cf. [E]). Denote the weight filtration and the Hodge filtration on $H^i(X_\infty)$ by W and F, respectively. Let $\hat{W}$ denote the filtration on $H^i(X_\infty)$ induced from the weight filtration of $H^i(X_\tau)$ $\tau \in D^*$. Then $dim Gr_F^p Gr_j^{\hat{W}} H^n(X_\infty) = dim Gr_F^p Gr_j^W H^n(X_\tau)$ for $\tau \in D^*$ [E,II, Proposition 2.1, (3)], where the right hand side is known to be zero for $p > j$ as well as $p < 0$. Hence $F^p \cap \hat{W}_j = 0$ for $p > j$ which yields $\bar{F}^p \cap \hat{W}_j = 0$ for $p > j$. We obtain

$$Gr_F^p Gr_{p+q}^W Gr_j^{\hat{W}} H^n(X_\infty) = 0 \text{ for } q > j.$$

In particular,

$$Gr_F^0 Gr_q^W Gr_j^{\hat{W}} H^n(X_\infty) = 0 \text{ for } q > j,$$

which implies that

$$dim Gr_F^0 Gr_j^{\hat{W}} H^n(X_\infty) = \sum_{k \leq j} dim Gr_F^0 Gr_k^W Gr_j^{\hat{W}} H^n(X_\infty).$$

Here the left hand side coincides with $dim Gr_F^0 Gr_j^W H^n(X_\tau)$, which turns out to be 1 in virtue of (2.2.1). Hence $Gr_F^0 Gr_k^W Gr_j^{\hat{W}} H^n(X_\infty) = \mathbf{C}$ for a suitable $k \leq j$. Since $\hat{W}$ is a filtration of $H^n(X_\infty)$ in the category of the mixed Hodge structure [E,II, Proposition 2.1, (2)], we have

(2.2.2) $$Gr_F^0 Gr_k^W H^n(X_\infty) \neq 0.$$

On the other hand, we can prove the isomorphism $Gr_F^0 H^n(X_0) \simeq Gr_F^0 H^n(X_\infty)$ of the Hodge structure by the same method as in the proof of [St, Theorem 2]. While Steenbrink states the theorem under the condition that X_τ are non-singular for all $\tau \in D^*$ and X_0 has only Du Bois singularities, his proof also works for our case that the singularities on X_τ $(\tau \in D)$ are all Du Bois singularities. Thus we have $Gr_F^0 Gr_k^W H^n(X_0) \neq 0$ by (2.2.2). We are going to show that it implies that the singularity on X_0 is of type (0, k). Let $g: \tilde{X}_0 \to X_0$ be a good resolution of the singularity on X_0 and C be the reduced exceptional divisor on $\tilde{X}_0$. From the exact sequence:

$$H^{n-1}(\tilde{X}_0, \mathbf{C}) \to H^{n-1}(C, \mathbf{C}) \to H^n(X_0, \mathbf{C}) \to H^n(\tilde{X}_0, \mathbf{C}),$$

we have an isomorphism

(2.2.3) $$Gr_F^0 Gr_k^W H^{n-1}(C) \simeq Gr_F^0 Gr_k^W H^n(X_0) \neq 0$$

which can be proved in the same way as we did for (2.2.1). The non-vanishing of the left hand side of (2.2.3) means that $H^{n-1}(C)$ has the (0, k)-Hodge component, which implies that the singularity of X_0 is of type (0, k) for $k \leq j$.

Remark 2.3. For a flat family $\pi: X \to D$ of normal isolated singularities on the unit disk, it is not yet known whether there exists a relative projective compactification $\bar{\pi}: \bar{X} \to D$ of π as in Proposition 2.2 (ii). However, if a family $\pi : X \to D$ satisfies certain conditions, it has a relative projective compactification. For example, if $\pi : X \to D$ is a family of hypersurface singularities, then it has a relative projective compactifications. In fact, let X be defined by a polynomial equation $f(x_0, x_1, .., x_n, t) = 0$ in $\mathbf{C}^{n+2}$, π be the projection $(x_0, .., x_n, t) \mapsto t$ and the singular locus of π be $\{(0, .., 0)\} \times \mathbf{C}^1$. Then we can take a suitable polynomial g in $x_0, .., x_n$ of higher degree such that $X' = \{f + g = 0\}$ gives the same family as X at the origins and the colsure of X' in $\mathbf{P}^{n+1} \times \mathbf{C}^1$ gives the relative compactification of π with the singular locus $\{(0, .., 0)\} \times \mathbf{C}^1$.

Proposition 2.4. *Let $\pi: X \to D$ be a flat family of Gorenstein purely elliptic singularities of dimension $n \geq 2$. Let $F: \tilde{X} \to X$ be a good resolution of X such that the restriction $F|_{[X_0]}$: $[X_0] \to X_0$ of F to the proper transform $[X_0]$ of X_0 in $\tilde{X}$ is a good resolution of X_0. Denote by ω the nowhere vanishing holomorphic (n+1)-form on the outside the singular locus of X. Then the pole divisor of ω on $\tilde{X}$ are all horizontal (i,e, they are all dominating D), and the order of poles are all one.*

Proof. First of all, for a general $\tau \in D$, $F_\tau : \tilde{X}_\tau \to X_\tau$ is a good resolution of a purely elliptic singularity. So, by Proposition 2.1, we get that a horizontal pole divisor of ω has the pole order one. Decompose the reduced exceptional divisor E into two parts E_h and E_v which consist of horizontal components and vertical components, respectively. By shrinking D and X, we may assume that $F(E_v) = x \in X_0$. Decompose the total fiber $(\pi \circ F)^{-1}(0)$ into $[X_0]$ and the other part E'_v . Recall the proof of the upper semi-continuity of δ_m in [I2]. The following relations are clear:

(2.4.1)
$$\delta_m(X_\tau) = dim(\omega_X^{[m]} \Big/ F_* i_* \omega^m_{\tilde{X}-E_v}((m-1)E_h)) \otimes \mathbf{C}(\tau)$$

$$\leq dim(\omega_X^{[m]} \Big/ F_* i_* \omega^m_{\tilde{X}-E_v}((m-1)E_h)) \otimes \mathbf{C}(0) \quad \text{(by Nakayama's Lemma)}$$

$$\leq dim(\omega_X^{[m]} \Big/ F_* \omega^m_{\tilde{X}}((m-1)E - mE'_v)) \otimes \mathbf{C}(0),$$

for general $\tau \in D^* = D - \{0\}$. The essential point of the proof in [I2,Theorem 2.1] is to show the existence of the inclusion:

$$\text{(2.4.2)} \qquad \omega_X^{[m]} \Big/ F_*\omega_{\tilde{X}}^m((m-1)E - mE_v') \otimes \mathbf{C}(0) \hookrightarrow \omega_{X_0}^{[m]} \Big/ F_*\omega_{[X_0]}^m((m-1)E|_{[X_0]}),$$

where the dimension of the right hand side gives $\delta_m(X_0)$. Now by the assumption of the proposition, $\delta_m(X_\tau)$ $(\tau \in D)$ is constantly 1 for every $m \in \mathbf{N}$. So the equalities should hold at every stage in (2.4.1) and (2.4.2). Therefore from the equality between the second and third terms in (2.4.1), we have:

$$\frac{F_*i_*\omega_{\tilde{X}-E_v}^m((m-1)E_h)}{F_*i_*\omega_{\tilde{X}-E_v}^m((m-1)E_h) \cap t\omega_X^{[m]}} = \frac{F_*\omega_{\tilde{X}}^m((m-1)E - mE_v')}{F_*\omega_{\tilde{X}}^m((m-1)E - mE_v') \cap t\omega_X^{[m]}},$$

which implies

$$F_*i_*\omega_{\tilde{X}-E_v}^m((m-1)E_h) = F_*i_*\omega_{\tilde{X}-E_v}^m((m-1)E_h) \cap t\omega_X^{[m]} + F_*\omega_{\tilde{X}}^m((m-1)E - mE_v'),$$

where t is the local parameter on D at 0. Take a holomorphic function g on X which vanishes on the singular locus of π and attains the minimal values $\nu_{E_{(i)}}(g)$ $(E_{(i)} < E_v)$ among such functions. Since g vanishes at E_h and the order of the poles of ω^m on the components of E_h are less than or equal to m, $g\omega^m$ belongs to the left hand side of the above equality. Therefore, we can write

$$g\omega^m = t \cdot h\omega^m + h'\omega^m,$$

where $t \cdot h\omega^m \in F_*i_*\omega_{\tilde{X}-E_v}^m((m-1)E_h) \cap t\omega_X^{[m]}$ and $h'\omega^m \in F_*\omega_{\tilde{X}}^m((m-1)E - mE_v')$. Then h vanishes on the singular locus in X. In fact, h vanishes at a horizontal exceptional component $E_{(1)}$ on $\tilde{X}$, because ω^m has a pole at a horizontal component $E_{(1)}$ of order m and t does not vanish at it. For an arbitray vertical component $E_{(i)}$, $\nu_{E_{(i)}}(th) = \nu_{E_{(i)}}(t) + \nu_{E_{(i)}}(h) > \nu_{E_{(i)}}(h)$. Here the last term is bounded below by $\nu_{E_{(i)}}(g)$ by the definition of g. Therefore $\nu_{E_{(i)}}(h'\omega^m) = \nu_{E_{(i)}}((g - th)\omega^m) = \nu_{E_{(i)}}(g\omega^m) = \nu_{E_{(i)}}(g) + m\nu_{E_{(i)}}(\omega)$. If $\nu_{E_{(i)}}(\omega) < 0$, the last term becomes negative for $m \gg 0$, while the first term is positive by the definition of h'. Hence $\nu_{E_{(i)}}(\omega) \geq 0$ for every vertical component $E_{(i)}$.

Proposition 2.5. *Let $\pi : X \to D$ be an (FG)-family of Gorenstein purely elliptic singularities, and $F : Y \to X$ be the canonical model of X. Then every fiber Y_τ $(\tau \in D)$ is irredicible.*

Proof. Let ω be a nowhere vanishing holomorphic (n+1)-form on the outside of the singular locus of X. Take a suitable blowing-up $G : \tilde{X} \to Y$ such that the composition $F \circ G : \tilde{X} \to X$ is a good resolution of X. By Proposition 2.4, every pole divisor of ω on $\tilde{X}$ is horizontal. Therefore by [I4, Lemma 2], G contracts all vertical components of $\tilde{X}$, which provides the irreducibility of the fibers Y_τ ($\tau \in D$).

Theorem 2. *Let $\pi : X \to D$ be an (FG)-family of n-dimensional Gorenstein purely elliptic singularities of type (0, n-1), where $n \geq 2$. Then π admits the simultaneous canonical model.*

Proof. Let $F : Y \to X$ be the canonical model of X and E be the reduced inverse image of the singular locus of π. Since the singularities on X are Gorenstein singularities and not canonical singularities, E is a Weil divisor and ω has poles at E by [I4, Lemma 2]. Moreover E is horizontal by Proposition 2.5. For a general $\tau \in D$, $F_\tau : Y_\tau \to X_\tau$ is the canonical model of X_τ. Here, E_τ is irreducible, since X_τ has a purely elliptic singularity of type (0, n-1). Therefore by [I4, Lemma 2], E is irreducible and $\omega_Y \simeq \mathcal{O}_Y(-E)$. For the theorem, it is sufficient to prove that the restriction $G_0 : Y_0 \to X_0$ is the canonical model of X_0. First we have $\omega_{Y_0} \simeq \mathcal{O}_{Y_0}(-E_0)$ and it is relatively ample $\mathbf{Q}$-Cartier divisor with respect to F_0. Because $\omega_{Y_0} = \omega_Y \otimes_{\mathcal{O}_Y} \mathcal{O}_{Y_0}$ by the fact that Y is Cohen-Macaulay and ω_Y is relatively ample with respect to F. Next claim that Y_0 is normal. Let $\sigma : \tilde{Y}_0 \to Y_0$ be the normarization of Y_0 and $\tilde{E}_0$ be $\sigma^{-1}(E_0)_{red}$. Since $F_0 \circ \sigma : \tilde{Y}_0 \to X_0$ is a partial resolution of a purely elliptic singularity, we have $\omega_{\tilde{Y}_0} \simeq \mathcal{O}_{\tilde{Y}_0}(-\tilde{E}_0 + C)$, where C is zero or a positive divisor with the support on $\tilde{E}_0$. Hence we have

(1) $$\sigma_* \mathcal{O}_{\tilde{Y}_0}(-\tilde{E}_0) \subset \sigma_* \omega_{\tilde{Y}_0}.$$

On the other hand, as σ is finite,

(2) $$\sigma_* \omega_{\tilde{Y}_0} \subset \omega_{Y_0} = \mathcal{J}_{E_0}.$$

Since $\sigma_* \mathcal{O}_{\tilde{Y}_0}(-\tilde{E}_0)$ is a reduced ideal and $Supp\ \mathcal{O}_{X_0} / \sigma_* \mathcal{O}_{\tilde{Y}}(-\tilde{E}_0) = E_0$, we obtain $\sigma_* \mathcal{O}_{\tilde{Y}_0}(-\tilde{E}_0) = \mathcal{J}_{E_0}$ by (1) and (2). Therefore the inclusion in (2) becomes the identity, which yields that Y_0 is normal. Finally we are going to prove that the singularities on Y_0 are terminal. Let $h : Y_0' \to Y_0$ be a resolution of Y_0.

Then $K_{Y_0'} = h^*K_{Y_0} + \sum_{i=1}^{r} a_i D_i$, where $a_i \in \mathbf{Q}$ and D_i's are the exceptional divisors for h. Since $h^*K_{Y_0}$ is represented as $-[E_0] - \sum_{i=1}^{r} b_i D_i$ with $b_i > 0$ for every exceptional component D_i, where $[E_0]$ is the proper transform of E_0 on Y_0'. Since $F_0 \circ h : Y_0' \to X_0$ is a resolution of a singularity of type (0, n-1), the essential divisor is irreducible (so it should be $[E_0]$). Therefore $a_i > a_i - b_i \geq 0$ for every i. This implies the singularities on Y_0 are all terminal.

Corollary 2.6. *Let $\pi : X \to D$ be an arbitrary flat family of simple elliptic singularitic singularities. Then π admits the simultaneous canonical model.*

Example 2.7. Yonemura classified quasi-homogeneous hypersurface simple K3- singularities into 95-classes by the weights. These classes have all different series $\{\gamma_m\}$ from each other. In fact, the series $\{\gamma_m\}$ are determined faithfully by the weights (cf 4.2, 4.8, 4.10 of [T]). Therefore, by Theorem 2 and Corollary 1.11, there exists no (FG)-families of simple K3-singularities with fibres belonging to different classes.

References.

[E] El Zein, F.: Dégenénérescence diagonale. I, II, C.R.Acad Sci. Paris, t296 51-55, 199-202, (1983)

[F] Fletcher, A.R.: Plurigenera of 3-folds and weighted hypersurfaces. Thesis submitted to the Unversity of Warwick for the degree of Ph.D (1988)

[I1] Ishii, S.: On isolated Gorenstein singularities. Math. Ann. 270, 541- 554 (1985)

[I2] Ishii, S.: Small deformations of normal singularities. Math. Ann. 275, 139-148 (1986)

[I3] Ishii, S.: The asymptotic behavior of plurigenera for a normal isolated singularity. Math. Ann. 286, 803-812 (1990)

[I4] Ishii, S.: Quasi-Gorenstein Fano 3-folds with isolated non-rational loci. to appear in Compositio. Math.

[I-W] Ishii, S. & Watanabe, K.: A geometric characterization of a simple K3-singularity. preprint 1990

[K] Knöller, F.W.: 2-dimensionale Singularitäten und Differentialformen. Math. Ann. 206, 205-213 (1973)

[KMM] Kawamata,Y.,Matsuda,K.& Matsuki, K.: Introduction to the minimal model problem. Algebraic Geometry in Sendai 1985 edited by Oda, Advanced Studies in Pure Math. 10, 283-360 Kinokuniya, Tokyo and North-Holland, Amsterdam,New York, Oxford (1987)

[K-SB] Kollár, J.& Shepherd-Barron, N.I.: Threefolds and deformations of surface singularities. Inv. Math. 91, 299-338 (1988)

[L] Laufer, H.: Weak simultaneous resolution for deformations of Gorenstein surface singularities. Proc. of Symposia in Pure Math. 40 Part 2, 1-29 (1983)

[Mo] Morales, M.: Calcul de quelques invariants des singularités de surface normale. Comptes rendes de séminaire tenu aux Plans-sur-Bex, 191-203 (1982)

[M] Mori, S.: Flip theorem and the existence of minimal models for 3-folds. J. Amer. Math. Soc. 1, 117-253 (1988)

[N] Nakayama, N.: Invariance of the plurigenera of algebraic varieties under minimal model conjecture. Toplogy 25, No.2 ,237-251 (1986)

[P] Pinkham, H.C.: Deformation of Algebraic Varietiees with $\mathbf{G}_m$ Action. astérisque 20 (1974)

[R] Reid, M.: Canonical 3-folds. Proc. Algebraic Geometry Anger 1979. Sijthoff and Nordhoff 273-310

[S] Saito,K.: Einfach elliptische Singularitäten. Inv. Math. 23, 289-325 (1974)

[St] Steenbrink, J.: Cohomologically insignificant degenerations. Compositio Math. 42, 315-320 (1981)

[T] Tomari, M.: The canonical filtration of higher dimensional purely ellptic singularity of a special type. to appear in Inv. Math.

[T-W] Tomari, M.& Watanabe, K.: On L^2-plurigenera of not-log -canonical Gorenstein isolated singularities. Amer. Math. Soc. 109, 931-935 (1990)

[W1] Watanabe, K.: On plurigenera of normal isolated singularities I, Math. Ann. 250, 65-94 (1980)

[W2] Watanabe, K.: Distribution formula of terminal singularities of a minimal resolution for simple K3-singularities. preprint 1989

[Y] Yonemura, T.: Hypersurface simple K3-singularities. Tohoku Math. J. The second series 42, 351-380 (1990)

Department of Mathematics, Tokyo Institute of Technology, Oh-Okayama, Meguro-ku, Tokyo 152, Japan. email address shihoko@math.titech. ac.jp

CONE THEOREMS AND CYCLIC COVERS

BY

JÁNOS KOLLÁR

This note is a continuation of [Kollár91b]. The aim is again to prove various forms of the Cone Theorem using methods similar to the original geometric arguments of [Mori79;82]. The basic idea is the same as in [Kollár91b]. Instead of deforming a curve $C \subset X$ directly, we construct a covering $Y \to X$ and deform another morphism $D \to Y$ where D is a suitable covering of C. In [ibid], Y was a bug-eyed cover of X, therefore a nonseparated algebraic space. The method of this article is to replace X by its formal completion along C and then find a suitable cyclic covering Y of this formal scheme.

The new method has both disadvantages and advantages. It can handle only quotients by cyclic groups and not by arbitrary groups. The method however is still strong enogh to recover the most general Cone Theorem for normal threefolds. The main advantage is that it can be used to investigate deformations of curves that are contained in the singular locus of X. In fact this method shows that the deformation problem of $C \to X$ where the canonical class of X is $\mathbb{Q}$-Cartier can be reduced to another deformation problem $D \to Y$ where the canonical class of Y is Cartier.

The following effective estimate on extremal rational curves is a combination of [Kollár91b,5.1,6.2] and (4.2). Presumably the 6 in the statement can be replaced by 4.

Theorem. *Let X be a projective threefold over a field of characteristic zero. Assume that K_X is $\mathbb{Q}$-Cartier and that X has only log-terminal singularities with the exception of finitely many points. Then*

(1) The set of extremal rays is locally finite in the open subset

$$\{z \in N_1(X) | z \cdot K_X < 0\}.$$

(2) For every extremal ray R the locus of R is covered by rational curves L_R such that $[L_R] \in R$ and

$$-6 \leq L_R \cdot K_X < 0.$$

The locus of a ray R is defined to be the set theoretic union of the closed points of all curves $D \subset X$ such that $[D] \in R$.

Partial financial support was provided by the NSF under grant numbers DMS-8707320 and DMS-8946082 and by an A. P. Sloan Research Fellowship. Typeset by $\mathcal{A}\mathcal{M}\mathcal{S}$-TEX.

1. Cyclic Covers

1.1 Lemma - Definition. *Let X be a scheme (or formal scheme) over a field k of characteristic $p \geq 0$ and let E be a $\mathbb{Q}$-Cartier divisor. Assume that mE is Cartier and that there is a homomorphism $s : \mathcal{O}_X(mE) \to \mathcal{O}_X$. Then*

(1.1.1) The natural multiplication and s define an algebra structure on

$$\sum_{i=0}^{m-1} \mathcal{O}_X(iE);$$

let

$$X[\sqrt[m]{s}] = Spec_X \sum_{i=0}^{m-1} \mathcal{O}_X(iE),$$

and let $\pi : X[\sqrt[m]{s}] \to X$ be the projection.

(1.1.2) If $p \nmid m$ then π is étale over $x \in X$ iff E is Cartier at x and s is an isomorphism at x.

(1.1.3) If X is strictly henselian, $p \nmid m$ and $u \in \Gamma(X, \mathcal{O}_X^*)$ then

$$X[\sqrt[m]{s}] \cong X[\sqrt[m]{us}].$$

(1.1.4) Assume that X is local, strictly henselian and that K_X is $\mathbb{Q}$-Cartier. Let $m > 0$ be the smallest integer such that mK_X is Cartier. Assume that $p \nmid m$. Let $s : \mathcal{O}_X(mK_X) \to \mathcal{O}_X$ be any isomorphism. Then $X[\sqrt[m]{s}]$ is independent of s. It is called the index one cover of X. The canonical divisor of $X[\sqrt[m]{s}]$ is Cartier.

(1.1.5) Let X be a three dimensional terminal singularitiy over a field of characteristic zero. Then its index one cover is a hypersurface singularity [Reid80,2.12]. □

1.2 Construction. Let X be a scheme and let $C \subset X$ be a proper and irreducible curve. Let p be the characteristic of the field of definition of C. Let $\hat{X}$ be the formal completion of X along C. Let furthermore E be a Weil divisor on $\hat{X}$ such that mE is Cartier for some fixed $m > 0$.

Let

$$d \equiv -mC \cdot E \mod m \quad \text{such that} \quad 0 \leq d < m.$$

Let $D \subset \hat{X}$ be a general Cartier divisor intersecting C transversally.

1.2.1 Claim. *If $p \nmid m$ then $\mathcal{O}_{\hat{X}}(-dD - mE)$ is divisible by m in $Pic(\hat{X})$.*

Proof. Let X_n be the nth order neighborhood of C in X. Then

$$Pic(\hat{X}) = \varprojlim Pic(X_n),$$

thus it is sufficient to show that $\mathcal{O}_{\hat{X}}(-dD - mE)$ is divisible by m in every $Pic(X_n)$. Let $b : \tilde{C} \to C$ be the normalization. By assumption $b^*\mathcal{O}_{\hat{X}}(-dD - mE)$ has degree 0, thus it is divisible by m in $Pic(\tilde{C})$. The kernel of the natural morphism

$$Pic(X_n) \to Pic(\tilde{C})$$

is a successive extension of the multiplicative and additive groups of k [Grothendieck62, no.232,6.5]. Thus, if $p \nmid m$ then the kernel is divisible by m. Therefore $\mathcal{O}_{\hat{X}}(-dD-mE)$ is divisible by m in $Pic(X_n)$. Let N be an mth-root. □

The map

$$s : N^{\otimes m} \otimes \mathcal{O}_{\hat{X}}(mE) \cong \mathcal{O}_{\hat{X}}(-dD) \to \mathcal{O}_{\hat{X}}$$

is an isomorphism outside *supp* D. Let $p : \bar{X} \to \hat{X}$ be the normalisation of $\hat{X}[\sqrt[m]{s}]$. By (1.1.4) the ramification locus of p consists of D and the set of points where E is not Cartier.

1.2.2 Claim. *If $\hat{X}$ has only log-terminal singularities then $\bar{X}$ also has only log-terminal singularities.*

Proof. p is étale in codimension one away from D so away from D this follows from [Reid80,1.7]. Let $Y \subset \hat{X}$ be an affine neighborhood of $C \cap D$ and let $Y' \subset \bar{X}$ be the corresponding cyclic cover of Y. Let $g : \tilde{Y} \to Y$ be a finite cover, étale in codimension one such that g^*K_Y and g^*E are Cartier. Let $\tilde{D} \subset \tilde{Y}$ be the pull back of D. Since $\bar{X}$ is log-terminal, by [Reid80,1.13] D is also log-terminal. Therefore $\tilde{D}$ is also log-terminal and it has index one.

Let $\tilde{Y}' = \tilde{Y} \times_Y Y'$. By definition $\tilde{Y}'$ is the normalization of

$$\sum_{i=0}^{m-1} \mathcal{O}_{\tilde{Y}}(ig^*E) \cong \sum_{i=0}^{m-1} \mathcal{O}_{\tilde{Y}}$$

where $s : (\mathcal{O}_{\tilde{Y}})^{\otimes m} \to \mathcal{O}_{\tilde{Y}}$ is multiplication by a local equation of $d\tilde{D}$. Therefore $\tilde{D}$ is a Cartier divisor on $\tilde{Y}'$. Since $\tilde{D}$ is rational and has index one the same holds for $\tilde{Y}'$. Thus $\tilde{Y}'$ is log-terminal. Since the natural morphism $\tilde{Y}' \to Y'$ is étale in codimension one, Y' is also log-terminal. □

1.2.3 Claim. *Let $\bar{C} \subset \bar{X}$ by any irreducible component of the preimage of C.*

(1.2.3.1) If $C \cdot K_{\hat{X}} \leq -2$ then

$$\bar{C} \cdot K_{\bar{X}} < \frac{1}{2} C \cdot K_{\hat{X}},$$

in particular $\bar{C} \cdot K_{\bar{X}} < -1$.

(1.2.3.2) If E is generically Cartier along C then

$$C \cdot K_{\hat{X}} < -1 \Rightarrow \bar{C} \cdot K_{\bar{X}} \leq -2.$$

Proof. Let

$$e = \frac{m}{gcd(d,m)}.$$

Then $C \cdot K_{\hat{X}}$ is an integral mutiple of $1/e$. In particular,

$$C \cdot K_{\hat{X}} < -1 \Rightarrow C \cdot K_{\hat{X}} \leq -1 - \frac{1}{e}.$$

The canonical divisor of $\bar{X}$ is computed by the adjunction formula:

$$K_{\bar{X}} \equiv p^*\left(K_{\hat{X}} + \left(1 - \frac{1}{e}\right)[D]\right).$$

If $C \cdot K_{\hat{X}} \leq -2$ then

$$\bar{C} \cdot K_{\bar{X}} \leq C \cdot \left(\frac{1}{2}K_{\hat{X}}\right) + deg(\bar{C}/C) \cdot \left(C \cdot \left(\frac{1}{2}K_{\hat{X}}\right) + 1 - \frac{1}{e}\right) < \frac{1}{2}C \cdot K_{\hat{X}}.$$

This shows (1.2.3.1).

If E is generically Cartier along C then every point of $\bar{C}$ lying over $C \cap D$ has ramification index precisely e. In particular, $deg(\bar{C}/C) \geq e$. Therefore

$$\bar{C} \cdot K_{\bar{X}} \leq e\left(-1 - \frac{1}{e} + 1 - \frac{1}{e}\right) = -2. \quad \square$$

2. The Cone Theorem for Normal Threefolds

The aim of this section is to present another proof of the Cone Theorem for normal threefolds in characteristic zero [Kollár91b,5.3]. The main steps of the argument are the same as in [Kollár91b]. First we state a variant of [Kollár91b,3.3] which asserts that if sufficienty many curves on a variety X have the expected deformation theory then the Cone Theorem holds for X.

The main difference is in the second part of the proof. In order to find deformations we use the cyclic coverings constructed in the previous section instead of bug-eyed covers.

We need to recall some definitions:

2.1 Definition. ([Kollár91b,3.1]) Let X be a scheme, C an irreducible curve and let $f : C \to X$ be a morphism. Let $\Sigma \in NS(X) \otimes \mathbb{Q}$.

(2.1.1) Assume first that everything is defined over a field of positive characteristic p.

We say that $f : C \to X$ satisfies the condition $ED_\Sigma(C, X)$ (for "Enough Deformations") if the following holds. There is a smooth, proper, irreducible curve D and a surjective morphism $g : D \to C$ with the following property. Let $F^m : D_m \to D$ be the Frobenius morphism of degree p^m. Let $g_m : D_m \to D \to C$ be the composite morphism. Then

$$\lim_{m\to\infty} \frac{dim_{[f \circ g_m]} Hom(D_m, X)}{p^m \cdot (-D \cdot_{f \circ g} \Sigma)} \geq 1.$$

(2.1.2) Now assume that X is defined over a field k of characteristic zero. We say that $f : C \to X$ satisfies condition $ED_\Sigma(C, X)$ if the following holds:

For every finitely generated subring $R \subset k$ such that $f : C \to X$ and Σ are definable over R, there is a Zariski dense subset $U \subset Specmax\ R$ such that for every $m \in U$ the reduction mod m of $f : C \to X$ satisfies the condition $ED_\Sigma(C, X)$.

(2.1.3) Let N be a real number. We say that the condition $ED_\Sigma(\geq N, X - Z)$ is satisfied if for every $f : C \to X$ such that

$$f(C) \not\subset Z \quad \text{and} \quad C \cdot_f \Sigma \geq N \cdot deg(C/f(C))$$

the condition $ED_\Sigma(C, X)$ is satisfied.

2.2 Theorem. *Let X be a projective variety over a field k and let $\Sigma \in NS(X) \otimes \mathbb{Q}$. Let $Z \subset X$ be a closed subvariety. Assume that the condition $ED_\Sigma(\geq N, X - Z)$ is satisfied. Let $R \subset \overline{NE}(X)$ be an extremal ray. Then*

(2.2.1) R satisfies at least one of the following conditions:

(2.2.1.1) $R \cdot \Sigma \geq 0$;

(2.2.1.2) $R \subset im[\overline{NE}(Z) \to \overline{NE}(X)]$;

(2.2.1.3) there is a rational curve L_R such that $[L_R] \in R$ and $-\max(2n+2, N) \leq L_R \cdot \Sigma < 0$.

(2.2.2) The set of extremal rays not contained in $im[\overline{NE}(Z) \to \overline{NE}(X)]$ is locally finite in the open subset

$$\{z \in N_1(X) | z \cdot \Sigma < 0\}.$$

Proof. This can be proved the same way as [Kollár91b,3.3]. Observe that in that proof the condition $ED_\Sigma(C, X)$ was used only for the sequence of curves C_i and then at the end for the curve $L_{j,k(j)}$. The proof still works if we know $ED_\Sigma(C_i, X)$ for infinitely many values of i. □

2.3 Theorem. *[Kollár91b,5.3] Let X be a normal projective threefold over a field of characteristic zero. Assume that $c_1(K_X) \in NS(X) \otimes \mathbb{Q}$ exists (e.g. K_X is $\mathbb{Q}$-Cartier). Then the set of extremal rays is locally finite in the open subset*

$$\{z \in N_1(X) | z \cdot K_X < 0\}.$$

Proof. By [Mori88] there is a projective threefold Y with only $\mathbb{Q}$-factorial terminal sigularities and a projective morphism $g : Y \to X$ such that K_Y is g-nef. Let $E_i \subset Y$ be the g-exceptional divisors. Let $K_Y \equiv g^*K_X + \sum a_i E_i$.

Since $\sum a_i E_i$ is g-nef a standard result (see e.g. [Kollár91a,5.2.5.3]) implies that $a_i \leq 0$ for every i. If C is a smooth, proper, irreducible curve and $f : C \to X$ is a morphism such that $f(C) \not\subset Sing\ X$ then there is a lifting $f' : C \to Y$. Furthermore,

$$C \cdot_{f'} K_Y = C \cdot_{f'} (g^*K_X + \sum a_i E_i) \leq C \cdot_f K_X.$$

Therefore

$$ED_{\frac{1}{2}K_Y}(C, Y) \Rightarrow ED_{\frac{1}{2}K_X}(C, X).$$

In order to use (2.2) I intend to show that the condition $ED_{\frac{1}{2}K_X}(\geq 1, X)$ is satisfied if X is a threefold with terminal singularities over a field of characteristic zero. This involves working also in positive characteristic, therefore we work with a more general class of singularities which behaves well with respect to reduction modulo p.

2.4 Theorem. *Let X be a normal projective variety over a field of characteristic $p \geq 0$. Assume that*

(2.4.1) X has isolated singularities,

(2.4.2) mK_X is Cartier and $p \nmid m$,

(2.4.3) for every singular point the index one cover is a hypersurface singularity.

Then the condition $ED_{\frac{1}{2}K_X}(\geq 1, X)$ is satisfied.

Proof. With these assumptions the reduction to positive characteristic goes automatically. Thus assume from now on that $p > 0$.

Using the notation of (2.1.1) we need to prove that certain maps $D_m \to X$ whose image is C have a large family of deformations. It is sufficient to prove that there is a large family of formal deformations. For this purpose we can replace X with the formal completion $\hat{X}$ of X along C. Let $E = K_{\hat{X}}$ and let $p : \bar{X} \to \hat{X}$ be as in (1.2). By (2.4.3) $\bar{X}$ has only hypersurface singularities. Therefore, if $g : D \to C$ factors through $D \xrightarrow{g'} \bar{C} \to C$ and $-C \cdot K_X \geq 2$ then by [Kollár91b,2.10] and (1.2.3.1)

$$\begin{aligned} dim_{[g_m]} Hom(D_m, X) &\geq dim_{[g'_m]} Hom(D_m, \bar{X}) \\ &\geq p^m \cdot (-D \cdot_{g'} K_{\bar{X}}) + 3\chi(\mathcal{O}_D) \\ &\geq \frac{1}{2} p^m \cdot (-D \cdot_g K_X) + 3\chi(\mathcal{O}_D). \end{aligned}$$

This proves $ED_{\frac{1}{2}K_X}(\geq 1, X)$. □

3. Further Cone Theorems

We start with some technical lemmas.

3.1 Lemma. *Let R be a local integral domain over a field k of characteristic $p \geq 0$. Assume that R is S_2 (e.g. normal). Assume that the cyclic group $\mathbb{Z}_m$ acts on R. If $p \nmid m$ then the action is completely reducible; let*

$$R = \sum_{i=0}^{m-1} R_i$$

be the decomposition into eigenspaces. Assume that every nonidentitity element of $\mathbb{Z}_m$ acts nontrivially.

Then every R_i is a torsionfree R_0 module of rank one and the mth power map

$$(R_1^{\otimes m})^{**} \to R_0 \qquad (^{**} \text{ denotes the double dual})$$

is an isomorphism.

Proof. By Galois theory the action on the quotient field of R decomposes into the sum of one dimensional eigenspaces over the quotient field of R_0. Thus every R_i is torsion free of rank one.

It is sufficent to prove that the mth power map is an isomorphism in codimension one. Let $P \subset R_0$ be a height one prime and let us localize at P. Then $R_P \supset (R_0)_P$ is étale, thus it is of the form

$$R_P \cong (R_0)_P[t]/(t^m - r) \qquad \text{for some } r \in (R_0)_P - P(R_0)_P.$$

The $\mathbb{Z}_m$-action is $t \mapsto \sqrt[m]{1} \cdot t$. Thus $(R_1)_P = (R_0)_P \cdot t$ and the mth power map becomes multiplication by r, which is a unit. □

3.2 Lemma. *Let R be a Noetherian local ring and let $I \subset R$ be an ideal. Assume that R is I-adically complete. Let $m \subset R$ be the maximal ideal and let $\hat{R}$ be the m-adic completion of R. Then*

$$J \mapsto \hat{R}J$$

provides a one-to-one correspondence between

$$\left\{\begin{matrix}\text{ideals } J \text{ of } R \text{ such that} \\ I + J \text{ is } m\text{-primary}\end{matrix}\right\} \quad \text{and} \quad \left\{\begin{matrix}\text{ideals } \hat{J} \text{ of } \hat{R} \text{ such that} \\ \hat{I} + \hat{J} \text{ is } \hat{m}\text{-primary}\end{matrix}\right\}$$

Proof. If I is nilpotent then J (resp. $\hat{J}$) is m (resp. $\hat{m}$) primary, hence the claim is clear.

In general we only have to check that

$$\hat{R}(\hat{J} \cap R) = \hat{J}.$$

The containement $\subset$ is clear, thus it is sufficient to prove that

$$\hat{R}(\hat{J} \cap R) + \hat{I}^n = \hat{J} + \hat{I}^n \quad \text{for every } n.$$

Since $\hat{R}/\hat{I}^n$ is the m-adic completion of R/I^n, the last equality follows from the already settled nilpotent case. □

3.3 Lemma. *Let R be a Noetherian local ring. Assume that R is S_2. Let $I \subset R$ be an ideal such that $dim\ R/I = 1$. Let M be a torsion free rank one module over R. Assume that M is locally free at every isolated prime of I. Then there is an ideal $J \subset R$ such that $J \cong M$ and $I + J$ is primary to the maximal ideal.*

Proof. Let M^* be the dual of M. Since $M^* \otimes (R/I)$ is locally free at every isolated prime of I, there is a map $R/I \to M^* \otimes (R/I)$ which is an isomorphism at every isolated prime of I. We can lift this to a map $R \to M^*$. Thus we get $M \to R^* \cong R$ which is an isomorphism at every isolated prime of I. Set $J = im(M \to R)$. □

3.4 Corollary. *Let $C \subset X$ be a curve in an S_2-scheme (e.g. X is normal). Let $\hat{X}$ be the formal completion of X along C. Assume that for finitely many $x_i \in C$ we are given rank one torsion free modules $\hat{F}_i$ over the completions of the local rings of $x_i \in X$. Then there is an ideal sheaf $\mathcal{J} \subset \mathcal{O}_{\hat{X}}$ such that*

(3.4.1) The completion of $\mathcal{J}$ at x_i is isomorphic to $\hat{F}_i$;

(3.4.2) $\mathcal{J} = \mathcal{O}_{\hat{X}}$ at every other point.

Proof. This follows from (3.2-3). □

3.5 Definition. Let Y be a normal scheme over a field k and let $y \in Y$ be a point. Let $\hat{Y}$ be the completion of Y along y. We say that Y is a solvable CIQ (complete intersection quotient) at y if there is a finite morphism $f : V \to \hat{Y}$ such that

(3.5.1) f is étale in codimension one;

(3.5.2) V is a complete intersection;

(3.5.3) $\hat{Y}$ is the quotient of V by a solvable group whose order is not divisible by the characteristic of k.

3.6 Theorem. *Let X be a projective variety over a field k. Assume that X is a solvable CIQ at every closed $x \in X$. Then*

(3.6.1) The set of extremal rays not contained in $im[\overline{NE}(Sing\ X) \to \overline{NE}(X)]$ is locally finite in the open subset

$$\{z \in N_1(X) | z \cdot K_X < 0\}.$$

(3.6.2) If a negative extremal ray R is not contained in $im[\overline{NE}(Sing\ X) \to \overline{NE}(X)]$ then there is a rational curve L_R such that $[L_R] \in R$.

Remark. The estimate on $L_R \cdot K_X$ depends on the length of the composition series of the groups.

Proof. Our aim is to show that the condition $ED_{\epsilon K_X}(\geq 1, X - Sing\ X)$ is satisfied for a suitable $\epsilon > 0$ which will be specified later. Let $C \subset X$ be a proper and irreducible curve such that $C \not\subset Sing\ X$. Let x_i be the singular points of X along C. Let S_0^i be the completion of the local ring of X at x_i. By assumption there is a sequence of complete local rings

$$S_0^i \subset S_1^i \subset \cdots \subset S_{k(i)}^i$$

such that every S_j^i is a the ring of invariants of S_{j+1}^i by a cyclic group action and that $S_{k(i)}^i$ is a complete intersection. Let $\hat{X}_0$ be the formal completion of X along C. We will construct a tower of cyclic covers

$$\hat{X}_0 \leftarrow \hat{X}_1 \leftarrow \cdots \leftarrow \hat{X}_{max\{k(i)\}}$$

corresponding to the above local covers. Assume that $\hat{X}_j$ is already constructed such that at its singular points the complete local rings are either complete intersections or one of the S_j^i. For every i we obtain a torsion free rank one sheaf F_j^i over S_j^i from the cyclic group action on S_{j+1}^i as in (3.1). By (3.4) there is an ideal sheaf $\mathcal{O}(-D_j)$ which is locally isomorphic to the sheaves F_j^i. Take $E = -D_j$ in the construction (1.2) to obtain $\hat{X}_{j+1}$.

At the end $\hat{X}_{max\{k(i)\}}$ has only complete intersection singularities. Estimating the intersection numbers as in the proof of (2.4) we conclude that $ED_{\epsilon K_X}(C, X)$ is satisfied for $\epsilon = 2^{-max\{k(i)\}}$ if $C \cdot K_X \geq \epsilon^{-1}$. □

4. Very Rigid Curves

4.1 Definition. Let X be an algebraic space and let $C \subset X$ be an irreducible proper curve. C is called very rigid if there is no flat family of one dimensional subschemes $D_t \subset X$ parametrised by a connected pointed curve $0 \in T$ such that

(4.1.1) *supp* $D_0 = C$,

(4.1.2) *supp* $D_t \not\subset C$ for $t \neq 0$, and

(4.1.3) D_t is purely one dimensional for $t \neq 0$.

[Kollár91b,section 6] considers very rigid curves that have negative intersection with the canonical class of X under the additional assuption that C is not contained in the singular locus of X. [ibid,5.5.2.2.2] shows that some restriction on the singularities along C is necessary. The next theorem gives the numerically best possible results assuming that the singularities along C are generically log-terminal.

4.2 Theorem. *Let X be a three dimensional normal algebraic space over a field of characteristic zero and let $C \subset X$ be a proper, reduced and irreducible curve. Assume that K_X is $\mathbb{Q}$-Cartier and that X is log-terminal at the generic point of C. If $C \cdot K_X < 0$ and C is very rigid then*

(4.2.1) C is rational, and

(4.2.2) $-2 < C \cdot K_X < 0$.

(4.2.3) If in addition X is canonical at the generic point of C then $-1 \leq C \cdot K_X < 0$.

Proof. We will use repeatedly the following simple observation:

Let $g : Y \to X$ be a morphism. Let $D \subset Y$ be a proper and irreducible curve. Assume that g is quasi-finite at the generic point of D and that $g(D) \subset X$ is very rigid. Then $D \subset Y$ is also very rigid.

Being very rigid depends only on the formal completion of the threefold along the curve, so it is sufficient to consider the case when X is a formal scheme along C. Since K_X is $\mathbb{Q}$-Cartier, we can take $E = K_X$ and construct $p : \bar{X} \to X$ as in (1.2). Let $\bar{C} \subset \bar{X}$ be as in (1.2.3). By the above remark $\bar{C} \subset \bar{X}$ is very rigid. Also, $\bar{X}$ is log-terminal and has index one at the generic point of $\bar{C}$, therefore it is canonical. Thus (4.2.3) implies (4.2.2) by (1.2.3.1).

Now assume that X is canonical at the generic point of C. Let $g : X' \to X$ be the canonical modification of X. Let $C' \subset X'$ be the proper transform of C. As in the proof of (2.3), it is sufficient to prove the theorem for $C' \subset X'$. We can replace X' with its formal completion along C'. $K_{X'}$ is $\mathbb{Q}$-Cartier hence again we can take the covering $p' : \bar{X}' \to X'$ as in (1.2). By construction $\bar{X}'$ has only index one canonical singularities. Also, by (1.2.3.2) it is sufficient to prove (4.2.1 and 3) for $\bar{X}'$.

We change notation again and assume this time that X has only index one canonical singularities along C. Let $f : Y \to X$ be a projective crepant terminalisation [Reid80,2.12]. Let $E \subset Y$ be the exceptional divisor. By [ibid] Y has only index one terminal singularities, in particular only isolated hypersurface singularities.

Let $C_1 \subset E$ be an irreducible curve such that $f(C_1) = C$. Let $n_1 : \bar{C}_1 \to Y$ be the normalization of C_1. Let us consider deformations of the pair $(n_1, \bar{C}_1)$, i.e. we deform both $\bar{C}_1$ and n_1. Let Z_1 be an irreducible component of $Def(n_1, \bar{C}_1)$. Let $\bar{\mathcal{C}}_1/Z_1$ be the universal curve and $\mathcal{N}_1 : \bar{\mathcal{C}}_1 \to Y$ be the universal morphism. Since C is very rigid, $\mathcal{N}_1(\bar{\mathcal{C}}_1) \subset E$. Let E_1 be the irreducible component of E containing this image. Let $F_1 \subset E_1$ be a general fiber of $E_1 \to C$ and let $d = deg(C_1/C)$. If $dim\ Z_1 \geq d+1$ then there is an at least one dimensional subfamily $W_1 \subset Z_1$ which corresponds to those deformations that keep the d points $F_1 \cap C_1$ fixed. By completing the family W_1 we obtain a curve $D_1 \subset E$ which is algebraically equivalent to C_1 and intersects F_1 in at least $d+1$ points. This is only possible if F_1 is an irreducible component of D_1. Let C_2 be an irreducible and reduced component of D_1 which dominates C.

Continuing in the above manner, finally we obtain a curve C_k such that the dimension of its deformation space is at most $deg(C_k/C)$. On the other hand, by [Kollár91b,2.14]

$$dim\ Def(n_k, \bar{C}_k) \geq -C_k \cdot K_Y = -C_k \cdot f^*K_X = deg(C_k/C)(-C \cdot K_X).$$

Thus $C \cdot K_X \geq -1$, which proves (4.2.3).

If $g(C) \geq 1$ then we have to show that $C \cdot K_X = -1$ is also impossible. Any étale cover $\tilde{C} \to C$ extends to an étale cover of the formal scheme $\tilde{X} \to X$. Since

$$\tilde{C} \cdot K_{\tilde{X}} = deg(\tilde{C}/C)(C \cdot K_X),$$

we conclude using (4.2.3) that $\tilde{C} \subset \tilde{X}$, and hence also $C \subset X$ are not very rigid. □

References

[Grothendieck62] A. Grothendieck, *Fondéments de la Géometrie Algébrique*, Sec. Math. Paris, 1962.

[Kollár91a] J. Kollár, *Flips, Flops, Minimal Models etc.*, to appear.

[Kollár91b] J. Kollár, *Cone Theorems and Bug-eyed Covers*, to appear.

[Mori79] S. Mori, *Projective Manifolds with Ample Tangent Bundles*, Ann. of Math. **110** (1979), 593-606.

[Mori82] S. Mori, *Threefolds whose Canonical Bundles are not Numerically Effective*, Ann. of Math. **116** (1982), 133-176.

[Mori88] S. Mori, *Flip theorem and the existence of minimal models for 3-folds*, Journal AMS **1** (1988), 117-253.

[Reid80] M. Reid, *Canonical Threefolds*, Géometrie Algébrique Angers, A. Beauville ed., Sijthoff & Noordhoff, 1980, pp. 273-310.

[SGAI71] A. Grothendieck, *Revêtements Etales et Groupes Fondémentales*, Springer LN 224, 1971.

University of Utah
Salt Lake City, UT 84112

Depth and Perversity

by Lê Dũng Tráng

In memory of J.L. Verdier

Introduction.

In the seminar on Algebraic Geometry (SGA 2) concerning the theorems of Lefschetz type ([G]) A. Grothendieck showed that the notion of depth introduced for rings and modules could be understood through the vanishing of adequate cohomologies. The topological counterpart of this notion of depth in homotopy and rational cohomology are related to the classical theorem of Lefschetz on hyperplane sections. The analogy between topology and commutative algebra led A. Grothendieck to formulate several conjectures which appear to be true (see [H-L]).

In this paper we pursue this analogy further and along the same lines. We show that the homotopy version of the classical theorem of Lefschetz is actually true under hypothesis weaker than the maximal homotopical depth. In fact it is true on complex analytic spaces for which local general fibers of complex analytic functions with isolated singularities have the homotopy type of bouquets of spheres of middle dimension. These spaces are called spaces with Milnor's property (see [L1]). These spaces have maximal rational homological depth and have the property that the constant sheaf $\mathbf{C}[d]$ shifted by the dimension d of the space is perverse in the sense of Beilinson-Bernstein-Deligne-Gabber ([B-B-D]). This remark leads naturally to a theorem of Lefschetz type for the hypercohomology of constructible complexes of sheaves satisfying a depth condition. This result was explicitly stated in the case of the intersection complex by Goresky and MacPherson in [G-M1] (§7) and follows from an argument due to Deligne which makes use of a result of M.Artin-A.Grothendieck on the direct images by affine maps of constructible complexes ([A] 3.1).

We are indebted to J.L. Verdier who suggested the relation between Grothendieck's depth and perversity, Z. Mebkhout who showed us this precise relation and P. Deligne for the proof of Theorem 4.10. This paper was written with the partial support of the NSF Grant DMS-9003498.

1. Rectified homotopical depth.

In SGA 2, A. Grothendieck introduced the notion of rectified homotopical depth of a complex analytic space. In this paragraph we recall its definition.

First we need to introduce the notion of good neighbourhoods of a point in the sense of D. Prill in [P].

Let X be a topological space and Y be a topological subspace of X. Consider a point $x \in X$.

1.1 Definition. *A neighbourhood U of x in X is called a good neighbourhood of x with respect to Y, if there are subsets $(U_\alpha)_{\alpha \in A}$ which satisfy the following conditions:*
i) The family $(U_\alpha)_{\alpha \in A}$ is a neighbourhood basis of the point x in X;
ii) Each set $U_\alpha - Y$ $(\alpha \in A)$ is a deformation retract of $U - Y$.

The interest of good neighbourhoods lies in the following (see [P]):

1.2 Lemma. *Let U and V be good neighbourhoods of x in X with respect to Y, then $U - Y$ and $V - Y$ have the same homotopy type.*

In the case of simplicial complexes, if Y is a subcomplex, the star of a point is a natural good neighbourhood of x in X. In the case of real analytic spaces (or more generally subanalytic sets) the following result gives us natural good neighbourhoods:

1.3 Proposition. *Let X be a subanalytic set and x be a point of X. Consider a subanalytic subset Y of X. Suppose that X is embedded in $\mathbf{R}^N$. There is $\epsilon_0 > 0$, such that, for any ϵ, $\epsilon_0 \geq \epsilon > 0$, the intersection $X \cap B_\epsilon(x)$ of X with the open ball $B_\epsilon(x)$ of $\mathbf{R}^N$ centered at x with radius $\epsilon > 0$, is a good neighbourhood of x in X with respect to Y.*

Remark. A subanalytic set can always be locally embedded in an affine space $\mathbf{R}^N$. Therefore the preceding proposition provides us with many good neighbourhoods which are easy to find. When we consider the local homotopy type of $X - Y$ at a point x of X, we mean the homotopy type of $U - Y$, where U is a good neighbourhood of x in X with respect to Y. Therefore we can consider in particular the homotopy type of $X \cap B_\epsilon(x) - Y$, where $B_\epsilon(x)$ is the open ball of an affine space $\mathbf{R}^N$ in which X is locally embedded in a neighbourhood of x.

Similarly if Z is a topological subspace of X, we can define:

1.4 Definition. *A neighbourhood U of Z in X is called a good neighbourhood of Z with respect to Y, if there are subsets $(U_\alpha)_{\alpha \in A}$ which satisfy the following conditions:*
i) The family $(U_\alpha)_{\alpha \in A}$ is a neighbourhood basis of the Z in X;
ii) Each set $U_\alpha - Y$ $(\alpha \in A)$ is a deformation retract of $U - Y$.

A result similar to 1.2 holds.

One gets the existence of good neighbourhoods of a subanalytic subset Z of a subanalytic set X with respect to a subanalytic set Y by using the triangulizability of subanalytic sets ([Hi] or [Har]).

1.5. *Definition. Let X be a complex analytic space. We say that the rectified homotopical depth (resp. rectified homological depth) of X at the point x is $\geq n$ if, for any locally closed complex analytic irreducible subspace Y of X which contains x there exists an open neighbourhood U of x in X such that, for any point $y \in Y \cap U$,* *there is a fundamental system of neighbourhoods* (U_α) *of* y *in* U *such that the topological pairs* $(U_\alpha, U_\alpha - Y)$ *are* $(n - dim_x Y - 1)$*-connected (resp. the integral homology* $H_k(U_\alpha, U_\alpha - Y))$ *vanishes for any* $0 \leq k \leq n - dim_x Y - 1)$.

Of course, as usual, one defines the rectified homotopical depth $rhd(X, x)$ (resp. rectified homological depth $rHd(X, \mathbf{Z}, x)$) of X at x as the maximum of the set of integers n for which $rhd(X, x) \geq n$ (resp. $rHd(X, \mathbf{Z}) \geq n$).

Remark. One checks easily that $dimX \geq rhd(X, x)$ by considering $Y = X$ in the definition.

In the above definition it is often better and more convenient to replace the fundamental system of neighbourhoods (U_α) by a good neighborhood as defined above in 1.1; e.g., if one locally embed X in some $\mathbf{C}^N$, one may consider open balls centered at x of sufficiently small radius.

If one considers the relative homology with coefficients in a field $\mathbf{F}$, one can define the rectified homological depth over $\mathbf{F}$ in a similar way and denote it by $rHd(X, \mathbf{F}, x)$.

The rectified homotopical depth $rhd(X)$ (resp. rectified homological depth $rHd(X, \mathbf{Z})$, resp. $rHd(X, \mathbf{F})$) is the infimum of the set of rectified homotopical depth (resp.rectified homological depth, resp.rectified homological depth over $\mathbf{F}$) of X at all the points x of X:

$$rhd(X) = \inf_{x \in X} rhd(X, x)$$

$$(resp.\ rHd(X, \mathbf{Z}) = \inf_{x \in X} rHd(X, \mathbf{Z}, x))$$

$$(resp.\ rHd(X, \mathbf{F}) = \inf_{x \in X} rHd(X, \mathbf{F}, x))$$

The remark above and Hurewicz Theorem imply:

$$dimX \geq rHd(X, \mathbf{F}) \geq rHd(X, \mathbf{Z}) \geq rhd(X)$$

In [H-L], H. Hamm and myself have studied the rectified homotopical depth of a complex analytic space and proved some conjectures formulated by Grothendieck in relation with theorems of Lefschetz type. Our first observation is that one can use the theory of Whitney stratifications to calculate the rectified homotopical (resp. homological depth). Namely we prove:

1.6 Theorem. *Let* X *be a reduced complex analytic space and* n *a positive integer. The following conditions are equivalent:*

a) $rhd(X) \geq n$ *(resp.* $rHd(X, \mathbf{Z}) \geq n$*, resp.* $rHd(X, \mathbf{F}) \geq n$*);*

b) *If* Y *is a locally closed irreducible complex analytic subspace of dimension* i*, there is an open dense analytic subset* Y_0 *of* Y *such that, for any point* y *in* Y_0*, there is a fundamental system of neighbourhoods* U_α *of* y *in* X *such that the pair* $(U_\alpha, U_\alpha - Y)$ *is* $(n - 1 - i)$*-connected (resp.* $H_k(U_\alpha, U_\alpha - Y, \mathbf{Z}) = 0$*, resp.* $H_k(U_\alpha, U_\alpha - Y, \mathbf{F}) = 0$*, for any* $0 \leq k \leq n - 1 - i$*);*

c) *If* $\mathcal{S}$ *is a complex analytic Whitney stratification of* X*, for any stratum* Y *of* $\mathcal{S}$ *of dimension* i*, for any point* y *in* Y *there is a fundamental system of* *neighbourhoods U_α of y in X such that the pair $(U_\alpha, U_\alpha - Y)$ is $(n-1-i)$-connected (resp. $H_k(U_\alpha, U_\alpha - Y, \mathbf{Z}) = 0$, resp. $H_k(U_\alpha, U_\alpha - Y, \mathbf{F}) = 0$, for any $0 \le k \le n-1-i$).*

This theorem shows that it is enough to check the vanishing of homotopy (resp. homology) groups at one point of each stratum of a given Whitney stratification of X to calculate the rectified homotopical (resp. homological depth). Thus, for instance, one has to consider a finite number of points in the case of algebraic varieties.
An obvious consequence of this theorem is that the rectified homotopical (resp. homological) depth of a non-singular space is equal to the complex dimension of the space. In fact one can show that this equality of the rectified homotopical depth and the complex dimension also holds when the space is locally a complete intersection (see [H-L], Corollary 3.2.2). This latter result is new in the case of the rectified homotopical depth. But for the case of the rectified homological depth over the rational field $\mathbf{Q}$, Z. Mebkhout showed that it is actually a consequence of a theorem of Grothendieck, as we shall show below (see §4).

2. Spaces with maximal depth.

To study a singular complex analytic space endowed with a Whitney stratification, the stratified Morse theory of M. Goresky and R. MacPherson shows that the homotopy type of the complex links of the strata are key invariants to study the topology of the space. In fact we can define the rectified homotopical depth using the connectivity of these complex links.

First let us define the complex link of a point x in a complex analytic space X embedded in $\mathbf{C}^N$ (see [G-M2] Part II, §2.3):

2.1 Theorem-Definition. *There is a dense Zariski open set Ω in the space of complex linear forms on $\mathbf{C}^N$, such that, for any $\ell \in \Omega$, there is $\epsilon_\ell > 0$, such that, for any ϵ, $\epsilon_\ell > \epsilon > 0$, there is $\tau_{\epsilon,\ell}$, such that, for any t, $\tau_{\epsilon,\ell} > |t| > 0$, the homotopy type of the space*

$$\mathcal{L} := B_\epsilon(x) \cap X \cap \ell^{-1}(\ell(x) + t)$$

is an analytic invariant of the germ (X, x). Such a space $\mathcal{L}$ is called a complex link of X at x.

In particular the definition of the complex link of x in X does not depend on the local embedding at x of X in $\mathbf{C}^N$ as it is an analytic invariant of the complex analytic germ (X, x).

Examples. i) If X is a complex analytic curve, the complex link of a point is a finite number of points and this number of points equals the multiplicity of the point.
ii) If X is a normal surface, the complex link of a point is a Riemann surface and, in [GS], G. Gonzalez-Sprinberg proved that this complex link was contractible if and only if the point is non-singular. Furthermore G. Gonzalez-Sprinberg proved that, for normal surfaces, the Euler characteristic of the complex link equals MacPherson's Euler

obstruction. Actually, when, more generally, the singularity of the complex analytic space is isolated, A. Dubson ([Du]) showed that the Euler characteristic of the complex link equals MacPherson's Euler obstruction.
iii) When X is a local complete intersection, in [L2] the author showed that the complex link of a point has the homotopy type of a bouquet of spheres of (real) dimension equal to the complex dimension $dim X$.
iv) In fact, for a general complex analytic space X, points of X where the complex link is not contractible are isolated. More precisely if a point x belongs to a stratum of a Whitney stratification which has dimension greater than or equal to 1, the complex link of this point in X is contractible.

We can now define the complex link of a Whitney stratum of X.

2.2 Theorem-Definition. *Let x be a point on the stratum S of a complex analytic stratification of X which satisfies Whitney conditions. Let $U \subset \mathbf{C}^N$ be an embedding of an open neighbourhood of x in X into $\mathbf{C}^N$. Let Z be a complex submanifold of $\mathbf{C}^N$ which intersects transversally $S \cap U$ in $\mathbf{C}^N$ at the point x. The homotopy type of a complex link of the germ $(Z \cap X, x)$ does not depend on the choice of x in the stratum S. Such a complex link is called a complex link of the stratum S in X.*

In terms of complex links we can give an estimation of the rectified homotopical (resp. homological) depth. As noticed by M. Goresky and R. MacPherson, it is often convenient to introduce the notion of transversal slice of a Whitney stratum. Let x be a point in a Whitney stratum S of X. As in 2.2, consider a local embedding $U \subset \mathbf{C}^N$ of an open neighbourhood U of x in X into some non-singular space $\mathbf{C}^N$ and consider a non-singular complex manifold Z which intersects transversally $S \cap U$ in $\mathbf{C}^N$ at the point x.

2.3 Definition. *A sufficiently small representative of the germ $(Z \cap X, x)$ which contains a closed complex link of s in X and which is contractible is called a local slice of X at x transversal to the stratum S.*

Now we can formulate (cf [H-L] Theorem 4.1.2):

2.4 Theorem. *Let X be a complex analytic space and x be a point in X. Let $\mathcal{S}$ be a Whitney stratification of X. The following conditions are equivalent:*

a) *$rhd(X, x) \geq n$ (resp. $rHd(X, x) \geq n$);*

b) *for any stratum S which contains x in its closure, the pair $(\mathcal{N}, \mathcal{L})$ of a normal slice $\mathcal{N}$ of S in X and a complex link $\mathcal{L}$ of S in X is $(n - dimS - 1)$-connected (resp. $H_k(\mathcal{N}, \mathcal{L}) = 0$, for any k, $0 \leq k \leq n - dimS - 1$).*

From this theorem we deduce a characterization of complex analytic spaces with maximal rectified homotopical (resp. homological) depth:

2.5 Corollary. *Let X be a complex analytic space and x be a point in X. Let $\mathcal{S}$ be a Whitney stratification of X. The following conditions are equivalent:*

a) $rhd(X,x) = dim_x X$ (resp. $rHd(X,x) = dim_x X$);

b) for any stratum S which contains x in its closure, a complex link $\mathcal{L}$ of S in X has the homotopy type (resp. the homology) of a bouquet of spheres of real dimension $dim_x X - dim S - 1$.

The proof of this corollary is based on the following lemma (cf [Mil], Proof of Theorem 6.5):

2.6 Lemma. *A $(d-1)$-connected CW-complex E of dimension d has the homotopy type of a bouquet of spheres of (real) dimension d.*

and the following theorem proved by H.Hamm in the case of singular spaces (cf [Ha] or [G-M2], Part II §1.1):

2.7 Theorem. *Stein spaces have the homotopy type of a CW-complex of real dimension less than their complex dimension.*

Our corollary is then obvious, once we observe that a complex link of a Whitney stratum in X is a Stein space, because it is isomorphic to a closed analytic subspace of an open ball in $\mathbf{C}^N$.

3. Theorems of Lefschetz type on singular spaces.

In SGA 2, A. Grothendieck conjectured that the rectified homotopical depth gives the level of homotopy comparison between a complex projective variety and one of its hyperplane sections. This fact was proved in [H-L]. This level is maximal and equal to the complex dimension, as in the case of non-singular spaces, if the rectified homotopical depth is maximal. In fact we are going to show that this level is maximal for spaces with Milnor's property.

Let us recall the definition of complex analytic spaces with Milnor's property:

3.1 Definition. *We shall say that a complex analytic space X has Milnor's Property at the point x if the space X is equidimensional in a neighbourhood of x in X, the complex link of $\{x\}$ in X has the homotopy type of a bouquet of spheres of middle dimension and the complex links of the strata of a Whitney stratification of X of dimension ≥ 1 which contain x in their closures have the homology type of a bouquet of spheres of middle dimension.*

In this definition, we say to simplify that a complex space has the homotopy type of a bouquet of spheres of middle dimension if the (real) dimensions of the spheres of the bouquet equals the complex dimension of the space.

In [L1] we showed that the general fiber of the germ of a complex analytic function which has an isolated singularity in the sense of [L3] on a equidimensional complex analytic space has the homotopy type of a bouquet of spheres of middle dimension. This property is a generalization of the fact proven by J. Milnor in [Mil] that the general fiber of the germ of a complex analytic function which has an isolated critical point in $\mathbf{C}^N$ has the homotopy type of a bouquet of spheres of (real) dimension $N-1$.

In this paragraph we shall prove that the level of homotopy comparison in a theorem of Lefschetz type is maximal if the space satisfies Milnor's property.

3.2 Theorem. *Let V be a complex projective variety of dimension d in the complex projective space $\mathbf{P}^N$. We suppose that V satisfies Milnor's property. Let H be any hyperplane of $\mathbf{P}^N$. Then:*

$$\pi_k(V, V\cap H, x) = 0 \qquad \textit{for any } k < d$$

Proof: The proof of the theorem proceeds as in [G-M2] (Part II §5.1). We imitate the original topological proof of A. Andreotti and T. Frankel in [A-F]. We identify $\mathbf{P}^N - H$ with $\mathbf{C}^N$. We stratify V with the Whitney conditions. If z is a sufficiently general point of $\mathbf{C}^N$:

i) The distance to the point z induces a real analytic Morse function δ on $V-H$ and according to a result of A. Andreotti and T. Frankel in [A-F], the index of a critical point of the restriction of δ to a stratum S is $\leq dimS$;
ii) On the stratified space V, the function δ is a stratified Morse function in the sense of [G-M2].

Because the variety V is projective, the number of strata is finite. So, if the number R is large enough, the critical points of the restrictions of δ to the strata of the Whitney stratification lie in the domain where $\delta < R$. We shall apply the stratified Morse theory on the stratified variety V using the restriction of the function $-\delta$ to V and starting from the subspace V_R of V where $\delta \geq R$.

Let c be a Morse singular point of the restriction of $-\delta$ to the stratified space V. In particular it is a Morse critical point of the restriction of $-\delta$ to the stratum S which contains c. From the results of [G-M2] (see Part I §1.2), when ϵ is small enough the pair

$$(V\cap\{-\delta\leq -\delta(c)+\epsilon\}, V\cap\{-\delta\leq-\delta(c)-\epsilon\})$$

is k-connected if the local Morse data at c of the restriction of $-\delta$ to V is k-connected. But according to [G-M2] (loc. cit.), the local Morse data at c is homotopic to the product of the tangential Morse data at c by the normal Morse data at c. The tangential Morse data of $-\delta$ at c is a pair $(B, \partial B)$ where B is a ball of dimension equal to the index of the restriction of $-\delta$ to the stratum S. Now the result of A. Andreotti and T. Frankel implies that the index of the restriction of $-\delta$ to S is $\geq dimS$, so that the dimension of B is $\geq dimS$. On the other hand the normal Morse data at c has the homotopy type of $(\mathcal{N}, \mathcal{L})$, where $\mathcal{N}$ is the contractible slice of V at c transversal to S and $\mathcal{L}\subset\mathcal{N}$ is a complex link of S in V. Therefore, to prove that the pair

$$(V\cap\{-\delta\leq -\delta(c)+\epsilon\}, V\cap\{-\delta\leq-\delta(c)-\epsilon\})$$

is $(dimV-1)$-connected, it is enough to prove that if V satisfies Milnor's property, the pair $(B\times\mathcal{N}, \partial B\times\mathcal{N}\cup B\times\mathcal{L})$ is k-connected with $k\geq dimV-1$.

In [H-L] (Lemma 1.8) we proved the following lemma:

3.3 Lemma. *Let (E, E') be a pair of locally trivial topological fibrations over a connected CW-complex B, let $\pi: E \to B$ be the corresponding projection. Let (N, N') be a pair of fibers of these fibrations. Consider B' a subcomplex of B. Assume that the pair (B, B') is $(r-1)$-connected, $r \geq 1$, the space N is contractible and $H_i(N, N'; \mathbf{Z}) = 0$, for $i \leq m - r - 1$, then the pair $(E, E' \cup \pi^{-1}(B'))$ is $(m-1)$-connected.*

We apply this lemma to estimate the connectivity of the pair of spaces

$$(B \times \mathcal{N}, \partial B \times \mathcal{N} \cup B \times \mathcal{L}).$$

If $dimS \geq 1$, the pair $(B, \partial B)$ is $(l-1)$-connected with $l \geq dimS \geq 1$, and because the space V has Milnor's property, we have

$$H_i(\mathcal{N}, \mathcal{L}; \mathbf{Z}) = 0,$$

for $i \leq dimV - dimS - 1$. Therefore the lemma implies that, when $dimS \geq 1$, the pair

$$(B \times \mathcal{N}, \partial B \times \mathcal{N} \cup B \times \mathcal{L})$$

is $dimV - dimS - 1 + l$-connected and $dimV - dimS - 1 + l \geq dimV - 1$. When $dimS = 0$, the pair

$$(B \times \mathcal{N}, \partial B \times \mathcal{N} \cup B \times \mathcal{L})$$

coincides with $(\mathcal{N}, \mathcal{L})$ which is $(dimV - 1)$-connected because the variety V satisfies Milnor's property.

In conclusion, for any stratified Morse point c of $-\delta$, the pair

$$(V \cap \{-\delta \leq -\delta(c) + \epsilon\}, V \cap \{-\delta \leq -\delta(c) - \epsilon\})$$

is $(dimV - 1)$-connected. On the other hand, if the interval $[a, b]$ does not contain any critical value of the stratified function $-\delta$, Thom-Mather first isotopy theorem (cf [T] and [Ma], see [G-M2] Part II §5.A Appendix) shows that the spaces $V \cap \{\delta \geq b\}$ and $V \cap \{\delta \geq a\})$ are homeomorphic. Finally the same argument used in the "classical" Morse theory (see [Mi2]) gives that the pair $(V, V \cap \{\delta \geq R\}) = (V, V_R)$ is $(dimV - 1)$-connected. To end the proof of Theorem 3.2, it remains to prove that $V \cap \{\delta \geq R\} = V_R$ retracts by deformation onto $V \cap H$ (see [G-M2] Part II §5.A Appendix).

In the preceding proof the main observation is that to have a Theorem of Lefschetz type on V with a level of comparison of the homotopy groups given by the dimension, as in the theorem for non-singular varieties, it is not needed that the space V has maximal rectified homotopical depth as the results of [H-L] and the conjectures of A. Grothendieck in [G] have suggested: it is sufficient to assume that the variety V satisfies Milnor's property.

4. Perverse Sheaves.

Instead of considering the homotopical depth, we could have considered the homological depth or the rational homological depth. Spaces with maximal rectified homological depth are characterized by:

4.1 Theorem. *Let X be a complex analytic space and x be a point of X. The following conditions are equivalent:*

i) $rHd(X,x) = dim_x X$ (resp. $rHd(X,x,\mathbf{Q}) = dim_x X$);

ii) There is an open neighbourhood U of x in X such that, for any y in U, the general fiber of a complex analytic function f defined in a neighbourhood of y having at y an isolated singularity has the homology (resp. rational homology) of a bouquet of spheres of real dimension $dim_x X - 1$;

iii) The space X is equidimensional at any point of a neighbourhood of x in X and the complex links of the strata of a Whitney stratification of X which contain x in their closures have the homology (resp. rational homology) of a bouquet of spheres of middle dimension.

This theorem is a consequence of Theorem 5.4 of [L1] and the characterization of spaces with the maximal rectified homological depth given above in Corollary 2.5. We have the following theorem of Lefschetz type which is proved in a similar way as Theorem 3.2 above:

4.2 Theorem. *Let V be a complex projective variety in complex projective space $\mathbf{P}^N$. We suppose that V has maximal rectified homological (resp. rational homological) depth. Let H be any hyperplane of $\mathbf{P}^N$. Then:*

$$H_k(V, V \cap H) = 0 \qquad (resp. H_k(V, V \cap H, \mathbf{Q}) = 0) \qquad \text{for any } k < d$$

Now let X be a complex analytic space and x be a point of X. Consider a closed complex analytic subspace Y of X which contains x. Let U be an open good neighbourhood of x in X with respect to Y. The condition:

$$H_k(U, U - Y, \mathbf{Q}) = 0 \qquad \text{for any } 0 \leq k \leq n - dim_x Y - 1$$

is equivalent to:

$$H^k(U, U - Y, \mathbf{Q}) = 0 \qquad \text{for any } 0 \leq k \leq n - dim_x Y - 1.$$

On the other hand the cohomology $H^k(U, U - Y, \mathbf{Q})$ is isomorphic to the support cohomology $H^k_{U \cap Y}(U, \mathbf{Q})$. Therefore we can assert:

4.3 Lemma. *Let X be a complex analytic space. The rectified rational homological depth $rHd(X, \mathbf{Q}, x)$ of X at a point x is $\geq n$ if and only if for any closed complex analytic subspace Y of X which contains x, there is an open neighbourhood U of x in X such that, for any point $y \in Y \cap U$, there is a fundamental system of good neighbourhoods (U_α) of y in U with respect to $U \cap Y$ such that:*

$$H^k_{U_\alpha \cap Y}(U_\alpha, \mathbf{Q}) = 0 \qquad \text{for any } 0 \leq k \leq n - dim_x Y - 1$$

In other words the rectified rational homological depth $rHd(X, \mathbf{Q}, x)$ of X at a point x is $\geq n$ if and only if for any closed complex analytic subspace Y of X which contains x, there is an open neighbourhood U of x in X such that the restriction to U of the k-th cohomology sheaf of sections of the constant sheaf $\mathbf{Q}_X$ with support in Y is the trivial sheaf:

$$h^k(\mathbf{R}\Gamma_Y(\mathbf{Q}_X))|U = 0 \qquad \text{for any } 0 \leq k \leq n - dim_x Y - 1.$$

This observation immediately leads to the following result (cf [H-L] Corollary 1.10):

4.4 Proposition. *Let X be a complex analytic space, equidimensional at x, the rectified rational homological depth $rHd(X, \mathbf{Q}, x)$ of X at a point x is equal to the dimension $dim_x X$ of X at x if and only if there is an open neighbourhood U of x in X such that the restriction of $\mathbf{C}_X[dimX]|U$ of the constant sheaf $\mathbf{C}_X[dimX]$ on U is perverse (for the autodual perversity).*

This proposition is an immediate consequence of the definition of the autodual perversity as it is defined in [B-B-D].

Let us recall the definition of the autodual perversity. As we shall only deal with the autodual perversity, we shall not mention it in the sequel. We adapt the definition of a perverse sheaf given in [B-B-D] (Introduction p. 9) to the complex analytic case.

Let $\mathbf{K}^*$ be a bounded complex of sheaves of complex vector spaces on X. We assume that the complex $\mathbf{K}^*$ is constructible, i.e. there is a complex analytic stratification of X such that the restrictions of the cohomology sheaves of this complex to the strata of this stratification are locally constant sheaves. We shall say that a complex analytic stratification $\mathcal{S}$ of X is adapted to the complex $\mathbf{K}^*$ if the restrictions to the strata of $\mathcal{S}$ of the cohomology sheaves of this complex are locally constant sheaves.

4.5 Definition. *We say that a complex $\mathbf{K}^*$ is perverse if, for any point x of X and any locally closed irreducible complex analytic subspace Y which contains x, there is an open neighbourhood U of x in X and an open dense analytic subset Y_0 of $Y \cap U$ such that we have:*
(a) $h^k(s^*\mathbf{K}^*_U) = 0$ for $k > -dimY$,
(b) $h^k(s^!\mathbf{K}^*_U) = 0$ for $k < -dimY$,

where $\mathbf{K}^*_U$ is the restriction of $\mathbf{K}^*$ to U and $s: Y_0 \to U$ is the inclusion.

Recall that the cohomology sheaf $h^k(s^!\mathbf{K}^*_U)$ is the restriction to Y_0 of the local cohomology sheaf $\mathcal{H}^k_{Y_0}(\mathbf{K}^*_U)$ with supports in Y_0 ([G]).
The condition (a) in the definition above is equivalent to the condition:

$$\dim Supp(h^k(\mathbf{K}^*)) \leq -k$$

(cf [A] §2-4, see [B-B-D] (4.0.1')) and is called the support condition for the complex $\mathbf{K}^*$. Then it is obvious that, for any complex analytic space X, the constant sheaf $\mathbf{C}_X[dimX]$ satisfies the support condition.

The condition (b) is called the cosupport condition for the complex $\mathbf{K}^*$. This terminology comes from the fact that the cosupport condition is nothing else, but the support condition for the Verdier dual complex $\mathcal{D}(\mathbf{K}^*)$ of $\mathbf{K}^*$.

The complex of sheaves $\mathbf{C}_X[dimX]$ satisfies condition (b) if and only if for any point x of X and any locally closed irreducible complex analytic subspace Y which contains x, there is an open neighbourhood U of x in X and an open dense analytic subset Y_0 of $Y \cap U$ such that

$$\mathcal{H}^k_{Y_0}(\mathbf{C}_X[dimX]) = 0$$

for $k < -dimY$. Because of the translation in the indices this is equivalent to the vanishing

$$\mathcal{H}^{k+dimX}_{Y_0}(\mathbf{C}_X) = 0$$

for $k < -dimY$. Theorem 1.6 shows that this is equivalent to $rHd(X, \mathbf{Q}) = dimX$. This obviously leads to proposition (4.4).

Remark. In fact, using the comparison theorem of Grothendieck as formulated by Mebkhout for $\mathcal{D}$-modules in [Me], the shifted constant sheaf $\underline{\mathbf{C}}[d]$ of complex numbers on a variety of dimension d which is locally a complete intersection is perverse. But Proposition 4.4 asserts that this is equivalent to $rHd(X, \mathbf{Q}) = dimX$. This explains the statement at the end of §1 above.

Recall that we have a stronger homotopy statement, because, from [L2] and Corollary 2.5 (see [H-L], Corollary 3.2.2 also), we know that, if the complex analytic space X is locally a complete intersection, the rectified homotopical depth equals the complex dimension of X.

In view of Proposition 4.4, it is natural to observe that a theorem of Lefschetz type with a depth condition is obtained for constructible complexes satisfying this depth condition. Namely we first define:

4.6 Definition. *Let X be a complex analytic space and let $\mathbf{K}^*$ be a constructible complex on X. We say that the depth $rHd(\mathbf{K}^*)$ is $\geq n$ if for any point x of X and any locally closed irreducible complex analytic subspace Y which contains x, there is an open neighbourhood U of x in X and an open dense analytic subset Y_0 of $Y \cap U$ such that $h^k(s^!\mathbf{K}^*_U) = 0$ for $k < -dimY + n - dimX$, where $\mathbf{K}^*_U$ is the restriction of $\mathbf{K}^*$ to U and $s: Y_0 \to U$ is the inclusion.*

Of course, we define the integer $rHd(\mathbf{K}^*)$ to be the maximum of integers n such that $rHd(\mathbf{K}^*) \geq n$. This definition gives by duality:

4.7 Lemma. *Let X be a complex analytic space. The following conditions are equivalent:*

a) *The depth $rHd(\mathbf{K}^*)$ is $\geq n$;*
b) *For any point x of X and any locally closed irreducible complex analytic subspace Y which contains x, there is an open neighbourhood U of x in X and an open dense analytic subset Y_0 of $Y \cap U$ such that*

$$h^k(s^* D(\mathbf{K}^*_U)) = 0 \text{ for } k > -dimY - n + dimX,$$

where $\mathbf{K}^*_U$ *is the restriction of* $\mathbf{K}^*$ *to* U *and* $s\colon Y_0 \to U$ *is the inclusion.*

We have immediately:

4.8 Lemma. *Let* X *be a complex analytic space. The following conditions are equivalent:*
a) $rHd(X, \mathbf{Q}) \geq n$;
b) $rHd(\mathbf{C}_X[dimX]) \geq n$.

4.9 Remark. In [B-B-D] (2.1.7), when X is a complex analytic space, the objects of the category ${}^pD^{\leq dimX-n}(X)$ are the constructible complexes $\mathbf{K}^*$ such that, for any point x of X and any locally closed irreducible complex analytic subspace Y which contains x, there is an open neighbourhood U of x in X and an open dense analytic subset Y_0 of $Y \cap U$ such that $h^k(s^*\mathbf{K}^*_U) = 0$ for $k > -dimY - n + dimX$, where $\mathbf{K}^*_U$ is the restriction of $\mathbf{K}^*$ to U and $s\colon Y_0 \to U$ is the inclusion. Therefore the complex $\mathbf{K}^*$ has a depth $rHd(\mathbf{K}^*)$ is $\geq n$ if and only if its dual $\mathcal{D}(\mathbf{K}^*)$ belongs to ${}^pD^{\leq dimX-n}(X)$.

Now we can state a theorem of Lefschetz type for constructible sheaves satisfying a depth condition:

4.10 Theorem. *Let* V *be a complex projective variety of dimension* d *in the complex projective space* $\mathbf{P}^N$. *Let* H *be a complex projective hyperplane in* $\mathbf{P}^N$. *Let* $\mathbf{K}^*$ *be a constructible complex on* V *and assume that* $rHd(\mathbf{K}^*|(V - V \cap H)) \geq n$, *then the homomorphism:*

$$\mathbf{H}^i(V, \mathbf{K}^*) \to \mathbf{H}^i(V \cap H, \mathbf{K}^*)$$

induced between the hypercohomologies by the inclusion of $V \cap H$ *into* V, *is an isomorphism for any* $i \leq n - d - 2$ *and is an injection for* $i = n - d - 1$.

Proof: The proof is very similar to the proof of the weak Lefschetz theorem in [D] (4.1.6) and uses a result of M.Artin-A.Grothendieck in [A] (3.1).

We have a distinguished triangle (see [B-B-D] (1.4.1.1)):

$$\to j_!j^*\mathbf{K}^* \to \mathbf{K}^* \to i_*i^*\mathbf{K}^* \to$$

where $j\colon V - V \cap H \to V$ is the inclusion of $V - V \cap H$ in V, and $i\colon V \cap H \to V$ is the inclusion of $V \cap H$ in V. The hypercohomology functor with compact support on V give the long exact sequence:

$$\to \mathbf{H}^i_c(V - V \cap H, \mathbf{K}^*) \to \mathbf{H}^i_c(V, \mathbf{K}^*) \to \mathbf{H}^i_c(V \cap H, \mathbf{K}^*)$$
$$\to \mathbf{H}^{i+1}_c(V - V \cap H, \mathbf{K}^*) \to$$

Because V and $V \cap H$ are compact we have:

$$\mathbf{H}^i_c(V, \mathbf{K}^*) = \mathbf{H}^i(V, \mathbf{K}^*)$$

$$\mathbf{H}^i_c(V \cap H, \mathbf{K}^*) = \mathbf{H}^i(V \cap H, \mathbf{K}^*)$$

and Verdier duality (see [V] §3, Théorème) gives an isomorphism:

$$\mathbf{H}^i_c(V - V \cap H, \mathbf{K}^*) \cong \mathbf{H}^{-i}(V - V \cap H, D(\mathbf{K}^*))$$

where $D(\mathbf{K}^*)$ is the dual complex of $\mathbf{K}^*$. The space $V - V \cap H$ being an affine variety, we can apply a theorem of M.Artin-A.Grothendieck (cf [A] (3.1)), as it is mentioned in [B-B-D] (Theorem 4.1.1). We use the notations of the remark 4.9 above to recall this theorem:

4.11 Theorem. *Let $f: X \to Y$ an affine morphism of complex algebraic varieties. Let $\mathbf{K}^*$ be a constructible complex on X which belongs to the category ${}^pD^{\leq dimX-n}(X)$. Then the derived direct image $f_*(\mathbf{K}^*)$ is in the category ${}^pD^{\leq dimX-n}(Y)$.*

Now we apply this theorem to prove Theorem 4.10. We consider the case $X = V - V \cap H$ and the space Y is reduced to a point $*$. The complex $D(\mathbf{K}^*)$, dual of the complex $\mathbf{K}^*$ considered in the hypothesis of Theorem 4.10, belongs to the category

$$ {}^pD^{\leq dimX-n}(V - V \cap H) $$

because we assume that $rHd(\mathbf{K}^*|(V - V \cap H) \geq n$ and we have the statement of Lemma 4.7. Therefore the Theorem of Artin-Grothendieck gives that the derived direct image $f_*(D(\mathbf{K}^*))$ is in the category ${}^pD^{\leq dimX-n}(*)$. By definition, the i-th cohomology of the complex $f_*(D(\mathbf{K}^*))$ is the hypercohomology $\mathbf{H}^i(V - V \cap H, D(\mathbf{K}^*))$. As $f_*(D(\mathbf{K}^*))$ belongs to the category ${}^pD^{\leq dimX-n}(*)$, we have:

$$ \mathbf{H}^i(V - V \cap H, D(\mathbf{K}^*)) = 0 $$

for any $i > -dimY - n + dimX = -n + d$, because $dimY = 0$ and $dimX = d$. Therefore:

$$ \mathbf{H}^{-i}(V - V \cap H, D(\mathbf{K}^*)) = 0 $$

for any $-i > -dimY - n + dimX = -n + d$ or $i < n - d$. This yields:

$$ \mathbf{H}^i_c(V - V \cap H, \mathbf{K}^*) = 0 $$

for any $i < n - d$ and the homomorphism:

$$ \mathbf{H}^i(V, \mathbf{K}^*) \to \mathbf{H}^i(V \cap H, \mathbf{K}^*) $$

induced between the hypercohomologies by the inclusion of $V \cap H$ into V, is an isomorphism for any $i \leq n - d - 2$ and is an injection for $i = n - d - 1$. This proves Theorem 4.10.

Remark. Another version of the Theorem of Artin-Grothendieck was proved by M. Goresky and R. MacPherson in [G-M1] §7.2 and is used to get a Theorem of Lefschetz type for the hypercohomology of intersection cohomology complex on a projective variety. Actually it can give a Theorem of Lefschetz type for perverse constructible complexes for a more general perversity than the autodual perversity. In fact Theorem 4.10 tells that to have a Theorem of Lefschetz type for the hypercohomology on a projective variety it is enough to consider constructible complexes satisfying a reasonable depth condition.

A Theorem like the Theorem of Artin-Grothendieck allows us to globalize to the hypercohomology of constructible complexes on an affine variety vanishing conditions which appear to be local with respect to these complexes. Such a Theorem for Stein maps should be true and this should imply a theorem of Lefschetz type for constructible complexes satisfying depth conditions on pairs of spaces (X, Y) for which $X - Y$ is Stein (compare to [H-L] Theorem 3.4.1).

Bibliography

[A] M. Artin, in Séminaire de Géométrie Algébrique 1963-1964 (SGA 4), Exposé XIV, *Théorème de finitude pour un morphisme propre; dimension cohomologique des schémas affines*, Théorie des Topos et Cohomologie étale des Schémas, tome 3, Springer Lect. Notes 305, 1973.

[A-F] A. Andreotti - T. Frankel, *The Lefschetz theorem on hyperplane sections*, Ann. of Math. (2) **69** (1959), 713-717.

[B-B-D] A.A. Beilinson, J. Bernstein, P.Deligne, *Faisceaux pervers*, Astérisque **100** (1982), 1-172.

[D] P. Deligne, *La Conjecture de Weil II*, Pub. I.H.E.S **52** (1980), 313-428.

[Du] A. Dubson, *Classes caractéristiques des variétés singulières*, C. R. Acad. Sc. **287** (1978), 237-239.

[GS] G. Gonzalez-Sprinberg, *Cycle Maximal et invariant local d'Euler des singularités isolées de surfaces*, Topology **21** (1982), 401-408.

[G-M1] M. Goresky - R. MacPherson, *Intersection Homology II*, Inv.Math. **71** (1983), 77-129.

[G-M2] M. Goresky - R. MacPherson, Stratified Morse Theory, Springer Ed., Berlin - Heidelberg - New York 1987.

[G] A. Grothendieck, Cohomologie locale des faisceaux cohérents et théorèmes de Lefschetz locaux et globaux (SGA2), Masson & North-Holland Ed., Paris, Amsterdam, 1968.

[Ha] H. A. Hamm, *Zum Homotopietyp Steinscher Raüme*, J. reine angew. Math. **338** (1983), 121-135.

[H-L] H. A. Hamm - Lê D. T., *Rectified homotopical depth and Grothendieck conjectures*, in The Grothendieck Festschrift, Volume II, Progress in Mathematics **87**, Birkhäuser 1990.

[Har] R.M. Hardt, *Stratifications of subanalytic sets and proper light subanalytic maps*, Inv. Math. **38** (1977), 207-217.

[Hi] H. Hironaka, *Triangulation of algebraic sets*, Proc.Symp.Pure Math. **29** (1975), Providence, 165-185.

[L1] Lê D. T., *Complex Analytic Functions with Isolated Singularities*, to be published in Jour. Alg. Geom. and Sing.

[L2] Lê D. T., *Sur les cycles évanouissants des espaces analytiques*, C. R. Acad. Sc. **288** (1979), 283-285.

[L3] Lê D. T., *Le concept de singularité isolée de fonction analytique*, Advanced Stud. in Pure Math. **8** (1986), 215-227.

[Ma] J. Mather, Notes on Topological Stability, Mimeographed Notes, Harvard University, July 1970.

[Me] Z. Mebkhout, *Local cohomology of analytic spaces*, Pub. Res. Inst. Math. Sc. **12**, (1977), 247-256.

[Mi1] J. Milnor, Singular Points of complex Hypersurfaces, Ann. of Math. Stud. **61**, Princeton, 1968.

[Mi2] J. Milnor, Morse Theory, Ann. of Math. Stud. **51**, Princeton, 1963.

[P] D. Prill, *Local classification of quotients of complex manifolds by discontinuous groups*, Duke Math. J. **34** (1967), 375-386.

[T] R. Thom, *Ensembles et morphismes stratifiés*, Bull.Amer.Math.Soc. **75** (1969), 240-284.

[V] J.L. Verdier, *Dualité dans la cohomologie des espaces localement compacts*, in Séminaire Bourbaki, Exposé **300**, Publications de l'IHP and Benjamin Pub., 1965-1966.

Department of Mathematics, Northeastern University, Boston, MA 02115, USA.

Elliptic Fibrations over Surfaces I

Noboru Nakayama*
Department of Math., Faculty of Science, Univ. of Tokyo
Hongo Tokyo 113 JAPAN

1 Introduction.

The structure of a degeneration of elliptic curves is completely clarified by the work of Kodaira [7] and this is very important in the classification theory of compact complex surfaces. We are now interested in the classification of threefolds. In this paper, we shall study proper surjective morphisms over surfaces: $f : X \to S$ whose general fibers are elliptic curves. These f are called elliptic fibrations over surfaces S. We want to treat in general situation, but here we assume that S is a nonsingular surface, f is smooth outside of a normal crossing divisor of S and that f is a locally projective morphism. If $f : X \to S$ does not satisfy this assumption, after suitable blow-up of S and X, the result fibration satisfies the assumption. Note that f may not be a flat morphism. Further we restrict ourselves to the study of local structure of the fibration. Namely we assume that the base surface S is the two-dimensional unit polydisc $\Delta^2 := \{(t_1, t_2) \in \mathbf{C}^2 | |t_1| < 1, |t_2| < 1\}$ or more precisely, the germ $(\Delta^2, 0)$. Here 0 denotes the origin $(0,0) \in \Delta^2$. And also we assume that f is a projective morphism and is smooth outside of $\{t_1 = 0\} \cup \{t_2 = 0\}$.

By the recent progress of minimal model theory, especially by three-dimensional Flip Theorem([9]), we have the following:

Minimal Model Theorem ([13], [4], [9]) Let $g : M \to (\Delta^2, 0)$ be a germ of projective surjective morphism from a nonsingular threefold M, then there exists a germ of projective surjective morphism $f : X \to (\Delta^2, 0)$ such that

(1) f and g are bimeromorphically equivalent over $(\Delta^2, 0)$,

(2) X has only terminal singularities and is $\mathbf{Q}$-factorial,

(3) the canonical divisor K_X is f-nef.

*Partially supported by Yukawa Foundation.

Here two projective surjective morphisms $g : M \to U$ and $g' : M' \to U'$ defined over open neighborhoods U and U', respectively, of the origin 0, are same as a germ of morphisms over $(\Delta^2, 0)$, if $g|_{g^{-1}(U'')}$ and $g'|_{g'^{-1}(U'')}$ are isomorphic over an open neighborhood $U'' \subset U \cap U'$ of 0. For the definition of terminal singularity, $\mathbf{Q}$-factoriality, e.t.c., the reader should refer [13],[4].

A projective elliptic fibration should be a projective surjective morphism having elliptic curves as general fibers. A germ of projective elliptic fibration $f : X \to (\Delta^2, 0)$ is called *minimal*, if X satisfies the conditions (2) and (3) in Minimal Model Theorem. By considering Minimal Model Theorem, it is very important to classify all the germs of minimal elliptic fibrations over $(\Delta^2, 0)$ which are smooth outside of $\{t_1 = 0\} \cup \{t_2 = 0\}$. It is not so easy because, for example, minimal elliptic fibtrations are not unique in its bimeromorphic class in general. But by a result of [2], we know that any two bimeromorphically equivalent minimal elliptic fibrations are connected by a finite times of flops. If f admits a section $(\Delta^2, 0) \to X$, then we have a Weierstrass model W of f as in [14]. This W has only canonical singularities and its canonical divisor K_W is trivial. So we obtain a minimal model of f by taking a partial crepant resolution of W. But usually W has one dimensional singular locus. In this case, there is a work of Miranda [8] where he obtains a flat model of $W \to (\Delta^2, 0)$ by resolving the singularities of W after blowing-up the base surface. It is also important to decide when f has a section. In the case of elliptic surfaces, i.e., elliptic fibration over curves, if the type of the singular fiber is not ${}_mI_a$ with $m > 1$, then there exists a section.

In this part I of this paper, we treat only the case the general singular fibers over $\ell_1 := \{t_1 = 0\}$ and $\ell_2 := \{t_2 = 0\}$ are of type I_a and of type I_b, respectively. For the definition of type I_a, see [7] or Notation below.

Main Theorem Let $f : X \to \Delta^2_t$ be a germ of minimal elliptic fibration whose discriminant locus is contained in $\ell_1 \cup \ell_2$. Assume that fibers over general points of ℓ_1 and ℓ_2 are of type I_a and I_b, respectively. Then the following four conditions are satisfied:

(A) X is nonsingular and has trivial canonical bundle,

(B) f is a flat morphism,

(C) the central fiber $f^{-1}(0)$ is of type I_{a+b},

(D) the pull-back $f^*(\ell_1 + \ell_2)$ is a simple normal crossing divisor with $(a+b)$-components.

In section 2, we shall prove this by using Hodge theory and obtained a nonsingular minimal model by resolving the singularities of the Weierstrass model. In section 3, we shall construct this minimal elliptic fibration by using the toric geometry.

In part II of this paper, we shall treat other cases, for example we can prove that if the types of singular fibers over ℓ_1 and ℓ_2 are ${}_mI_a$ and ${}_nI_b$, then there exist positive integers p and q such that $(p,m)=1$, $(q,n)=1$ and $qma \equiv pnb \mod mn$.

1.1 Notation.

Minimal Model: We use the same notation as in [13] and [4].

Normal Crossing Divisor: An effective divisor D in a complex manifold M is said to be a *normal crossing divisor* , if for any point $P \in \mathrm{Supp}(D)$, there exist an open neighborhood U in M and coordinate functions $\{u_1, u_2, \ldots, u_k\}$ on U such that $D|_U = \mathrm{div}(u_1 \cdots u_k)$. A *simple normal crossing divisor* should be a normal crossing divisor whose components are all smooth divisors.

Elliptic Fibration: Let $f : X \to Y$ be a proper surjective morphism between normal complex varieties X and Y. f is called a *fibration*, if every fibers are connected. This is equivalent to the surjectivity of the canonical homomorphism $\mathcal{O}_Y \to f_*\mathcal{O}_X$. In this paper, a fibration $f : X \to Y$ is called an *elliptic fibration*, if f is a *projective* morphism whose general fibers are elliptic curves. The locus $\{y \in Y | f \text{ is not smooth along } f^{-1}(y)\}$ is an analytic subset which is called the *discriminant locus* of f.

Unit Polydisc: Let $\Delta_t^2 := \{(t_1,t_2) \in \mathbf{C}^2 | |t_1| < 1, |t_2| < 1\}$ be the two dimensional unit disc with the coordinate $t = (t_1, t_2)$. For simplicity, we denote the above Δ_t^2 by Δ^2. Also the lines $\{t_i = 0\}$ are denoted by ℓ_i for $i = 1, 2$, respectively.

Types of Singular fibers of Elliptic Surface: We shall use the same notation as in [7]. Especially, the singular fiber of type I_b is a cycle of smooth rational curves with b components, for $b \geq 2$. I_1 is a rational curve with one node and I_0 is a smooth elliptic curve. In the case of type I_b, the monodromy matrix is writen as $\begin{pmatrix} 1 & b \\ 0 & 1 \end{pmatrix}$. Therefore the J-function has a pole of order b at the origin.

Hodge Theory: Let $f : X \to S$ be a locally projective fibration between complex manifolds X and S. Assume that f is smooth outside of a normal crossing divisor E of S. Then for every $i \geq 0$, the higher direct image sheaf $H^i := R^i f_* \mathbf{Z}_X|_U$ is locally constant on $U := S \setminus E$. Further H^i becomes a variation of Hodge structures. Since E is a normal crossing divisor, we have the upper and lower canonical extensions ${}^u\mathcal{H}_S^i$ and ${}^\ell\mathcal{H}_S^i$ of $H^i \otimes \mathcal{O}_U$, respectively (cf. [5] and [10]). When the local monodromies are unipotent, the lower and upper canonical extensions coinside, which we denote simply by $\mathcal{H}_S^i$. For $* = u$, or ℓ, let $F^p({}^*\mathcal{H}^i)$ be the induced p-th filtration from the p-th Hodge filtration of H^i. Let us denote by $\omega_{X/S} := \omega_X \otimes f^*(\omega_S)^{-1}$ the relative dualizing sheaf of f and let $d = \dim X - \dim S$. Then we have the following:

Theorem 1.1 ([10], cf. [1], [5], [12])) *There exist canonical isomorphisms*

(1) $R^i f_* \omega_{X/S} \simeq F^d({}^u\mathcal{H}_S^{d+i})$ *and,*

(2) $R^i f_* \mathcal{O}_X \simeq Gr_F^0({}^l\mathcal{H}_S^i)$.

2 Proof of Main Theorem.

Let $f : X \to \Delta^2$ be a minimal elliptic fibration as in the Main Theorem.

Lemma 2.1 *ω_X is trivial.*

Proof. By Theorem 1.1, the direct image $f_*\omega_X$ is an invertible sheaf. Thus from the canonical bundle formula of minimal elliptic surface, we see that $\omega_X \simeq \mathcal{O}_X(E)$ for some effective Weil divisor E with $\mathrm{Supp}(E) \subset f^{-1}(0)$. Since $f : X \to \Delta^2$ is minimal, K_X is f-nef. Therefore, E must be zero, hence ω_X is trivial. Q.E.D.

Take a general smooth curve $C \subset \Delta^2$ such that $C \cap (\ell_1 \cup \ell_2) = \{0\}$ and denote the fiber product $X \times_{\Delta^2} C$ by F. Then F is an effective Cartier divisor of X.

Lemma 2.2 *F is an irreducible surface with only rational double points.*

Proof. Let $\mu : S \to F$ be the minimal desingularization of the main component of F. Then $f \circ \mu : S \to C$ is an elliptic surface. It is enough to show that the trace map $\mu_*\omega_S \to \omega_F$ is an isomorphism. Since $\omega_F \simeq (\omega_X \otimes \mathcal{O}(F)) \otimes \mathcal{O}_F \simeq \mathcal{O}_F$, we have only to show that $f_*\mu_*\omega_{S/C} \to f_*\omega_{F/C}$ is isomorphic. By Theorem 1.1, we have the following canonical isomorphisms:

(1) $f_*\omega_{X/\Delta^2} \simeq F^1(\mathcal{H}^1_{\Delta^2})$,

(2) $f_*\mu_*\omega_{S/C} \simeq F^1(\mathcal{H}^1_C)$ and

(3) $R^1 f_*\omega_{X/\Delta^2} \simeq \mathcal{O}_{\Delta^2}$.

Since the monodromy matrices are unipotent, we have $\mathcal{H}_{\Delta^2} \otimes \mathcal{O}_C \simeq \mathcal{H}_C$ and hence $F^1(\mathcal{H}_{\Delta^2}) \otimes \mathcal{O}_C \simeq F^1(\mathcal{H}_C)$. On the other hand, from the exact sequence

$$0 \to \omega_{X/\Delta^2}(-F) \to \omega_{X/\Delta^2} \to \omega_{F/C} \to 0,$$

we obtain $f_*\omega_{X/\Delta^2} \otimes \mathcal{O}_C \simeq f_*\omega_{F/C}$. By considering the commutative diagram:

$$\begin{array}{ccc} (f_*\omega_{X/\Delta^2}) \otimes \mathcal{O}_C & \to & f_*\omega_{F/C} \\ \downarrow & & \uparrow \\ F^1(\mathcal{H}^1_{\Delta^2}) \otimes \mathcal{O}_C & & f_*\mu_*\omega_{S/C} \\ \searrow & & \swarrow \\ & F^1(\mathcal{H}_C) & \end{array},$$

we see that the homomorphism $f_*\mu_*\omega_{S/C} \to f_*\omega_{F/C}$ must be an isomorphism. Q.E.D.

Proposition 2.3 *The central fiber $f^{-1}(0)$ is a reduced curve.*

Proof. Since F has only rational double points and trivial dualizing sheaf, S has also trivial canonical bundle. Thus the singular fiber of $f \circ \mu$ is of type I_{a+b}. The desingularization μ may contract some components of the singular fiber. Therefore the central fiber $f^{-1}(0)$ is a reduced curve. Q.E.D.

By the above Proposition 2.3, we have a section $\Delta^2 \to X$ of f, if we replace the disc Δ^2 by small one. Thus by [14], there exist holomorphic functions α and β on Δ^2 and a bimeromorphic morphism $\phi : X \to W(\alpha, \beta)$. Here $W(\alpha, \beta)$ is called the Weierstrass model of type (α, β) and defined to be the divisor $\{y^2 z = x^3 + \alpha x z^2 + \beta z^3\} \subset \mathbf{P}^2 \times \Delta^2$, where $(x : y : z)$ is a homogeneous coordinate of $\mathbf{P}^2$. Let $p : W := W(\alpha, \beta) \to \Delta^2$ be the induced morphism. Then fibers of p are irreducible cubic curves and a fiber $p^{-1}(t)$ is smooth if and only if $4\alpha(t)^3 + 27\beta(t)^2 \neq 0$. Note that W has only rational Gorenstein singularities and has trivial dualizng sheaf, since X has the same property. Thus $\omega_X \simeq \phi^*\omega_W$. From the assumption of the Main Theorem, there is a unit function ε on Δ^2 such that $4\alpha^3 + 27\beta^2 = \varepsilon t_1^a t_2^b$. Hence if $a = b = 0$, then p is a smooth morphism. Therefore ϕ must be an isomorphism and Main Theorem is proved in this case. In what follows, we assume that $(a, b) \neq (0, 0)$. Next we shall resolve the singularities of W.

Lemma 2.4 **(1)** *If $(a, b) \neq (0,0)$, then α and β are unit functions over Δ^2.*

(2) *For given $(a, b) \neq (0, 0)$, the Weierstrass models $W(\alpha, \beta)$ are unique up to isomorphisms over the germ $(\Delta^2, 0)$.*

(3) *The singularity of W is analytically isomorphic to the hypersurface singularity $\{uv = t_1^a t_2^b\} \subset \mathbf{C}^4$, where (u, v, t_1, t_2) is a coordinate of $\mathbf{C}^4$.*

Proof. (1). Since S in Lemma 2.2 has a singular fiber of type I_{a+b}, $\alpha(0)$ and $\beta(0)$ are not zero.

(2). As in [14], we can replace (α, β) by $(\alpha\lambda^4, \beta\lambda^6)$ for unit functions λ over Δ^2. Therefore over the germ $(\Delta^2, 0)$, we can normalize $\alpha = -3$ and $27(\beta^2 - 4) = t_1^a t_2^b$. If we take $\lambda = \sqrt{-1}$, then we can replace β by $-\beta$.

(3). (cf. [8]) The singular locus of W is contained in $\{z \neq 0\}$. Thus we use the affine coordinate (x, y, t_1, t_2). Then W is defined to be $\{y^2 = x^3 - 3x + \beta\}$, where $27(\beta^2 - 4) = t_1^a t_2^b$. We may assume that $\beta(0) = 2$. Therefore $x^3 - 3x + \beta = (x-1)^2(x + 2) + (unit)t_1^a t_2^b$. Thus the singular locus of W is contained in $\{x = 1, y = 0\}$. Since $x + 2$ is unit near the singular locus, the singularity is analytically isomorphic to $\{uv = t_1^a t_2^b\}$. Q.E.D.

Proposition 2.5 *There exists a resolution of singularities $\rho : \widetilde{W} \to W$ such that*

(1) *$\widetilde{W}$ has trivial canonical bundle, i.e., ρ is crepant,*

(2) *the central fiber of $\tilde{p} := p \circ \rho : \widetilde{W} \to \Delta^2$ is a cycle of smooth rational curves with $(a+b)$-components and that*

(3) *the pull back $\tilde{p}^*(\ell_1 + \ell_2)$ is a simple normal crossing divisor with $(a+b)$-components.*

Proof. If $a = 0$, then the singularity is analytically isomorphic to $\ell_2 \times (A_{b-1}$- sing.). Thus we obtained desired resolution $\rho : \widetilde{W} \to W$ by successive blow-ups along smooth centers. Thus we may assume that a and b are positive integers. First blow-up the variety $\{uv = t_1^a t_2^b\}$ along the ideal $(u, t_1{}^a)$. Then the exceptional set is $\mathbf{P}^1$ which is mapped to the origin. Thus we can extend this blow-up to W and obtained a crepant modification $\rho_1 : \widehat{W} \to W$. The singularity of $\widehat{W}$ is isomorphic to a disjoint union of $\ell_1 \times (A_{a-1}$- sing.) and $\ell_2 \times (A_{b-1}$- sing.). (cf. FIGURE $\widehat{W}$). Hence we have a crepant resolution of singularities of $\rho_2 : \widetilde{W} \to \widehat{W}$ by successive blow-ups along smooth centers. Thus the composition $\rho := \rho_1 \circ \rho_2$ is also crepant. It is easy to see the properties (2) and (3). (cf. FIGURE $\widetilde{W}$). Q.E.D.

FIGURE $\widehat{W}$

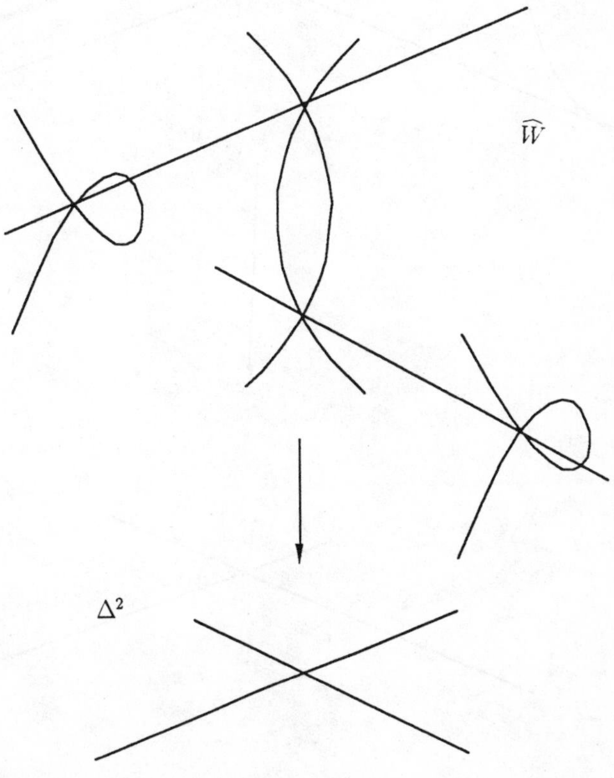

FIGURE $\widetilde{W}$

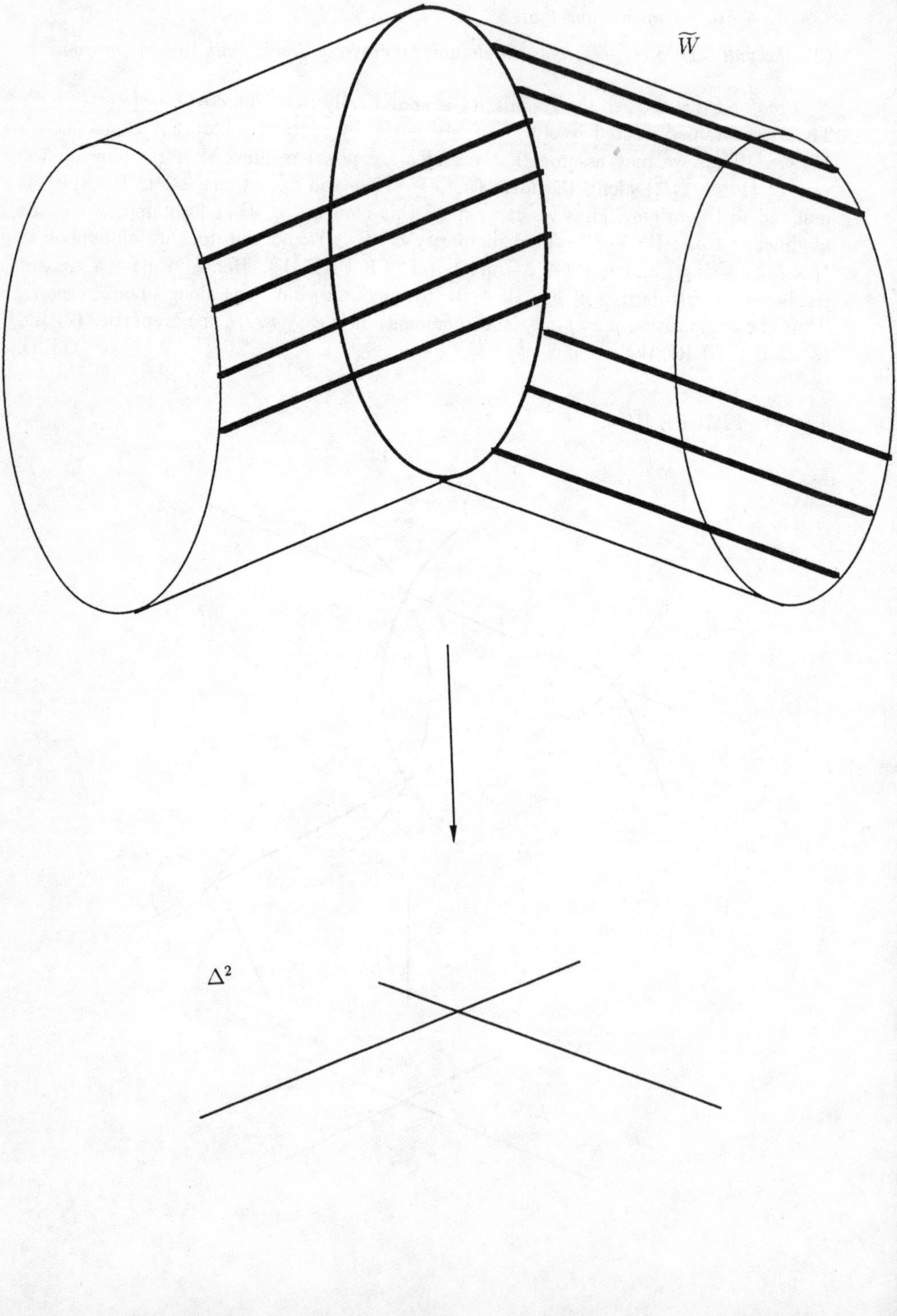

Therefore $\tilde{p} : \widetilde{W} \to \Delta^2$ satisfies the conditions of Main Theorem. By the construction of $\rho_2 : \widetilde{W} \to \widehat{W}$, we see that if C is a component of the central fiber $\tilde{p}^{-1}(0)$, then one of the following conditions (I) or (II) is satisfied (cf. FIGURE I and II):

(I) there exist an open neighborhood U of C and smooth divisors D_1, D_2, E_1, E_2 in U such that

(I-1) C is the scheme-theoretic intersection of D_1 and E_1,

(I-2) $D_2 \cap E_2 = \emptyset$,

(I-3) the intersection numbers $D_2 \cdot C = E_2 \cdot C = 1$ and

(I-4) $D_1 + D_2$ and $E_1 + E_2$ are pull-backs of the divisors ℓ_1 and ℓ_2, respectively, or

(II) there exist an open neighborhood U of C and smooth divisors D_1, D_2, D_3, E in U such that

(II-1) C is the scheme-theoretic intersection of D_2 and E,

(II-2) $D_1 \cap D_3 = \emptyset$,

(II-3) the intersection numbers $D_1 \cdot C = D_3 \cdot C = 1$ and

(II-4) $D_1 + D_2 + D_3$ and E are pull-backs of ℓ_1 and ℓ_2 or ℓ_2 and ℓ_1, respectively.

FIGURE I

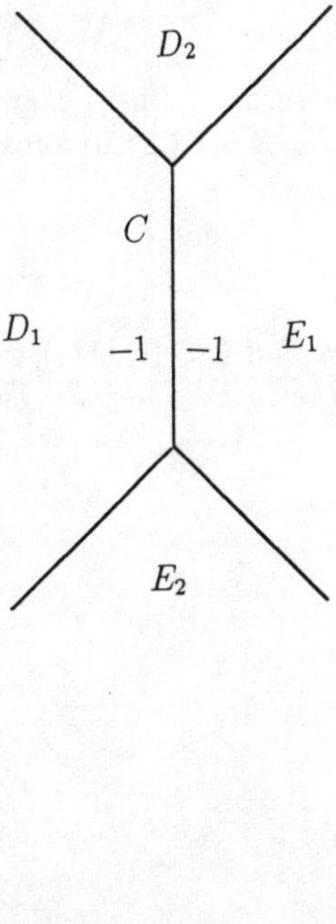

FIGURE II

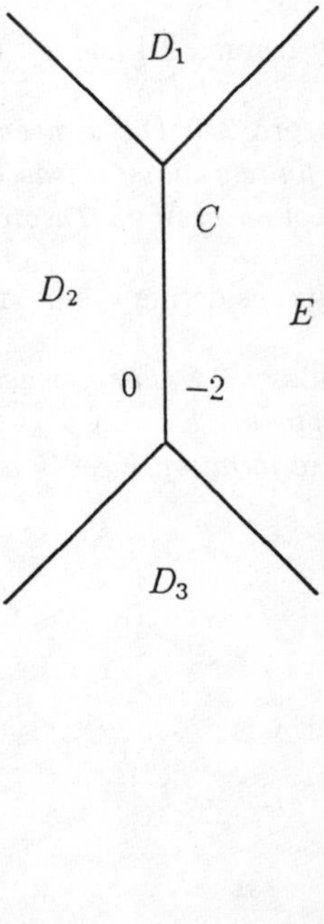

Claim 2.6 **(1)** *If C is a smooth rational curve satisfies the above condition (I), then the normal bundle is isomorphic to $\mathcal{O}_{\mathbf{P}^1}(-1)\oplus\mathcal{O}_{\mathbf{P}^1}(-1)$. If U' is a threefold obtained by the flop along $C \subset U$, D_1', D_2', E_1', E_2' are proper transforms of D_1, D_2, E_1, E_2, respectively, and if C' is the indeterminancy locus of the meromorphic map $U' \cdots \to U$, then $U', C', D_2', D_1', E_2', E_1'$ also satisfy the above condition (I).*

(2) *If C is a smooth rational curve satisfies the above condition (II), then the normal bundle is isomorphic to $\mathcal{O}_{\mathbf{P}^1}\oplus\mathcal{O}_{\mathbf{P}^1}(-2)$ and C cannot be an exceptional curve of any small contraction.*

Proof. (1). By the conditions (I-3) and (I-4), we have $D_1 \cdot C = -1$ and $E_1 \cdot C = -1$. Therefore by condition (I-1), the normal bundle is $\mathcal{O}_{\mathbf{P}^1}(-1)\oplus\mathcal{O}_{\mathbf{P}^1}(-1)$. The rest is clear.

(2). By the condition (II-3), we have $D_2 \cdot C = -2$ and $E \cdot C = 0$. Therefore by the condition (II-4), the normal bundle is $\mathcal{O}_{\mathbf{P}^1}\oplus\mathcal{O}_{\mathbf{P}^1}(-2)$ and C is not isolated. Q.E.D.

If $C \subset \tilde{p}^{-1}(0)$ is the proper transform in $\widetilde{W}$ of a component of the central fiber of $\widehat{W} \to \Delta^2$, then C satisfies the condition (I). Let $W' \cdots \to \widetilde{W}$ be the flop along C, then by the above Claim 2.6, $W' \to \Delta^2$ also satisfies the conditions of Main Theorem. Further if C' is a component of the central fiber of $W' \to \Delta^2$, then C' must satisfy the condition (I) or (II). If C' satisfies the condition (I), we have the flop $W'' \cdots \to W'$ along C' and $W'' \to \Delta^2$ also satisfies the conditions of Main Theorem by Claim 2.6. In this way, we see that any minimal elliptic fibrations obtained from $\widetilde{W} \to \Delta^2$ by a finite times of flops along components of central fibers satisfy the conditions of Main Theorem.

Proof of Main Theorem. By the construction, X and $\widetilde{W}$ are bimeromorphically equivalent over Δ^2. Further they are isomorphic in codimension one, since they are $\mathbf{Q}$-factorial and minimal over $(\Delta^2, 0)$. Therefore X and $\widetilde{W}$ are connected by flops (cf. [2]). Thus by the above argument, we are done. Q.E.D.

By Lemma 2.4 and by [3], we have:

Theorem 2.7 *For a given pair of non-negative integers $(a, b) \neq (0, 0)$, there exist at most finitely many germs of minimal elliptic fibrations $f : X \to (\Delta^2, 0)$ satisfying the assumption of Main Theorem.*

By considering the proof of Lemma 2.2, we have:

Corollary 2.8 *For a general smooth curve $C \subset \Delta^2$ such that $C \cap (\ell_1 \cup \ell_2) = \{0\}$, the fiber product $F := X \times_{\Delta^2} C$ is a nonsingular surface. The type of singular fiber of the induced filtration $F \to C$ is I_{a+b}.*

3 Toric construction.

In this section, we shall construct the minimal models of elliptic fibrations obtained in previous section, in the case a or b is positive, by using toric geometry.

Let $\delta : \mathbf{Z} \to \{1,2\}$ is a map such that

(1) $\delta(m+a+b) = \delta(m)$ for any $m \in \mathbf{Z}$,

(2) $\sharp\{0 \leq m < a+b \,|\, \delta(m) = 1\} = a$ and hence $\sharp\{0 \leq m < a+b \,|\, \delta(m) = 2\} = b$.

For a given pair of integers (p,q), we define inductively the functions i and j as follows:

$$i(0) = p, \qquad j(0) = q,$$

$$\begin{cases} i(m+1) = i(m)+1 \\ j(m+1) = j(m), \end{cases} \quad if \quad \delta(m) = 1 \quad \text{and,} \quad \begin{cases} i(m+1) = i(m) \\ j(m+1) = j(m)+1, \end{cases} \quad if \quad \delta(m) = 2$$

For any integer $k \in \mathbf{Z}$, let $\sigma_k \subset \mathbf{R}^3$ be the cone

$$\{(u,v,y) \in \mathbf{R}^3 |\, u \geq 0, v \geq 0 \quad \text{and} \quad i(k)u + j(k)v \leq y \leq i(k+1)u + j(k+1)v\},$$

$\sigma_k^* \cap \mathbf{Z}^3$ the semigroup $\{(i,j,\ell) \in \mathbf{Z}^3 |\, iu + jv + \ell y \geq 0, \quad \text{for any} \quad (u,v,y) \in \sigma_k\}$, $\mathbf{C}[\sigma_k^* \cap \mathbf{Z}^3]$ the associated semi-group ring and let $\widetilde{\mathcal{X}}_k^{p,q}$ be the affine scheme $\mathrm{Spec}\mathbf{C}[\sigma_k^* \cap \mathbf{Z}^3]$. By definition, $\mathbf{C}[\sigma_k^* \cap \mathbf{Z}^3]$ is isomorphic to the subring

$$\mathbf{C}[t_1, t_2, st_1^{-i(k)} t_2^{-j(k)}, s^{-1} t_1^{i(k+1)} t_2^{j(k+1)}] \subset \mathbf{C}[t_1^{\pm 1}, t_2^{\pm 1}, s^{\pm 1}].$$

Thus $\widetilde{\mathcal{X}}_k^{p,q}$ is isomorphic to $\mathbf{C}^3$. Let $\mathcal{X}_k^{p,q}$ be the fiber product $\widetilde{\mathcal{X}^{p,q}}_k \times_{\mathrm{Spec}\mathbf{C}[t_1,t_2]} \Delta_t^2$. By usual toric construction, we can patch these $\mathcal{X}_k^{p,q}$. Let us denote this union $\cup \mathcal{X}_k^{p,q}$ by $\mathcal{X}_\delta^{p,q}$. By the construction we have:

Lemma 3.1 *For all (p,q), there is an isomorphism $\mathcal{X}_\delta^{0,0} \to \mathcal{X}_\delta^{p,q}$ such that $(t_1,t_2,s) \mapsto (t_1, t_2, st_1^p t_2^q)$.*

Hence we can normalize $(p,q) = (0,0)$ and we denote $\mathcal{X}_\delta^{0,0}$ by $\mathcal{X}_\delta$. The general fiber of the induced fibration $\tilde{f} : \mathcal{X}_\delta \to \Delta_t^2$ is $\mathbf{C}^* = \mathbf{C} \setminus \{0\}$ and the central fiber is an infinite chains of smooth rational curves. Now we consider the isomorphism $g : \mathcal{X}_\delta \to \mathcal{X}_\delta$ such that $g : (t_1,t_2,s) \mapsto (t_1, t_2, st_1^a t_2^b)$. Then g gives an isomorphism $\mathcal{X}_k \to \mathcal{X}_{k+a+b}$ for all $k \in \mathbf{Z}$. By the same argument as in [11], we have:

Claim 3.2 *The action of g on $\mathcal{X}$ is fixed point free and properly discontinuous.*

Therefore we have the quotient manifold $X_\delta := \mathcal{X}_\delta / \langle g \rangle$ and an elliptic fibration $f_\delta : X_\delta \to \Delta^2$, which satisfies the conditions of Main Theorem .

Theorem 3.3 *Let $X \to \Delta^2$ be a germ of minimal elliptic fibration satisfying the assumption of Main Theorem. Then X is isomorphic to one of X_δ constructed above.*

Proof. Since the central fiber of $X_\delta \to \Delta^2$ is a cycle of smooth rational curves with $(a+b)$-components , we can take a section of the fibration, and thus X_δ is a minimal resolution of a Weierstrass model. But by Lemma 2.4, this Weierstrass model is unique. Thus X is obtained from X_δ by taking a finitely many times of flops. If X_δ is flopped along some $(-1,-1)$-curve, then the result is also isomorphic to $X_{\delta'}$ for some δ', since this flop is described by means of the changing the division $\{\sigma_k\}$ of $\{(u,v,y) \in \mathbf{R}^3 | u, v > 0\} \cup \{0\}$ (cf. [15]) and by Lemma 3.1. Therefore we are done. Q.E.D.

References

[1] Y. Kawamata, Characterization of abelian varieties, Composit. Math., **43** (1981), 253-276.

[2] Y. Kawamata, Crepant blowing-up of 3-dimensional canonical singularities and its application to degenerations of surfaces, Annals of Math., **127** (1988), 93-163.

[3] Y. Kawamata and K. Matsuki, The number of the minimal models for a 3-fold of general type is finite, Math. Ann., **276** (1987), 595-598.

[4] Y. Kawamata, K. Matsuda and K. Matsuki, Introduction to the minimal model problem, in *Algebraic Geometry Sendai 1985*, Adv. Studies in Pure Math., **10** (1987) Kinokuniya and North-Holland, 283-360.

[5] J. Kollár, Higher direct images of dualizing sheaves I, II, Annals of Math., **123** (1986), 11-42, ibid. **124** (1986), 171-202.

[6] J. Kollár, Flops, Nagoya Math. J., **113** (1989), 14-36.

[7] K. Kodaira, On complex analytic surfaces II, III, Annals of Math., **77** (1963), 563-626, ibid. **78** (1963), 1-40.

[8] R. Miranda, Smooth models for elliptic threefolds, in *The Birational Geometry of Degenerations*, Progress in Math., **29** (1983) Birkhäuser, 85-113.

[9] S. Mori, Flip theorem and the existence of minimal models for 3-folds, Journal of AMS., **1** (1988), 117-253.

[10] A. Moriwaki, Torsion freeness of higher direct images of canonical bundle, Math. Ann., **276** (1987), 385-398.

[11] I. Nakamura, Relative compactification of the Néron model and its application, in *Complex Analysis and Algebraic Geometry*, (1977) Iwanami Shoten and Cambridge Univ. Press, 207-225.

[12] N. Nakayama, Hodge filtrations and the higher direct images of canonical sheaves, Invent. Math., **85** (1986), 217-221.

[13] N. Nakayama, The lower semi-continuity of the plurigenera of complex varieties, in *Algebraic Geometry Sendai 1985*, Adv. Studies in Pure Math., **10** (1987) Kinokuniya and North-Holland, 551-590.

[14] N. Nakayama, On Weierstrass models, in *Algebraic Geometry and Commutative Algebra* in Honor of M. Nagata vol. II (1987) Kinokuniya and North-Holland, 405-431.

[15] M. Reid, Decomposition of toric morphisms, in *Arithmetic and Geometry II*, Progress in Math., **36** (1983) Birkhäuser, 395-418.

WEIL LINEAR SYSTEMS ON SINGULAR K3 SURFACES

Viacheslav V.Nikulin

Steklov Mathematical Institute,

ul.Vavilova 42, Moscow, GSP-1, 117966,USSR

§ 0. Introduction.

We recall that K3 surface is a smooth projective algebraic surface X over an algebraically closed field k with $K_X=0$ and $H^1(X,\mathcal{O}_X)=0$.

A normal projective algebraic surface Y is a singular K3 surface if for the minimal resolution of singularities $\sigma:X\longrightarrow Y$ the nonsingular surface X is a K3 one. In this case, all singularities of Y are Du Val singularities $\mathbf{A}_m$, $\mathbf{D}_m$, $\mathbf{E}_m$, and we get Y if we blow down trees of nonsingular rational curves of the type $\mathbf{A}_m$, $\mathbf{D}_m$, $\mathbf{E}_m$ on X.

V.A.Alekseev asked me: what one can say about a complete ample linear system $|\bar{D}|$ of integral Weil divisors $\bar{D}$ on singular K3 surface Y. For example, what one can say about the fixed part of the linear system, multiplicities of fixed components with respect to $\bar{D}^2$, $\dim|\bar{D}|$?

This problem is very important maybe from the viewpoint of a classification of Fano threefolds F with $\mathbb{Q}$-factorial terminal singularities. If the linear system $|-K_F|$ has a good member $Y\in|-K_F|$ then, by the adjunction formula, Y is a singular K3 surface and the restriction of the linear system $|-K_F|$ on Y is a complete ample linear system of Y. Thus, we can reduce a description of the $|-K_F|$ to a linear system on the surface Y. And a classification of Fano threefolds is very closely related with a description of linear systems on K3 surfaces.

Unfortunately, it is not proved yet that this good member does exist. Recently, V.A.Alekseev got some results in this direction, and it was the reason why he asked me about. On the other hand, as I think, one can consider results about linear systems on singular K3 surfaces as a good model for the system $|-K_F|$ on Fano threefolds with terminal singularities and can try to generalize these results for Fano threefolds with terminal singularities.

It was very strange to me that I did not see in literature some results devoted to linear systems on singular K3 surfaces. Except, of course, Saint-Donat's paper [S-D] devoted to nonsingular ones. It is required to construct some theory devoted to this problem.

At first, me and a little later Alekseev considered the case when rk Pic Y=1 (see § 3, **3.3** here). It was solved by different methods. I worked with nonsingular K3 surface X, and Alekseev used Riemann-Roch for singular K3 surface Y. Later, I considered the general case when rk Pic Y is arbitrary. The last case is much more complicated, and we will consider this case here. On the other hand, for the case rk Pic Y=1 we have a very precise answer. For an arbitrary rk Pic Y, we have a theory only. Using this theory, one can get the full description of all cases in principle.

At last, we recall some results about linear systems on nonsingular K3 surfaces. Here we have the

Proposition 0.1. *Let* $H\in\text{Pic } X$ *is* nef. *Then one of the cases* (i)-(iv) *below holds:*

(i) $H^2>0$, $|H|$ *contains an irreducible curve and has not fixed points,* $\dim |H|=H^2/2+1>0$;

(ii) $H^2=0$, $|H|=m|E|$, $m>0$, *where* $|E|$ *is an elliptic pencil* ($|H|$ *contains an irreducible curve for* $m=1$ *only*).

(iii) $H=0$, $|H|=\emptyset$.

(iv) $H^2>0$ *and* $|H|=m|E|+\Gamma$, $m>1$, *where* $|E|$ *is an elliptic pencil,* Γ *is an irreducible curve with* $\Gamma^2=-2$, *and* $E\cdot\Gamma=1$. *Here* $m=\dim|H|=H^2/2+1$, Γ *is the fixed part of* $|H|$.

Proof. It is well known to specialists and follows very easy from [S-D]. We will give a proof.

Let $H\neq 0$. Since H is nef, $H^2\ge 0$. Then, by Riemann-Roch theorem, $\dim|H|>0$. Let $|C|$ be the moving part of $|H|$ and Δ the fixed part. By [S-D], (i), or (ii) holds for $|C|$.

At first, let $|C|$ contains an irreducible curve C. By Riemann-Roch theorem, $(C+\Delta)^2\le C^2$. Thus, $\Delta\cdot(2C+\Delta)\le 0$. It follows $\Delta\cdot(C+\Delta)+\Delta\cdot C\le 0$. Since $C+\Delta$ and C are nef, $\Delta\cdot C=\Delta\cdot(C+\Delta)=0$. Then $\Delta^2=0$. If $\Delta=0$, we get the case (i). If $\Delta\neq 0$, by Riemann-Roch theorem, $\dim|\Delta|\ge 1$, and we get the contradiction.

Let $|C|=m|E|$ where $|E|$ is an elliptic pencil. By Riemann-Roch theorem, $(mE+\Delta)^2/2+1\le m$. Thus, $(mE+\Delta)\cdot\Delta+mE\cdot\Delta\le 2m-2$. Since $mE+\Delta$ is nef, either $E\cdot\Delta=0$ or $E\cdot\Delta=1$ and $\Delta^2\le -2$. We consider these possibilities.

Let $E\cdot\Delta=0$. By Hodge index theorem, $\Delta^2\le 0$. Since $E+\Delta$ is nef, $\Delta^2=0$. If $\Delta=0$, we get the case (ii). If $\Delta\neq 0$, we get the contradiction since $\dim|\Delta|\ge 1$.

Let $E\cdot\Delta=1$ and $\Delta^2\le -2$. Then $\Delta=\Gamma+\Delta'$ where Γ is an irreducible curve with $\Gamma^2=-2$, and $E\cdot\Gamma=1$, and $E\cdot\Delta'=0$, and Γ is not a component of the divisor Δ'. If $\Delta'=0$, we get (iv). Let $\Delta'\neq 0$. By Hodge index theorem and Riemann-Roch theorem, $(\Delta')^2<0$ and $(\Gamma+\Delta')^2=-2+2\Gamma\cdot\Delta'+(\Delta')^2<0$. Since Picard lattice of K3 surface is even, $2\Gamma\cdot\Delta'+(\Delta')^2=(\Gamma+\Delta')\cdot\Delta'+\Gamma\cdot\Delta'\leq 0$. Since $C+\Gamma+\Delta'$ is nef and $C\cdot\Delta'=0$, $(\Gamma+\Delta')\cdot\Delta'\geq 0$. Since Γ is not a component of Δ', $\Gamma\cdot\Delta'\geq 0$. It follows $(\Gamma+\Delta')\cdot\Delta'=\Gamma\cdot\Delta'=0$. Thus, $(\Delta')^2=0$. We get the contradiction. ∎

We want to get something similar for singular K3 surfaces. On the other hand, the Proposition 0.1 will be very important for us in the case of singular K3 surfaces also.

§ 1 Fixed part of linear system on nonsingular K3 surfaces.

Let X be a nonsingular K3 surface, H an effective divisor on X and $|H|$ the corresponding complete linear system. Let $|H|=|C|+\Delta$, where $|C|$ is the moving part and Δ is the fixed part of $|H|$. What one can say about the $|C|$ and Δ?

From the Proposition 0.1, it follows the following statement:

(*) $|C|$ satisfies the condition (i), (ii), or (iii) of the Proposition 0.1, and $\Delta=\sum k_i\Gamma_i$, where any Γ_i is an irreducible -2 curve and $k_i\in\mathbb{N}$. If $|C|=m|E|$ where E is an elliptic curve and $m\geq 2$ then there does not exist more than one irreducible component Γ_i of Δ such that $E\cdot\Gamma_i\geq 1$; if here $m\geq 4$, then the multiplicity k_i of the Γ_i is $k_i=1$.

Our question is: If (*) holds, when

$$|C+\Delta|=|C|+\Delta? \qquad (1.1)$$

We correspond to this situation a graph $G(C,\Delta)$ (and $G(\Delta)$) by the obvious way. The $G(C,\Delta)$ is the dual graph of intersections of the irreducible components C and Γ_i of $C+\Delta$. Here C is a general member of $|C|$ if $C^2>0$, and $C=mE$ where E is a general member of the pencil $|E|$ if $C^2=0$ and $|C|=m|E|$. The weight of the vertex C is equal to C^2, the weight of the vertex Γ_i is equal to -2. The multiplicity of C is equal to 1 if $C^2>0$, and is equal to m if $|C|=m|E|$ where $|E|$ is an elliptic pencil; the multiplicity of Γ_i is equal to k_i. For the case (i) let $C_{red}=C$, and for the case (ii) $C_{red}=E$ where $|C|=m|E|$. We denote by ∘ a vertex of the weight -2, and by C∘ (or C^2∘) a vertex of the weight C^2.

The question is: What are graphs of this kind possible? It is obvious that if $G(C,\Delta)$ is possible (has the property (1.1)) then an every subgraph of $G(C,\Delta)$ is possible. Here a subgraph corresponds to a divisor D such that $0 \le D \le C+\Delta$.

We prove the following basic theorem.

Theorem 1.1. *Let C and Δ are divisors on the nonsingular K3 surface X which satisfy the condition (*) above.*

Then $|C+\Delta| = |C| + \Delta$ if and only if $G(C,\Delta)$ *is a tree (particularly, all components of $C+\Delta$ are intersected transversely in no more than one point) and* $G(C,\Delta)$ *has no subtrees* $\tilde{D}_m$, $\tilde{E}_6$, $\tilde{E}_7$, $\tilde{E}_8$, $\tilde{D}_m(C)$, $\tilde{E}_6(C)$, $\tilde{E}_7(C)$, $\tilde{E}_8(C)$, $\tilde{B}_m(C)$ *or* $\tilde{G}_2(C)$ *below:*

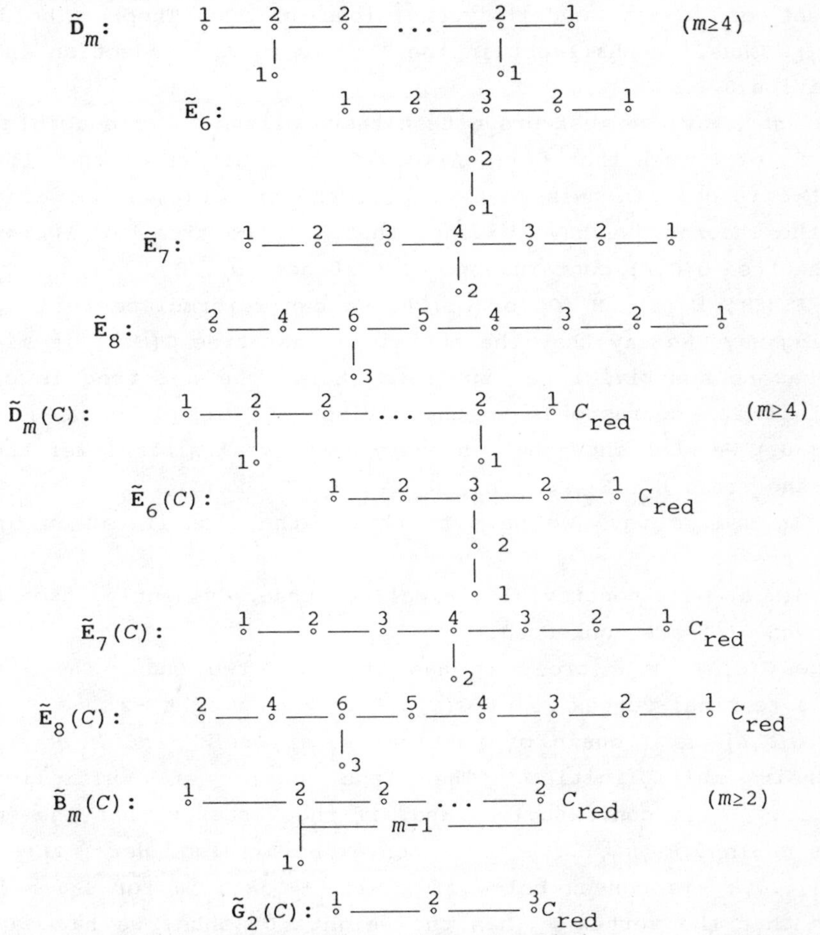

<u>Proof.</u> By the Proposition 0.1, these conditions are necessary: The divisors corresponding to subgraphs $\tilde{A}_m$, $\tilde{D}_m$, $\tilde{E}_6$, $\tilde{E}_7$, $\tilde{E}_8$, $\tilde{A}_m(C)$, $\tilde{D}_m(C)$, $\tilde{E}_6(C)$, $\tilde{E}_7(C)$, $\tilde{E}_8(C)$, $\tilde{B}_m(C)$, $\tilde{G}_2(C)$ are nef.

Let us prove the inverse statement which is much more difficult.

If $\Delta=0$, the statement is trivial. If $|C|=m|E|$, where $|E|$ is an elliptic pencil, $m\geq 2$ and $\Delta=\Gamma$ is an irreducible curve, then the statement holds by the Proposition 0.1 and the condition (*).

Let $\Delta\neq\emptyset$ and Δ is not an irreducible curve if $|C|=m|E|$, $m\geq 2$ and $|E|$ is an elliptic pencil. Let $G(C,\Delta)$ be a tree and it has no subtrees $\tilde{D}_m$, $\tilde{E}_6$, $\tilde{E}_7$, $\tilde{E}_8$, $\tilde{D}_m(C)$, $\tilde{E}_6(C)$, $\tilde{E}_7(C)$, $\tilde{E}_8(C)$, $\tilde{B}_m(C)$ or $\tilde{G}_2(C)$. We will show that then there exists an irreducible component Γ_i of Δ such that $\Gamma_i\cdot(C+\Delta)<0$. It follows the Theorem. Indeed, then Γ_i is a fixed component of $|C+\Delta|$, and the conditions of the Theorem hold for $C+(\Delta-\Gamma_i)$. Thus, we shall obtain the Theorem by the induction and the Proposition 0.1.

In such a way, we must prove that there exists an irreducible component Γ_i of Δ such that $\Gamma_i\cdot(C+\Delta)<0$. If it is not true, then the divisor $C+\Delta$ is nef. In this case we call the tree $G(C,\Delta)$ nef also. To prove the Theorem, we have to show that, if the tree $G(C,\Delta)$ is nef, then the tree $G(C,\Delta)$ contains one of subtrees $\tilde{D}_m$, $\tilde{E}_6$, $\tilde{E}_7$, $\tilde{E}_8$, $\tilde{D}_m(C)$, $\tilde{E}_6(C)$, $\tilde{E}_7(C)$, $\tilde{E}_8(C)$, $\tilde{B}_m(C)$ or $\tilde{G}_2(C)$. We can reformulate this by the following way. We say that the nontrivial nef tree $G(C,\Delta)$ is minimal if it has no nontrivial nef subtrees (Here, the nef tree is called trivial if it corresponds to the divisors C, or kE, or $kE+\Gamma$ where $k\geq 2$, or 0.) We must show that an every nontrivial minimal nef tree is one of the trees $\tilde{D}_m$, $\tilde{E}_6$, $\tilde{E}_7$, $\tilde{E}_8$, $\tilde{D}_m(C)$, $\tilde{E}_6(C)$, $\tilde{E}_7(C)$, $\tilde{E}_8(C)$, $\tilde{B}_m(C)$ or $\tilde{G}_2(C)$. In such a way, we have to obtain the classification of nef minimal trees.

Let $G(C,\Delta)$ be a nontrivial minimal nef tree. Evidently, then trees $G(C,\Delta)$ and $G(\Delta)$ are connected.

Since $G(C,\Delta)$ is a tree, it has at least two ends. Thus, there exists a terminal vertex v_1 of $G(C,\Delta)$ with the weight -2.

Let $G(C,\Delta)$ be a chain of vertices $v_1,v_2,\ldots,v_m$ and $k_1,k_2,\ldots,k_m$ are their multiplicities. Then the chain of multiplicities $0,k_1,k_2,\ldots,k_m$ is convex below, and, if the vertex v_m has the weight -2, the chain $0,k_1,k_2,\ldots,k_m,0$ is convex below also. Here, the chain $0,k_1,k_2,\ldots,k_m$ is convex below if $k_i-k_{i-1}+k_i-k_{i+1}\leq 0$ for $1\leq i\leq m-1$. It follows that the vertex v_m has the weight ≥ 0 (thus, we have a case (i) or (ii)) and the chain of multiplicities $0,k_1,k_2,\ldots,k_m$ is strongly increased. It follows very easy that $m=3$, $k_1=1$, $k_2=2$, $k_3=3$ and the vertex $v_3=E$ where $|E|$ is an elliptic pencil ($G(C,\Delta)$ is nontrivial minimal nef!). Thus, $G(C,\Delta)$ is the tree $\tilde{G}_2(C)$.

We recall that the valence of a vertex v of a tree is the number of edges of the tree which come out from v. Suppose that G(C,Δ) is not a chain. Then we can suppose that the chain $v_1,v_2,\ldots,v_m$ consists of vertices $v_2,\ldots,v_{m-1}$ of the valence 2, and the v_m has a valence ≥3. For the cases (i) and (ii), the vertex C is a terminal vertex of the tree G(C,Δ) since the tree G(Δ) is connected. Thus, the vertex v_m has the weight -2. The multiplicity k_m of the v_m is ≥2 since the chain of multiplicities $0,k_1,\ldots k_m$ is increased.

If the vertex v_m has the valence ≥4, then G(C,Δ) contains a subtree of type $\tilde{\mathbf{D}}_4$ or $\tilde{\mathbf{D}}_4(C)$ with the vertex v_m of the subtree of the valence 4. It follows that G(C,Δ) is this subtree, since G(C,Δ) is a minimal nef tree.

Thus, further, we can suppose that v_m has the valence 3. Let $\alpha_1,\alpha_2,\ldots\alpha_n=v_m$ and $\beta_1,\beta_2,\ldots,\beta_p=v_m$ be two other chains of vertices of G(C,Δ) which are different from the chain $v_1,v_2,\ldots,v_m$ and come out from v_m. Here we suppose that the valence of $\alpha_2,\ldots\alpha_{n-1}$ and $\beta_2,\ldots,\beta_{p-1}$ is 2 and the vertices a_1 and β_1 have valence 1 or ≥3.

Suppose that the vertex α_1 has the valence ≥3. Let $t_1,\ldots,t_n=k_m\geq 2$ are multiplicities of $\alpha_1,\alpha_2,\ldots,\alpha_n$. In this case, if all multiplicities $t_1,t_2,\ldots,t_n$ are strongly greater than 1, the tree G(C,Δ) contains a subtree $\tilde{\mathbf{D}}_{n+2}$ or $\tilde{\mathbf{D}}_{n+2}(C)$ with the vertices α_1 and v_m of the valence 3 in this subtree. Then G(C,Δ) is coincided with this subtree. Thus, we can suppose that there exists $i\geq 1$ such that $t_i=1$ and all $t_{i+1},\ldots,t_n=\mathrm{k}_m$ are strongly greater than 1.

It follows that we can find nef subtree T of G(C,Δ) with vertices z, $u_1,\ldots,u_{l-1}$, $v_1,\ldots,v_{m-1}$, $w_1,\ldots,w_{t-1}$ and with the form

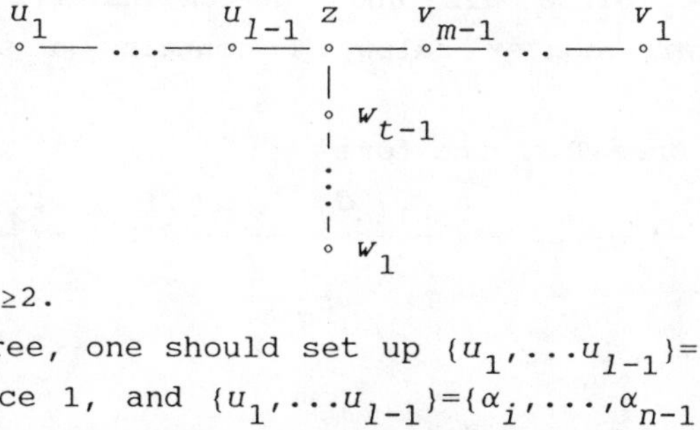

where $l\geq 2$, $m\geq 2$, $t\geq 2$.

To get this tree, one should set up $\{u_1,\ldots u_{l-1}\}=\{\alpha_1,\ldots,\alpha_{n-1}\}$ if α_1 has the valence 1, and $\{u_1,\ldots u_{l-1}\}=\{\alpha_i,\ldots,\alpha_{n-1}\}$ if α_1 has the valence ≥3. By the same way, one gets the chain $w_1,\ldots,w_{m-1}$ using the chain $\beta_1,\ldots,\beta_p$. Since G(C,Δ) is minimal, G(C,Δ)=T. We should prove that G(C,Δ)=$\tilde{\mathbf{E}}_6$, $\tilde{\mathbf{E}}_7$, $\tilde{\mathbf{E}}_8$, $\tilde{\mathbf{E}}_6(C)$, $\tilde{\mathbf{E}}_7(C)$, $\tilde{\mathbf{E}}_8(C)$, or $\tilde{\mathbf{B}}_m(C)$. We prove it in the Lemmas below. We denote by D the curve C in the case (i), the curve E in the case (ii), and one of the terminal vertices of $G(C,\Delta)=G(\Delta)$ in the case (iii). We denote by δ the multiplicity of D. Thus, $\delta=1$ in the case (i), $|C|=\delta|E|$ in the case (ii) where $|E|$ is an elliptic pencil, δ is equal to the multiplicity of D in the case (iii). Indexes near vertices on pictures are multiplicities of the vertices.

Lemma 1. *If a tree* T *of the form*

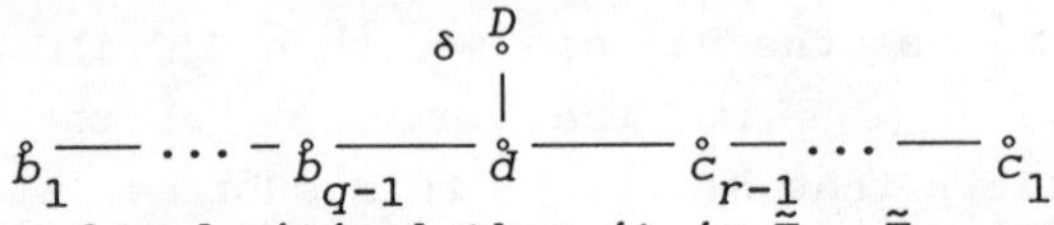

where $2\le q\le r$, *is* nef *and minimal then it is* $\tilde{E}_7$, $\tilde{E}_8$, *or* $\tilde{B}_3(C)$.

<u>Proof.</u> The chains $0,b_1,\ldots,b_{q-1},d$ and $0,c_1,\ldots,c_{r-1},d$ are convex below and $\delta\ge d-b_{q-1}+d-c_{r-1}$ where $d-b_{q-1}\ge 1$ and $d-c_{r-1}\ge 1$. It follows that $\delta\ge 2$ and $d\ge 2$. Thus, we have the case (ii) or (iii).

Let us consider the case (ii). Since $\delta\ge 2$, $d\ge 2$ and $2\le q\le r$, the tree T contains the subtree $\tilde{B}_3(C)$. It follows $T=\tilde{B}_3(C)$, since T is minimal.

Let us consider the case (iii). Then the chain $0,\delta,d$ is also convex below. It follows that we have an inequality

$$d/2\ge\delta\ge d/q+d/r. \qquad (1.2)$$

It follows that $3\le q\le r$. Let $q=3$. From (1.2), $1/2\ge 1/3+1/r$ and $r\ge 6$. It follows $c_i\ge i$ and $d\ge r\ge 6$ since the chain $0,c_1,\ldots,c_{r-1},d$ is convex below. From (1.2), then $\delta\ge 3$. If $b_1=1$, then $d/2\ge\delta\ge(d-1)/2+d/r$. It follows, $d/r\le 1/2$. We obtain the contradiction, since $d\ge r$. Thus, $b_1\ge 2$. Since the chain $0,b_1,b_2,d$ is convex below, $b_2\ge 4$. As a result, we prove that T contains a subtree $\tilde{E}_8$. Then $T=\tilde{E}_8$.

Suppose $4\le q\le r$. Since $\delta\ge 2$ and the chains $0,b_1,\ldots,b_{q-1},d$ and $0,c_1,\ldots,c_{r-1},d$ are convex below, T contains a subtree $\tilde{E}_7$. Then $T=\tilde{E}_7$. ∎

Lemma 2. *If a tree* T *of the form*

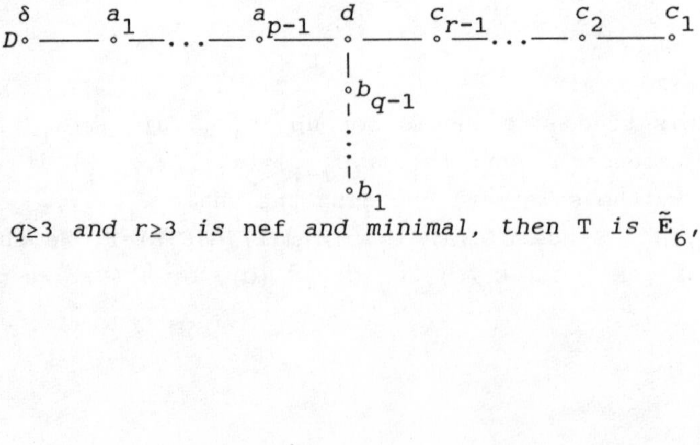

where $p\ge 1$, $q\ge 3$ *and* $r\ge 3$ *is* nef *and minimal, then* T *is* $\tilde{E}_6$, $\tilde{E}_7$, $\tilde{E}_8$ *or* $\tilde{E}_6(C)$.

Proof. Let us use an induction by p. For $p=1$ it was proved in the Lemma 1 that T is $\tilde{E}_7$ or $\tilde{E}_8$.

Now suppose that $p\geq 2$.

At first, suppose that $a_1=1$. Then, evidently, $D^2\geq 0$ and we have the case (i) or (ii). Let us set up $C'=C+\Gamma_1$ where Γ_1 is the component corresponding to the vertex with the multiplicity a_1 and the weight of C' is 1. Then we get the statement by induction: The case $p=2$ is impossible; if $p\geq 3$ then $p=3$ and T is $\tilde{E}_6(C')$. It follows that T contains the subtree $\tilde{E}_6$ and $T=\tilde{E}_6$. We get the contradiction.

Let $a_1\geq 2$. The chains $\delta,a_1,\ldots,a_{p-1},d$, and $0,b_1,\ldots,b_{q-1},d$, and $0,c_1,\ldots,c_{r-1},d$ are convex below. Since $q\geq 3$ and $r\geq 3$, we get $d\geq 3$, $b_{q-1}\geq 2$, and $c_{r-1}\geq 2$. If $\delta=1$, then also $a_{p-1}\geq 2$, since $a_1\geq 2$. It follows $T=\tilde{E}_6$ or $\tilde{E}_6(C)$. If $\delta=2$, then the chain $\delta=2,a_1,\ldots,a_{p-1},d$ is increased (may be not strongly), since $a_1\geq 2$. It follows T contains the subtree $\tilde{B}_{p+2}(C)$. Then $T=\tilde{B}_{p+2}(C)$, and we get the contradiction, since $q\geq 3$ and $r\geq 3$. If $\delta\geq 3$, then T contains the subtree $\tilde{G}_2(C)$ since $a_1\geq 2$ and $p\geq 2$. Then $T=\tilde{G}_2(C)$, and we get the contradiction, since $q\geq 3$ and $r\geq 3$. ∎

Lemma 3. *If a tree* T *of the form*

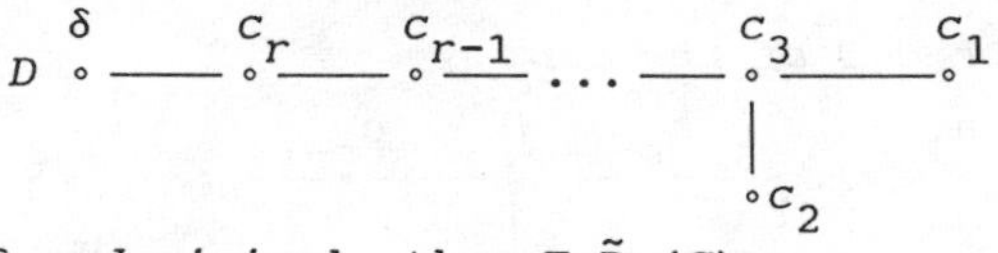

where $r\geq 4$, *is* nef *and minimal, then* $T=\tilde{B}_r(C)$.

Proof. The chains $0,c_1,c_3$, and $0,c_2,c_3$, and $\delta,c_r,\ldots,c_3$ are convex below and $c_4\geq c_3-c_1+c_3-c_2$. It follows $c_4\geq c_3\geq 2$, and the chain $\delta,c_r,\ldots,c_3$ is decreased. It follows, $D^2=0$ and we have the case (ii). Then T contains the subtree $\tilde{B}_r(C)$, hence $T=\tilde{B}_r(C)$. ∎

Lemma 4. *If a tree* T *of the form*

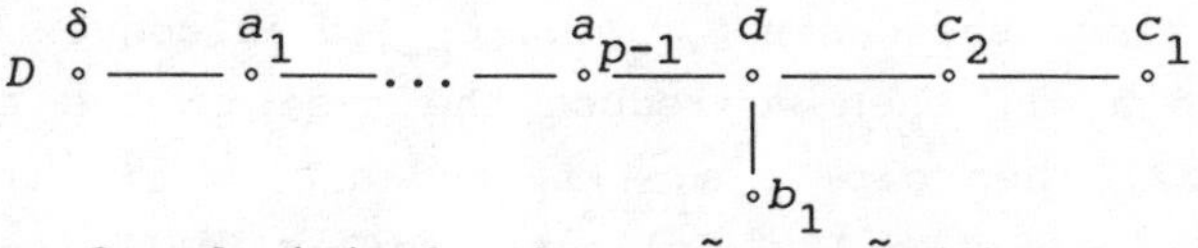

where $p\geq 1$, *is* nef *and minimal, then* $T=\tilde{E}_8$ *or* $\tilde{E}_8(C)$.

Proof. The case (i). Then $\delta=1$. The chains $1,a_1,\ldots,a_{p-1},d$, and $0,b_1,d$, and $0,c_1,c_2,d$ are convex below, and $b_1+c_2+a_{p-1}\geq 2d$. It follows that $d/2+2d/3+1+(d-1)(p-1)/p\geq 2d$. Thus, $d(2-1/2-2/3-(p-1)/p)\leq\leq 1-(p-1)/p$. Or $d(1/p-1/6)\leq 1/p$. Evidently, $d\geq 3$. It follows, $p\geq 4$. If $p=4$, we get $d/12\leq 1/4$. It follows $d=3$. One can see very easy that this case is impossible. If $p=5$, we get $d/30\leq 1/5$. If follows that $d\leq 6$. One can see very easy, that then $d=6$ and $T=\tilde{E}_8(C)$.

Let us suppose that $p\geq 6$. If $a_1=1$, we set up $C_1=C+\Gamma_1$ where Γ_1 corresponds to the vertex with the multiplicity a_1, and this case is reduced to the case $p-1$: we obtain that $p=6$ and $T=\tilde{E}_8(C_1)$. Then T contains $\tilde{E}_8$ and $T=\tilde{E}_8$. We get the contradiction. If $a_1\geq 2$, then $d\geq p+1\geq 7$ and $pd/(p+1)+2d/3+b_1\geq 2d$. Or $d(1/3+1/(p+1))\leq b_1$. Since $d\geq 7$, we get $b_1\geq 3$. If $c_1=1$, we get $pd/(p+1)+d/2+1+(d-1)/2\geq 2d$. It follows $d/(p+1)\leq 1/2$. This is impossible since $d\geq p+1$. It proves that T contains the subtree $\tilde{E}_8$ and $T=\tilde{E}_8$. We get the contradiction.

The case (ii). If $a_1=1$, we set up $C_1=C+\Gamma_1$ (like above). It reduces the case to the previous one, and we get that T contains the subtree $\tilde{E}_8$. Particularly, it holds if $\delta\geq 4$. Let $a_1\geq 2$. If $\delta=3$, then T contains the subtree $\tilde{G}_2(C)$ and $T=\tilde{G}_2(C)$. We get the contradiction. If $\delta=2$, we get that the chain $\delta,a_1,\dots,a_{p-1},d$ is increased since $a_1\geq 2$. It follows that T contains the subtree $\tilde{D}_{p+1}(C)$, and $T=\tilde{D}_{p+1}(C)$. We get the contradiction. If $\delta=1$, the proof is the same as for the case (i).

The case (iii). Then the chain $0,a_1,\dots,a_{p-1},d$ is convex below, and the proof is similar to the case (i). ■

Lemma 5. *If a tree* T *of the form*

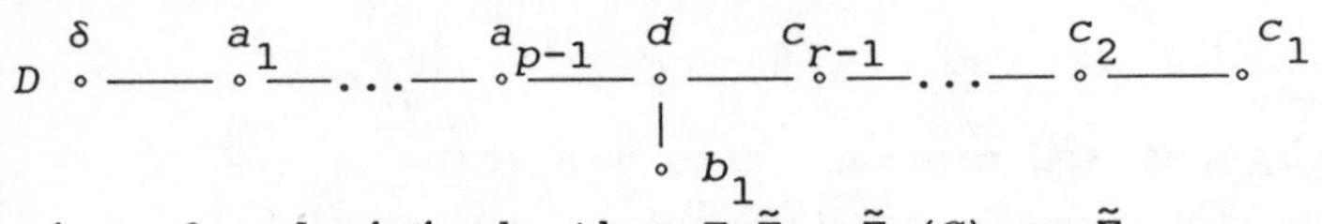

where $r\geq 4$, is nef *and minimal, then* $T=\tilde{E}_7$, $\tilde{E}_7(C)$ *or* $\tilde{E}_8$.

<u>Proof.</u> The case (i).

If $p=1$, we get the statement from the Lemma 1.

Let $p=2$. Then $d-b_1\geq d/2$, and $d-a_1\geq(d-1)/2$, and we get $c_{r-1}\geq d-b_1+d-a_1\geq d-1/2$. It follows that $c_{r-1}\geq d$. We get the contradiction since the chain $0,c_1,\dots,c_{r-1},d$ is convex below.

Let $p\geq 3$. If $a_1=1$, then we reduce the case to the case $p-1$ like above. Let $a_1\geq 2$. Then $d\geq p+1$, $a_i\geq 1+i$, and $d\geq r$, $c_i\geq i$. Let $b_1=1$. Then $dp/(p+1)+d(r-1)/r+1\geq 2d$. It follows, $d(2-p/(p+1)-(r-1)/r)\leq 1$. Or $d(1/(p+1)+1/r)\leq 1$. But $d\geq p+1$ and $d\geq r$. We get the contradiction. Thus, $b_1\geq 2$. It follows, T contains the subtree $\tilde{E}_7(C)$.

The case (ii). The proof is the same as for the Lemma 4.

The case (iii). We have the inequality $2d\leq d(p/(p+1)+1/2+(r-1)/r)$. Thus, $1/(p+1)+1/r\leq 1/2$ and $p\geq 2$. The case $p=2$ follows from the Lemma 4. Let $p\geq 3$. Then $\delta\geq 1$, $a_i\geq i+1$, $d\geq 4$, $c_j\geq j$.

Let $b_1=1$. Then $d/(p+1)+d/r\leq 1$. But $d\geq p+1$ and $d\geq r$. We get the contradiction, and $b_1 \geq 2$.

As a result, we proved that T contains a subtree $\tilde{E}_7$. Then $T=\tilde{E}_7$. It finishes the proof of the Lemma and the Theorem 1.1. ■

The basic Theorem 1.1 reduces a description of all possible graphs $G(C,\Delta)$ with the condition (1.1) to a description of nonsingular curves trees $\mathcal{T}$ on K3 surfaces which satisfy the condition

(*)′ $\mathcal{T}$ does not contain more than one curve C with a square $C^2 \geq 0$ (if $\mathcal{T}$ has not such a curve, we set up $C=0$); all other curves Γ_i, $i \in I$, of the $\mathcal{T}$ are nonsingular rational.

To obtain all possible graphs $G(C,\Delta)$, one should prescribe to the curves C and Γ_i of the trees $\mathcal{T}$ multiplicities m and k_i such that the condition (*) holds, and prove the condition of the Theorem 1.1. Here any tree $\mathcal{T}$ is possible if these multiplicities are equal to one:

Corollary 1.2. *If $\mathcal{T}$ is a tree satisfying to the condition* (*)′, *then for the divisor* $\Delta = \sum_{i \in I} \Gamma_i$ *holds that* $|C+\Delta| = |C| + \Delta$.

Proof. This follows from the theorem 1.1, or one can prove it independently (consider a terminal vertex with a weight -2 of $G(C)$). ■ Δ

§ 2. Trees of nonsingular curves on a nonsingular K3 surface.

2.1. General remarks. We consider here results on a classification of nonsingular curves trees $\mathcal{T}$ on K3 surfaces which satisfy to the condition (*)′. $G(C)$ is the graph of intersections of curves of $\mathcal{T}$ and G the graph of intersections of the curves Γ_i, $i \in I$.

To obtain this classification, we use the following reasons (I), (II), (III), (IV) below, which are purely algebraic.

(I). *Hodge index theorem:* A *tree* $G(C)$ *should not be more than hyperbolic - the corresponding intersection matrix has not more than one positive square.*

By (I), connected component G_i of G may be elliptic (with negative definite intersection matrix), parabolic (with semidefinite intersection matrix), and hyperbolic (with hyperbolic intersection matrix).

Proposition 2.1.1. (1) *An elliptic connected component of* G *is a tree* $\mathbf{A}_m, \mathbf{D}_m, \mathbf{E}_6, \mathbf{E}_7$ *or* $\mathbf{E}_8$.

(2) A *parabolic connected component of* G *is a tree* $\tilde{\mathbf{D}}_m, \tilde{\mathbf{E}}_6, \tilde{\mathbf{E}}_7$ *or* $\tilde{\mathbf{E}}_8$.

(3) *A hyperbolic connected component* G_{hyp} *of* G *is unique.*

(4) *If* $G_{hyp} \neq \emptyset$, *then all other components of* G *are elliptic.*

(5) *If* $C^2 \geq 0$ *and* $C \neq 0$ *and* $G_{hyp} \neq \emptyset$, *then* C *is joined to a vertex* Γ_{hyp}

of G_{hyp}. *If* $C^2>0$ *and* G_i *is a parabolic component of* G, *then* C *is joined to a vertex* Γ_i *of* G_i.

Proof. It is obvious. ■

For the matrix M we denote by $D(M)$ the determinant of M, and $\bar{D}(M)=D(-M)$. For the subgraph T of $G(C)$ we denote by the same letter the corresponding intersection matrix. It is obvious that

$$\bar{D}(T) \quad \text{is} \begin{cases} >0 \text{ if T is elliptic,} \\ =0 \text{ if T is parabolic,} \\ \leq 0 \text{ if T is hyperbolic,} \\ <0 \text{ if T is hypebolic and linearly independent.} \end{cases} \qquad (2.1)$$

We use the following simple formula: Let a tree G has a form:

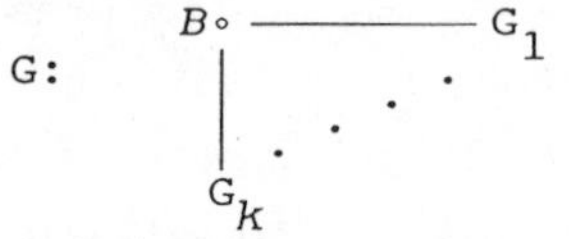

Let v_i be the vertex of G_i joined to B. Then

$$\bar{D}(G)=\bar{D}(G_1)\bar{D}(G_2)\cdots\bar{D}(G_k)(-B^2)-\bar{D}(G_1-v_1)\bar{D}(G_2)\cdots\bar{D}(G_k)-$$
$$-\bar{D}(G_1)\bar{D}(G_2-v_2)\bar{D}(G_3)\cdots\bar{D}(G_k)-\ldots-\bar{D}(G_1)\bar{D}(G_2)\cdots\bar{D}(G_{k-1})\bar{D}(G_k-v_k)=$$
$$=\bar{D}(G_1)\bar{D}(G_2)\cdots\bar{D}(G_k)(-B^2-\bar{D}(G_1-\{v_1\})/\bar{D}(G_1)-\ \ldots-\bar{D}(G_k-\{v_k\})/\bar{D}(G_k)). \qquad (2.2)$$

(II). *On a K3 surface, if* E *is an effective curve with* $E^2=0$, *then* $C\cdot E\geq 2$ *for any irreducible curve* C *with* $C^2\geq 0$.

We can use (II) by the following way.

Let we have a connected parabolic subtree $\mathcal{P}$ of $G(C)$: $\{C\}$, where $C^2=0$ and $C\neq 0$, $\tilde{D}_n$, $\tilde{E}_6$, $\tilde{E}_7$ or $\tilde{E}_8$. This tree corresponds to all components of an elliptic pencil fiber on a K3 surfaces. Thus, an every vertex v of $\mathcal{P}$ has the invariant $m(\mathcal{P},v)$ which is equal to the multiplicity of the corresponding to v irreducible component of the fiber. (This invariants are shown as the multiplicities of the vertices of the trees $\tilde{D}_n$, $\tilde{E}_6$, $\tilde{E}_7$, $\tilde{E}_8$ of the Theorem 1.1.) By (II), we have the

Proposition 2.1.2. (1) *Let* $C^2\geq 0$ *and* $C\neq 0$. *Let* $\mathcal{P}$ *be a connected parabolic subtree of* $G(C)$, *let* p *be a vertex of* $\mathcal{P}$ *joined to* C. *Then* $m(\mathcal{P},p)>1$.

(2) *Let* $\mathcal{P}$ *and* Q *be two connected parabolic subtrees of* $G(C)$ *which have not common vertices. Let* p *be a vertex of* $\mathcal{P}$ *and* q *of* Q *and* pq *an edge of* $G(C)$. Then *either* $m(\mathcal{P},p)>1$ *or* $m(Q,q)>1$. ■

For example, it follows that $G(C)$ has not subtrees (where $C^2\geq 0$):

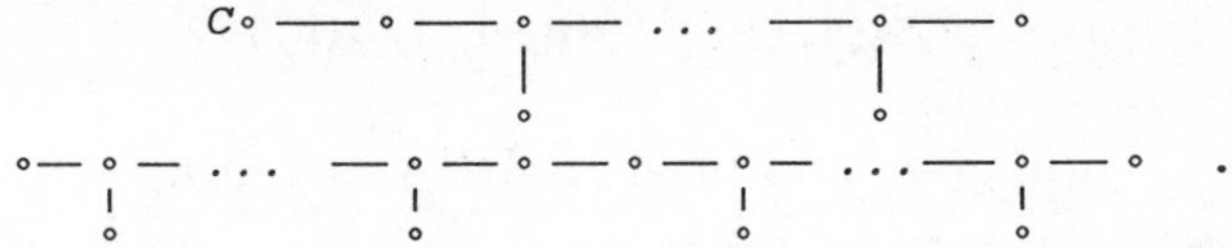

(III). *An elliptic pencil on a K3 surface has not multiple fibers.*

It follows the

Proposition 2.1.3. *Let a tree* $G(C)$ *has two disjoint connected parabolic subtrees* $\mathcal{P}$ *and* Q, *and a vertex* w *of* $G(C)-\mathcal{P}$ *is joined to a vertex* p *of the* $\mathcal{P}$. *Then* w *is joined to some vertex* q *of the* Q *and*

$$m(\mathcal{P},p)=m(Q,q). \blacksquare$$

Let $\mathcal{E}=\mathbf{A}_n$, $\mathbf{D}_n$ or $\mathbf{E}_n$ be an elliptic subtree of $G(C)$ and e be a vertex of $\mathcal{E}$. We can introduce the invariant $m(\mathcal{E},e)$ which is equal to the set of multiplicities of the vertex e under all possible embeddings of $\mathcal{E}$ into all parabolic connected graphs $\tilde{A}_m$, $\tilde{D}_m$, $\tilde{E}_m$.

Proposition 2.1.4. *Let* $G(C)$ *has two disjoint connected subtrees* $\mathcal{P}$ *and* $\mathcal{E}$ *where* $\mathcal{P}$ *is parabolic and* $\mathcal{E}$ *is elliptic. Let a vertex* w *of* $G(C)-\mathcal{P}$ *is joined to a a vertex* p *of* $\mathcal{P}$ *and to a vertex* e *of* $\mathcal{E}$.

Then $m(\mathcal{P},p)\geq \min m(\mathcal{E},e)$.

Proof. Vertices of $\mathcal{P}$ correspond to all components of a degenerate fiber of an elliptic pencil E on X. Vertices of $\mathcal{E}$ correspond to some components of an other degenerate fiber of E and also have multiplicities. By (III), we get the statement. $\blacksquare$

(IV). *The rank of Picard lattice of* K3 *surface* ≤ 22, *and it is* ≤ 20 *if a basic field has the characteristic* 0.

It follows the

Proposition 2.1.5. $\mathrm{rk}\, G(C)\leq 22$, *and* $\mathrm{rk}\, G(C)\leq 20$ *if* char=0.

We describe all possible trees $G(C)$ which satisfy the condition (I), the Proposition 2.1.2 of (II), the Propositions 2.1.3 and 2.1.4 of (III) and the Proposition 2.1.5 of (IV). It is a purely algebraic problem about sets of vectors in a linear space with a symmetric pairing.

2.2. The case $C^2>0$. For K3 surface $C^2\geq 2$, and we have the

Theorem 2.2.1. 1. *Let* $C^2\geq 2$ *and the* $G_{hyp}\neq\emptyset$, *let* Γ_{hyp} *be the vertex of the* G_{hyp} *joined to* C. *Then* $G_{hyp}-\Gamma_{hyp}$ *is elliptic.*

2. *Let* $H_1,\dots,H_t$ *be all connected components of* $G_{hyp}-\Gamma_{hyp}$ *and* Γ_i *be a vertex of* H_i *joined to* Γ_{hyp}. *Then*

$$\bar{D}(G_{hyp})=\bar{D}(H_1)\cdots\bar{D}(H_t)(2-\bar{D}(H_1-\Gamma_1)/\bar{D}(H_1)-\dots-\bar{D}(H_t-\Gamma_t)/\bar{D}(H_t))<0$$

where

$$\bar{D}(H_1-\Gamma_1)/\bar{D}(H_1)+\ldots+\bar{D}(H_t-\Gamma_t)/\bar{D}(H_t)>2.$$

3. *For* $G_{hyp}(C)$ *we have*

$$\bar{D}(G_{hyp}(C))=\bar{D}(H_1)\cdot\cdot\bar{D}(H_t)(-C^2(2-\bar{D}(H_1-\Gamma_1)/\bar{D}(H_1)-\ldots-\bar{D}(H_t-\Gamma_t)/\bar{D}(H_t))-1)\le 0$$

$$2\le C^2\le \bar{D}(G_{hyp}-\Gamma_{hyp})/(-\bar{D}(G_{hyp}))=1/(\bar{D}(H_1-\Gamma_1)/\bar{D}(H_1)+\ldots+\bar{D}(H_t-\Gamma_t)/\bar{D}(H_t)-2).$$

4. *It follows:* $2 < \bar{D}(H_1-\Gamma_1)/\bar{D}(H_1)+\ldots+\bar{D}(H_t-\Gamma_t)/\bar{D}(H_t) \le 5/2.$

5.

$$\operatorname{rk} G_{hyp}(C) = \begin{cases} \#G(C) \text{ if } C^2<1/(\bar{D}(H_1-\Gamma_1)/\bar{D}(H_1)+\ldots+\bar{D}(H_t-\Gamma_t)/\bar{D}(H_t)-2) \\ \#G(C)-1 \text{ if } C^2=1/(\bar{D}(H_1-\Gamma_1)/\bar{D}(H_1)+\ldots+\bar{D}(H_t-\Gamma_t)/\bar{D}(H_t)-2) \end{cases}.$$

6. *From 1-4 and the Proposition 2.1.2, it follows:*

If $C^2>42$ *then* $G_{hyp}=\emptyset$. *If* $G_{hyp}\neq\emptyset$, *then the tree* G_{hyp} *has three (only if* $C^2\le 42$*), four (only if* $C^2\le 6$*) or five (only if* $C^2=2$*) ends, and* $G_{hyp}(C)$ *is one of the following trees:*

G_{hyp} *has three ends:*

$G_1(C^2;p,q,r;a)$

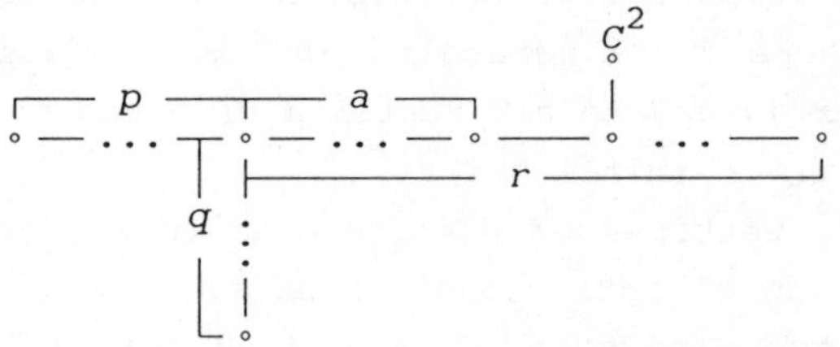

where $a\ge 0$, $2\le p\le q\le r$, *if* $p\ge 3$ *then* $a\le 1$, *if* $q\ge 3$ *then* $a\le 4$, *if* $q\ge 4$ *then* $a\le 2$; $1/p+1/q+1/r<1$, $1/p+1/q+1/a>1$,

$2\le C^2\le (r-a)(-pqa+pq+qa+ap)/(pqr-pq-qr-rp)\le 42$ *(the case* $C^2=42$ *corresponds to the case* $a=0$, $p=2$, $q=3$, $r=7$ *only).*

$\bar{D}(G_1(C^2;p,q,r;a))=C^2(pqr-pq-qr-rp)-(r-a)(pqa-pq-qa-ap)$,

$\bar{D}(G_{hyp})=-pqr+pq+qr+rp$, $\bar{D}(G_{hyp}-\Gamma_{hyp})=(r-a)(-pqa+pq+qa+ap)$.

G_{hyp} *has four ends:*

$G_2(C^2;2,q;a,b,):$

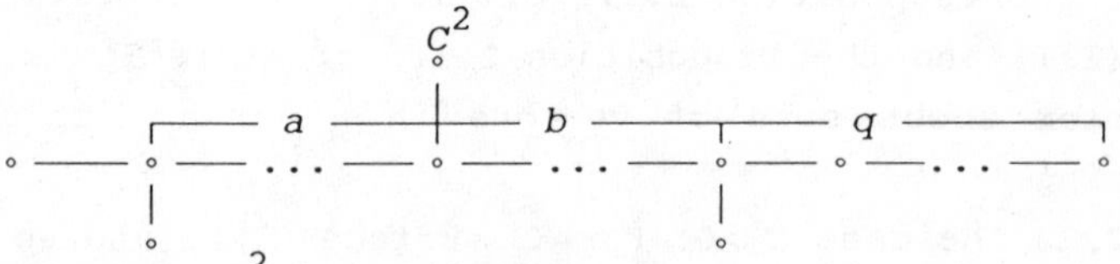

where $2\le C^2\le 6$, $a\ge 1$, $b\ge 1$, $q\ge 3$, $C^2+b\le 3+4/(q-2)$.

$\bar{D}(G_2(C^2;2,q;a,b))=4((q-2)C^2+((q-2)(b-3)-4))$, $\bar{D}(G_{hyp})=8-4q$,

$\bar{D}(G_{hyp}-\Gamma_{hyp})=4(4-(q-2)(b-3))$.

$G_2(2;3,q;1,2)$

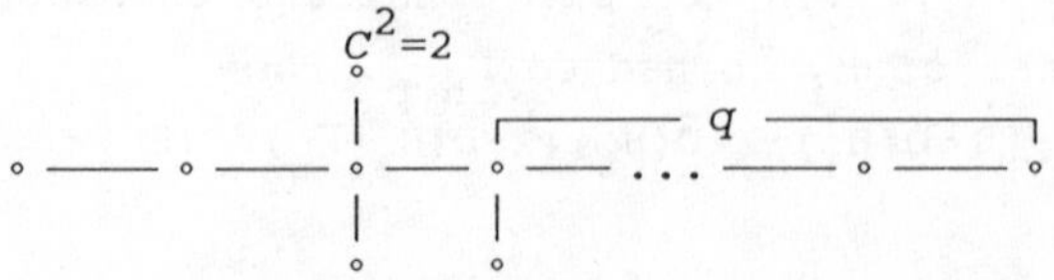

where $q=3,4$. $\bar{D}(G_2(2;3,q;1,2))=8(q-4)$, $\bar{D}(G_{hyp})=-7q+10$, $\bar{D}(G_{hyp}-\Gamma_{hyp})=6(q+2)$.

$G_2(2;p,3;1,2)$

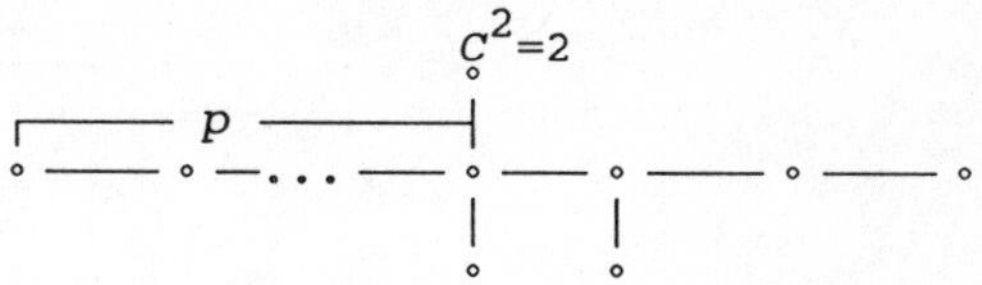

where $p=4,5$. $\bar{D}(G_2(2;p,3;1,2))=4(p-5)$, $\bar{D}(G_{hyp})=-7p+10$. $\bar{D}(G_{hyp}-\Gamma_{hyp})=10p$.

$G_2(2;p,3;1,3)$

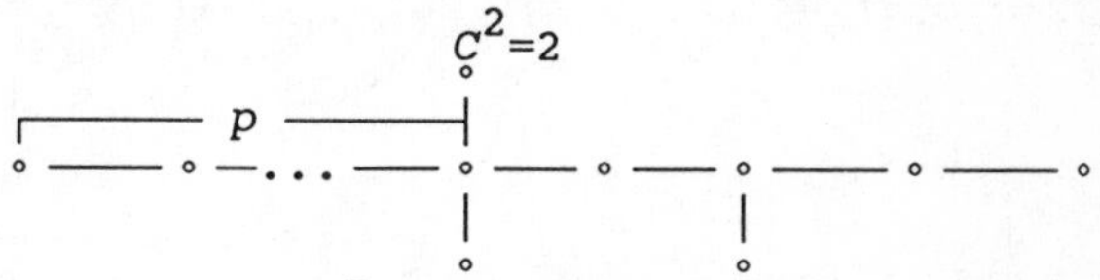

where $p=3,4$. $\bar{D}(G_2(2;p,3;1,3)=4p-16$, $\bar{D}(G_{hyp})=8-6p$, $\bar{D}(G_{hyp}-\Gamma_{hyp})=8p$.

$G_2(2;3,3;1,4)$

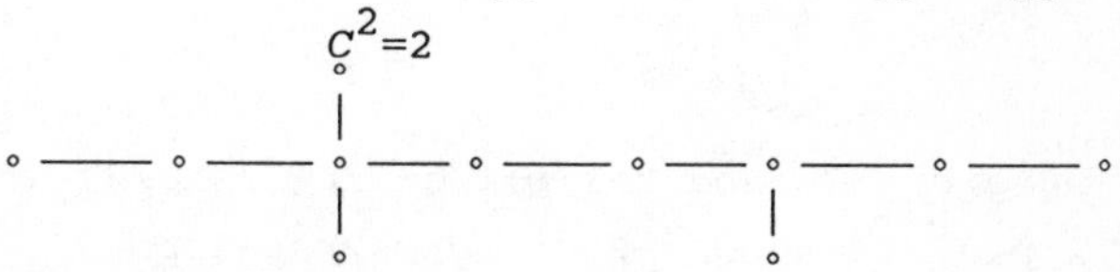

$\bar{D}(G_2(2;3,3;1,4)=0$, $\bar{D}(G_{hyp})=-9$, $\bar{D}(G_{hyp}-\Gamma_{hyp})=18$.

$G_2(2;3,3;2,2)$

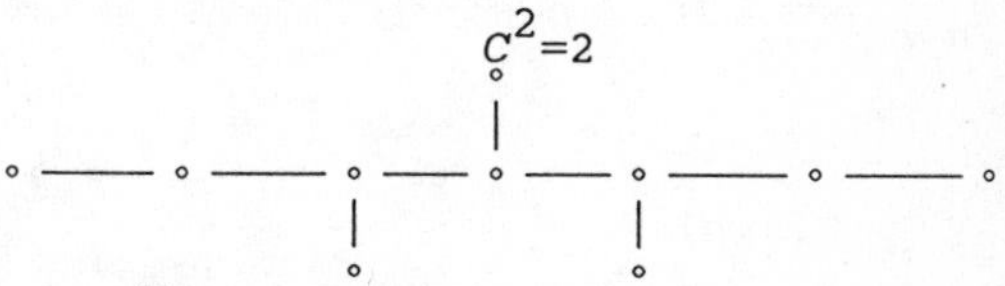

$\bar{D}(G_2(2;3,3;2,2)=-5$, $\bar{D}(G_{hyp})=-10$, $\bar{D}(G_{hyp}-\Gamma_{hyp})=25$.

$G_2(2;3,3;2,3)$

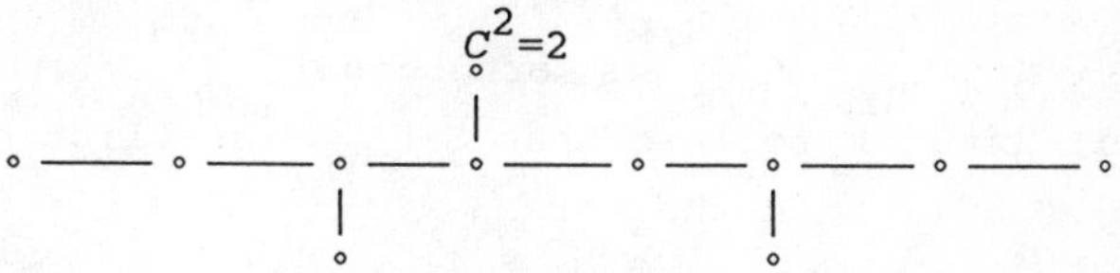

$\bar{D}(G_2(2;3,3;2,3)=-2$, $\bar{D}(G_{hyp})=-9$, $\bar{D}(G_{hyp}-\Gamma_{hyp})=20$.

$G_2(2;3,3;3,3)$

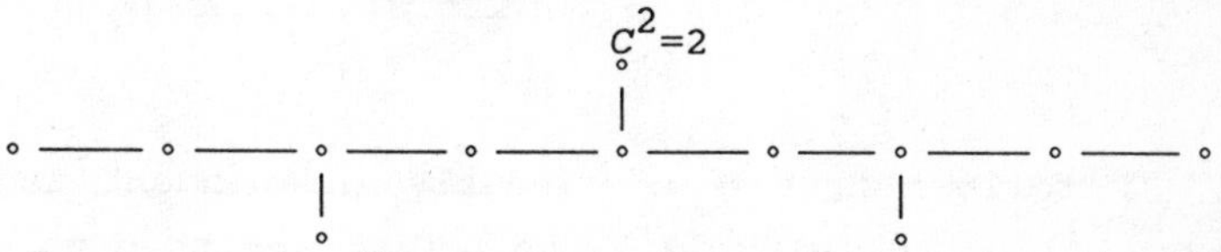

$\bar{D}(G_2(2;3,3;3,3)=0$, $\bar{D}(G_{hyp})=-8$, $\bar{D}(G_{hyp}-\Gamma_{hyp})=16$.

$G_3(2;a)$:

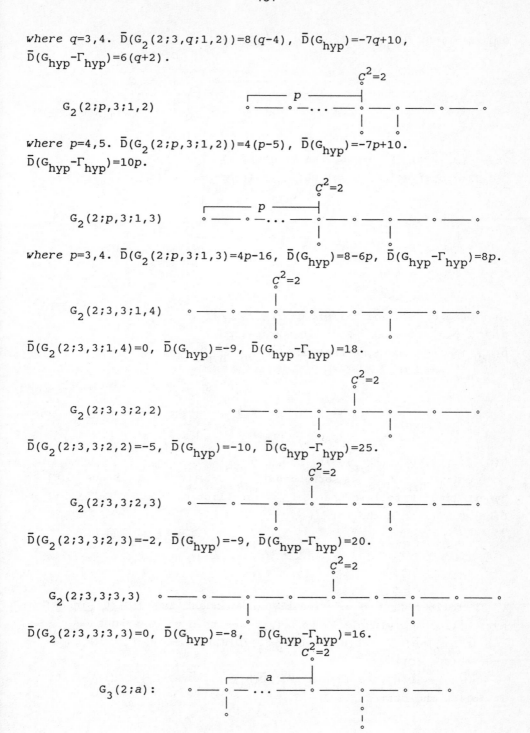

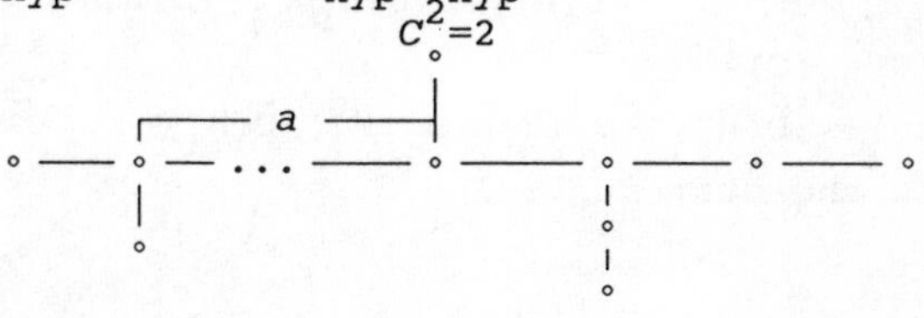

$\bar{D}(G_3(2;a))=0$, $\bar{D}(G_{hyp})=-12$, $\bar{D}(G_{hyp}-\Gamma_{hyp})=24$;

$G_4(2;a;p,q)$

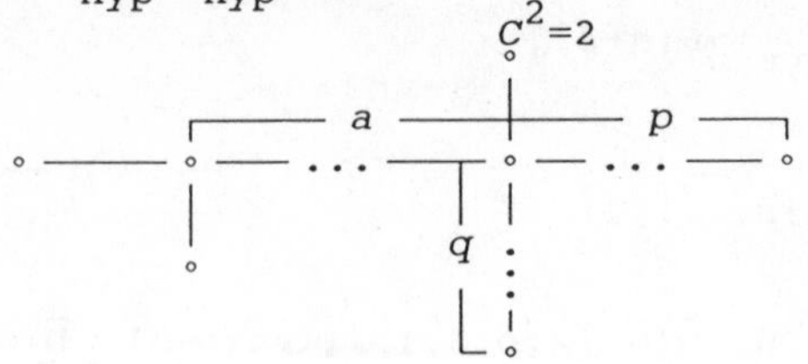

where $(p,q)=(3,q)$, $3\le q\le 6$, *and* $(p,q)=(4,4)$.

$\bar{D}(G_4(2;a;p,q))=4(pq-2p-2q)$, $\bar{D}(G_{hyp})=4(-pq+p+q)$, $\bar{D}(G_{hyp}-\Gamma_{hyp})=4pq$.

$G_5(2;p,q,r,s)$

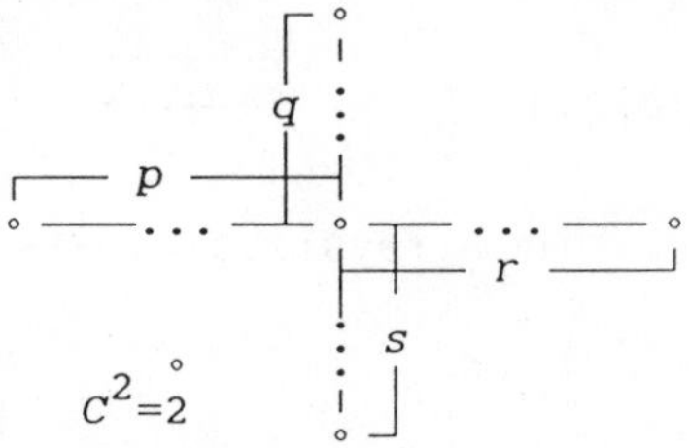

where $p\ge q\ge r\ge s\ge 2$, *and* $2>1/p+1/q+1/r+1/s\ge 3/2$;

$\bar{D}(G_5(2;p,q,r,s)=3pqrs-2pqr-2qrs-2rsp-2spq\le 0$,

$\bar{D}(G_{hyp})=-2pqrs+pqr+qrs+rsp+spq$, $\bar{D}(G_{hyp}-\Gamma_{hyp})=pqrs$.

G_{hyp} *has five ends:*

$G_6(2;a,b)$

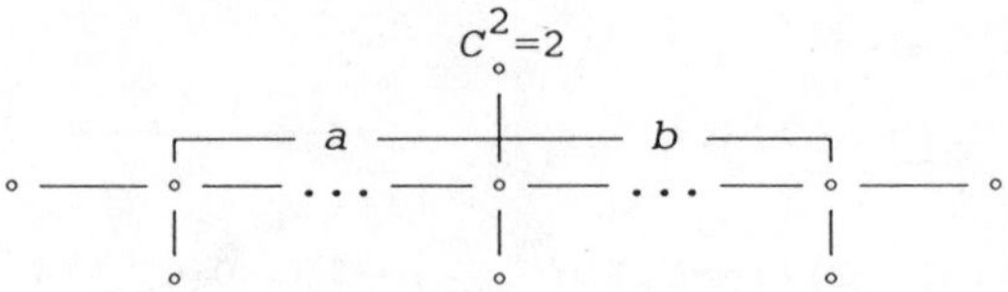

$\bar{D}(G_6(2;a,b))=0$, $\bar{D}(G_{hyp})=-16$, $\bar{D}(G_{hyp}-\Gamma_{hyp})=32$.

Proof. $G_{hyp}-\Gamma_{hyp}$ is orthogonal to C with $C^2>0$. Since $G(C)$ is hyperbolic, it follows that $G-\Gamma_{hyp}$ is elliptic. Thus, $\bar{D}(G-\Gamma_{hyp})>0$ and $\bar{D}(H_i)>0$. $G(C)$ has the form

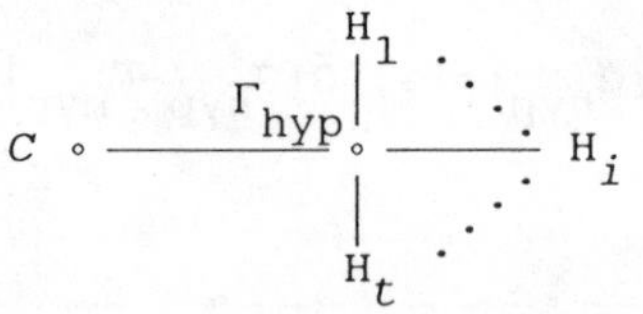

It follows that G is linearly independent and $\bar{D}(G)<0$ since G is hyperbolic. Applying (2.2) to $D=\Gamma_{hyp}$ and then to C, we get

$$\bar{D}(G_{hyp})=\bar{D}(H_1)\cdots\bar{D}(H_t)(2-\bar{D}(H_1-\Gamma_1)/\bar{D}(H_1)-\ldots-\bar{D}(H_t-\Gamma_t)/\bar{D}(H_t))<0$$

and $\bar{D}(G_{hyp}(C))=$

$$=\bar{D}(H_1)\cdots\bar{D}(H_t)(-C^2(2-\bar{D}(H_1-\Gamma_1)/\bar{D}(H_1)-\ldots-\bar{D}(H_t-\Gamma_t)/\bar{D}(H_t))-1)\le 0.$$

It follow the statements 1 - 5.

To get the last statement 6, we should find all possible sets $(H_1,\Gamma_1;\ldots;\ H_t,\Gamma_t)$ such that

$$2 < \bar{D}(H_1-\Gamma_1)/\bar{D}(H_1)+\ldots+\bar{D}(H_t-\Gamma_t)/\bar{D}(H_t) \le 5/2.$$

For an elliptic tree $H=\mathbf{A}_n$, $\mathbf{D}_m$, $\mathbf{E}_6$, $\mathbf{E}_7$ or $\mathbf{E}_8$ and its vertex Γ let

$$I(H,\Gamma)=\bar{D}(H-\Gamma)/\bar{D}(H).$$

One can find very easy the full list of such pairs with $I(H,\Gamma)\le 5/2$:

$(\mathbf{A}_q,\Gamma^{(1)})$: 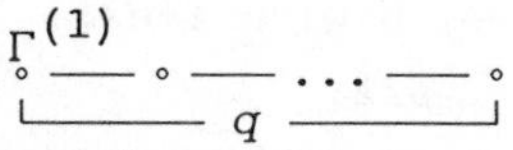

with $I(\mathbf{A}_q,\Gamma^{(1)})=1-1/(1+q)\ge 1/2$, where $q\ge 1$;

$(\mathbf{A}_{1+q},\Gamma^{(2)})$: 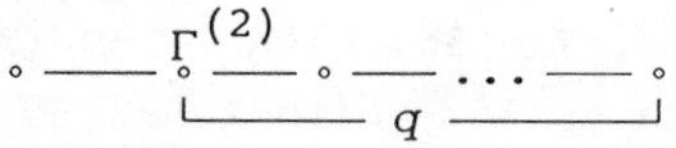

with $I(\mathbf{A}_{1+q},\Gamma^{(2)})=2-4/(2+q)\ge 1$, where $q\ge 2$;

$(\mathbf{A}_{2+q},\Gamma^{(3)})$: 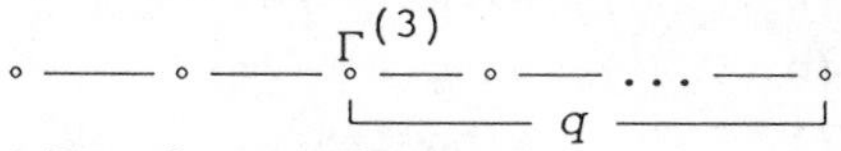

with $I(\mathbf{A}_{2+q},\Gamma^{(3)})=3-9/(3+q)\ge 3/2$, where $q\ge 3$;

$(\mathbf{A}_{3+q},\Gamma^{(4)})$: 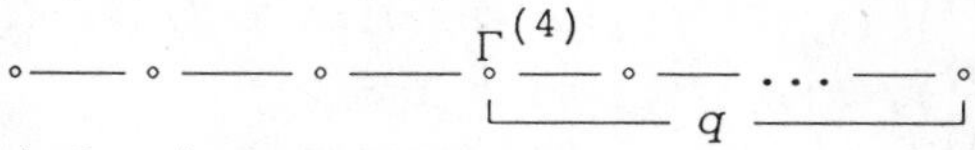

with $I(\mathbf{A}_{3+q},\Gamma^{(4)})=4-16/(4+q)\ge 2$, where $q\ge 4$;

$(\mathbf{A}_9,\Gamma^{(5)})$: 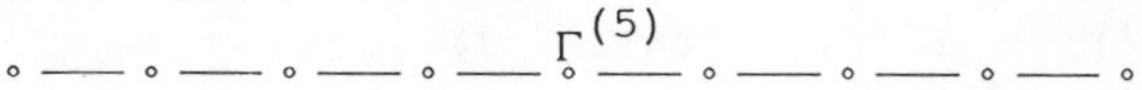

with $I(\mathbf{A}_9,\Gamma^{(5)})=5/2$.

$(\mathbf{D}_n,\Gamma^{(k)})$: 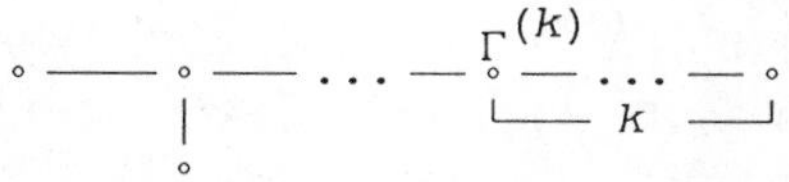

with $I(\mathbf{D}_n,\Gamma^{(k)})=k$, where $n-2\ge k\ge 1$;

$(\mathbf{D}_n,\Gamma^{(n-1)})$: 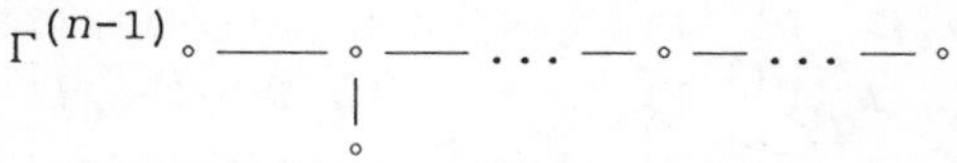

with $I(\mathbf{D}_n,\Gamma^{(n-1)})=n/4\ge 5/4$, where $n\ge 5$;

$(\mathbf{E}_6,\Gamma^{(1)})$: 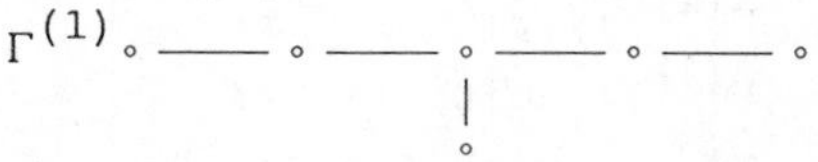

with $I(\mathbf{E}_6,\Gamma^{(1)})=4/3$;

$(\mathbf{E}_6,\Gamma^{(2)})$: 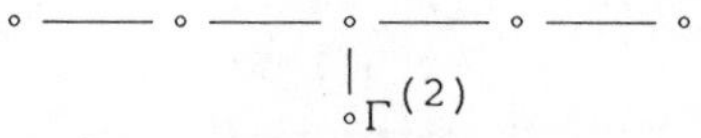

with $I(\mathbf{E}_6,\Gamma^{(2)})=2$;

$(\mathbf{E}_7,\Gamma^{(7)})$: 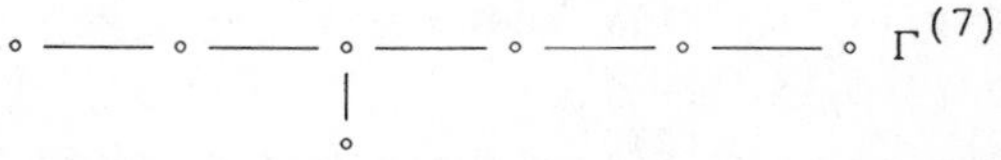

with $I(\mathbf{E}_7,\Gamma^{(7)})=3/2$;

$(\mathbf{E}_7,\Gamma^{(1)})$: 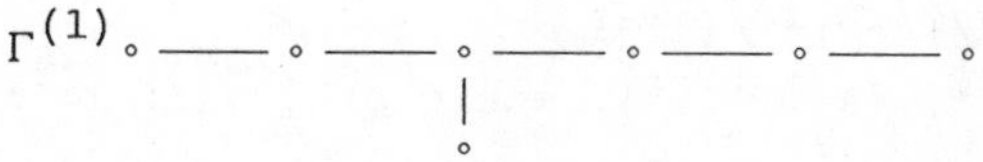

with $I(\mathbf{E}_7,\Gamma^{(1)})=2$;

$(\mathbf{E}_8,\Gamma^{(8)})$:

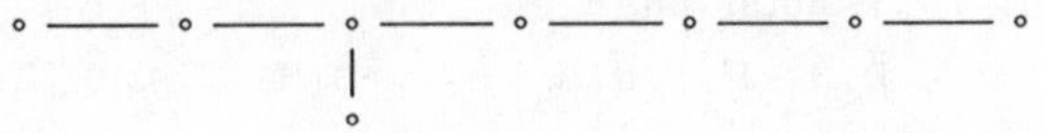

$\Gamma^{(8)}$

with $I(\mathbf{E}_8,\Gamma^{(8)})=2$.

It follows, we have the following and only following possibilities for $(H_1,\Gamma_1;\ldots;H_t,\Gamma_t)$ with $2<I=I(H_1)+\ldots+I(H_t)\le 5/2$:

$\{(\mathbf{D}_n,\Gamma^{(n-1)})\}$, $n=9,10$ with $I=n/4$;

$\{(\mathbf{A}_9,\Gamma^{(5)})\}$ with $I=5/2$;

$\{(\mathbf{A}_{3+q},\Gamma^{(4)})\}$, $q=5,6$ with $I=4-16/(4+q)$;

$\{(\mathbf{A}_{2+q},\Gamma^{(3)})\}$, $7\le q\le 15$, with $I=3-9/(3+q)$;

$\{(\mathbf{E}_8,\Gamma^{(8)}),(\mathbf{A}_1,\Gamma^{(1)})\}$ with $I=5/2$;

$\{(\mathbf{E}_7,\Gamma^{(1)}),(\mathbf{A}_1,\Gamma^{(1)})\}$ with $I=5/2$;

$\{(\mathbf{E}_7,\Gamma^{(7)}),(\mathbf{D}_n,\Gamma^{(1)})\}$ with $I=5/2$;

$\{(\mathbf{E}_7,\Gamma^{(7)}),(\mathbf{A}_3,\Gamma^{(2)})\}$ with $I=5/2$;

$\{(\mathbf{E}_7,\Gamma^{(7)}),(\mathbf{A}_q,\Gamma^{(1)})\}$, $q\ge 2$, with $I=5/2-1/(q+1)$;

$\{(\mathbf{E}_6,\Gamma^{(2)}),(\mathbf{A}_1,\Gamma^{(1)})\}$ with $I=5/2$;

$\{(\mathbf{E}_6,\Gamma^{(1)}),(\mathbf{D}_n,\Gamma^{(1)})\}$ with $I=7/3$;

$\{(\mathbf{E}_6,\Gamma^{(1)}),(\mathbf{A}_3,\Gamma^{(2)})\}$ with $I=7/3$;

$\{(\mathbf{E}_6,\Gamma^{(1)}),(\mathbf{A}_q,\Gamma^{(1)})\}$, $q\ge 3$, with $I=7/3-1/(1+q)$;

$\{(\mathbf{D}_5,\Gamma^{(4)}),(\mathbf{D}_5,\Gamma^{(4)})\}$ with $I=5/2$;

$\{(\mathbf{D}_m,\Gamma^{(m-1)}),(\mathbf{D}_n,\Gamma^{(1)})\}$, $m=5,6$, with $I=m/4+1$;

$\{(\mathbf{D}_5,\Gamma^{(4)}),(\mathbf{A}_{1+q},\Gamma^{(2)})\}$, $q=2,3$, with $I=13/4-4/(2+q)$;

$\{(\mathbf{D}_6,\Gamma^{(5)}),(\mathbf{A}_3,\Gamma^{(2)})\}$ with $I=5/2$;

$\{(\mathbf{D}_5,\Gamma^{(4)}),(\mathbf{A}_q,\Gamma^{(1)})\}$, $q\ge 4$, with $I=9/4-1/(1+q)$;

$\{(\mathbf{D}_6,\Gamma^{(5)}),(\mathbf{A}_q,\Gamma^{(1)})\}$, $q\ge 2$, with $I=5/2-1/(1+q)$;

$\{(\mathbf{D}_7,\Gamma^{(6)}),(\mathbf{A}_q,\Gamma^{(1)})\}$, $1\le q\le 3$, with $I=11/4-1/(1+q)$;

$\{(\mathbf{D}_8,\Gamma^{(7)}),(\mathbf{A}_1,\Gamma^{(1)})\}$, with $I=5/2$;

$\{(\mathbf{D}_n,\Gamma^{(2)}),(\mathbf{A}_1,\Gamma^{(1)})\}$, with $I=5/2$;

$\{(\mathbf{D}_n,\Gamma^{(1)}),(\mathbf{A}_5,\Gamma^{(3)})\}$ with $I=5/2$;

$\{(\mathbf{D}_n,\Gamma^{(1)}),(\mathbf{A}_{1+q},\Gamma^{(2)})\}$, $3\le q\le 6$, with $I=3-4/(2+q)$;

$\{(\mathbf{A}_7,\Gamma^{(4)}),(\mathbf{A}_1,\Gamma^{(1)})\}$ with $I=5/2$;

$\{(\mathbf{A}_5,\Gamma^{(3)}),(\mathbf{A}_3,\Gamma^{(2)})\}$ with $I=5/2$;

$\{(\mathbf{A}_{2+p},\Gamma^{(3)}),(\mathbf{A}_q,\Gamma^{(1)})\}$, where $p\ge 3$, $q\ge 1$, $2>9/(3+p)+1/(1+q)\ge 3/2$, with $I=4-9/(3+p)-1/(1+q)$;

$\{(\mathbf{A}_{1+p},\Gamma^{(2)}),(\mathbf{A}_{1+q},\Gamma^{(2)})\}$, where $2\le p\le q$, $2<q$, $1/(2+p)+1/(2+q)\ge 3/8$, with $I=4-4/(2+p)-4/(2+q)$

$\{(\mathbf{A}_{1+p},\Gamma^{(2)}),(\mathbf{A}_q,\Gamma^{(1)})\}$, where $p>2$, $q\ge 1$, $1>4/(2+p)+1/(1+q)\ge 1/2$,

with $I=3-4/(2+p)-1/(1+q)$;
$\{(\mathbf{E}_7,\Gamma^{(7)}),(\mathbf{A}_1,\Gamma^{(1)}),(\mathbf{A}_1,\Gamma^{(1)})\}$, with $I=5/2$.
$\{(\mathbf{E}_6,\Gamma^{(1)}),(\mathbf{A}_p,\Gamma^{(1)}),(\mathbf{A}_q,\Gamma^{(1)})\}$, where $1\le p\le q$, $1/(1+p)+1/(1+q)\ge 5/6$, with $I=10/3-1/(1+p)-1/(1+q)$;
$\{(\mathbf{D}_5,\Gamma^{(4)}),(\mathbf{A}_p,\Gamma^{(1)}),(\mathbf{A}_q,\Gamma^{(1)})\}$, where $1\le p\le q$, $1/(1+p)+1/(1+q)\ge 3/4$, with $I=13/4-1/(1+p)-1/(1+q)$;
$\{(\mathbf{D}_6,\Gamma^{(5)}),(\mathbf{A}_1,\Gamma^{(1)}),(\mathbf{A}_1,\Gamma^{(1)})\}$, with $I=5/2$;
$\{(\mathbf{D}_m,\Gamma^{(1)}),\ (\mathbf{D}_n,\Gamma^{(1)}),(\mathbf{A}_1,\Gamma^{(1)})\}$ with $I=5/2$;
$\{(\mathbf{D}_m,\Gamma^{(1)}),\ (\mathbf{A}_3,\Gamma^{(2)}),(\mathbf{A}_1,\Gamma^{(1)})\}$ with $I=5/2$;
$\{(\mathbf{D}_m,\Gamma^{(1)}),(\mathbf{A}_p,\Gamma^{(1)}),(\mathbf{A}_q,\Gamma^{(1)})\}$, where $1\le p\le q$, $q>1$, $1/(1+p)+1/(1+q)\ge 1/2$, with $I=3-1/(1+p)-1/(1+q)$;
$\{(\mathbf{A}_5,\Gamma^{(3)}),(\mathbf{A}_1,\Gamma^{(1)}),(\mathbf{A}_1,\Gamma^{(1)})\}$, with $I=5/2$;
$\{(\mathbf{A}_3,\Gamma^{(2)}),(\mathbf{A}_3,\Gamma^{(2)}),(\mathbf{A}_1,\Gamma^{(1)})\}$ with $I=5/2$;
$\{(\mathbf{A}_{1+p},\Gamma^{(2)}),(\mathbf{A}_q,\Gamma^{(1)}),(\mathbf{A}_r,\Gamma^{(1)})\}$ where $p\ge 2$, $1\le q\le r$, $3/2\le 4/(2+p)+1/(1+q)+1/(1+r)<2$, with $I=4-4/(2+p)-1/(1+q)-1/(1+r)$;
$\{(\mathbf{A}_p,\Gamma^{(1)}),(\mathbf{A}_q,\Gamma^{(1)}),(\mathbf{A}_r,\Gamma^{(1)})\}$ where $1\le p\le q\le r$, $1/2\le 1/(1+p)+1/(1+q)+1/(1+r)<1$, with $I=3-1/(1+p)-1/(1+q)-1/(1+r)$;
$\{(\mathbf{D}_n,\Gamma^{(1)}),(\mathbf{A}_1,\Gamma^{(1)}),(\mathbf{A}_1,\Gamma^{(1)}),(\mathbf{A}_1,\Gamma^{(1)})\}$, with $I=5/2$;
$\{(\mathbf{A}_3,\Gamma^{(2)}),(\mathbf{A}_1,\Gamma^{(1)}),(\mathbf{A}_1,\Gamma^{(1)}),(\mathbf{A}_1,\Gamma^{(1)})\}$, with $I=5/2$;
$\{(\mathbf{A}_p,\Gamma^{(1)}),(\mathbf{A}_q,\Gamma^{(1)}),(\mathbf{A}_r,\Gamma^{(1)}),(\mathbf{A}_s,\Gamma^{(1)})\}$ where $1\le p\le q\le r\le s$, $s>1$, $3/2\le 1/(1+p)+1/(1+q)+1/(1+r)+1/(1+s)$, with $I=4-1/(1+p)-1/(1+q)-1/(1+r)-1/(1+s)$;
$\{(\mathbf{A}_1,\Gamma^{(1)}),(\mathbf{A}_1,\Gamma^{(1)}),(\mathbf{A}_1,\Gamma^{(1)}),(\mathbf{A}_1,\Gamma^{(1)}),(\mathbf{A}_1,\Gamma^{(1)})\}$, with $I=5/2$.

If we draw the trees corresponding to all this possibilities, we get all trees of the Theorem and one additional tree corresponding to the case $\{(\mathbf{D}_n,\Gamma^{(2)}),(\mathbf{A}_1,\Gamma^{(1)})\}$. The last tree is impossible by the Proposition 2.1.2. The same Proposition gives the additional inequalities: if $p\ge 3$ then $a\le 1$, if $q\ge 3$ then $a\le 4$, if $q\ge 4$ then $a\le 2$ for the tree $G_1(C^2;p,q,r;a)$. ∎

Proposition 2.2.2. 1. *Let* $C^2\ge 2$ *and the hyperbolic connected component* $G_1=G_{hyp}\ne\emptyset$. *Let* G_i, $1\le i\le k$, *are all connected components of* G *which are connected by the edge* Cv_i, $v_i\in G_i$, *with* C, *and* G_j, $k<i\le l$ *are all other connected components of* G *(disconnected with* C*).*

Then all connected components G_i, $2\le i\le l$ *are elliptic and*

1. $\bar{D}(G(C))=\bar{D}(G_1)\bar{D}(G_2)\cdots\bar{D}(G_l)(-C^2-$
$-\bar{D}(G_1-v_1)/\bar{D}(G_1)-\bar{D}(G_2-v_2)/\bar{D}(G_2)-\ldots\ldots-\bar{D}(G_k-v_k)/\bar{D}(G_k))\le 0,$

where

$$2\le C^2\le\bar{D}(G_1-v_1)/(-\bar{D}(G_1))-\bar{D}(G_2-v_2)/\bar{D}(G_2)-\ldots\ldots-\bar{D}(G_k-v_k)/\bar{D}(G_k)).$$

2. rk $G(C)=\#G(C)$ *if the right inequality above is strong, and* rk $G(C)=\#G(C)-1$ *if this inequality is an equality.*

3. *If* $\mathcal{P}$ *is a parabolic subtree of the tree* $G_1=G_{hyp}$, *then*

$$m(\mathcal{P},v_1)\geq \min\, m(G_i,v_i),\ 2\leq i\leq k.$$

<u>Proof</u>. Use the formula (2.2) for $B=C$ and the Proposition 2.1.4. ∎

Remark 2.2.3. If $G_{hyp}=\varnothing$, then all restrictions for the tree $G(C)$ we can give here follow from the Propositions 2.1.1 - 2.1.5. We would like to emphasis the difference of this case from the case $G_{hyp}\neq\varnothing$. For the case $G_{hyp}=\varnothing$ the C^2 (equivalently, the $\dim|C|=C^2/2+1$) <u>may be arbitrary large.</u> ∎

2.3. The case $C^2=0$ and $C\neq 0$. Here we have the

Theorem 2.3.1. *Let* $C^2=0$ *and* $C\neq 0$, *and the* $G_{hyp}\neq\varnothing$, *let* Γ_{hyp} *be the vertex of the* G_{hyp} *joined to* C.

Then all connected components $H_1,\dots,H_t$ *of the* $G_{hyp}-\Gamma_{hyp}$ *are parabolic or elliptic. Let* Γ_i *be the vertex of* H_i *joined to* Γ_{hyp}. *If all components* $H_1,\dots,H_t$ *are elliptic, then hyperbolicity of* G_{hyp} *is equivalent to inequality*

$$\bar{D}(G_{hyp})=\bar{D}(H_1)\cdots\bar{D}(H_t)(2-\bar{D}(H_1-\Gamma_1)/\bar{D}(H_1)-\dots-\bar{D}(H_t-\Gamma_t)/\bar{D}(H_t))<0$$

or

$$\bar{D}(H_1-\Gamma_1)/\bar{D}(H_1)+\dots+\bar{D}(H_t-\Gamma_t)/\bar{D}(H_t)>2.$$

Moreover, $m(H_i,\Gamma_i)=1$ *if the component* H_i *is parabolic, and* min $m(H_i,\Gamma_i)=1$ *if the component* H_i *is elliptic. The* rk $G_{hyp}(C)=\#G_{hyp}(C)-p$ *where* p *is the number of parabolic components from* $H_1,\dots,H_t$.

<u>Proof</u>. Use the formula (2.2) and the Propositions 2.1.3 and 2.1.4. ∎

The following proposition is an analog of the Proposition 2.2.2.

Proposition 2.3.2. *Under the conditions and notations of the Theorem 2.3.1, suppose that there are other connected components* $G_2,\dots,G_k$, $k\geq 2$, *of* G *different from the hyperbolic component* $G_1=G_{hyp}\neq\varnothing$, *such that there exists* $\Gamma_j\in G_j$, $k\geq j\geq 2$, *which is joined to* C. *Let* G_j, $l\geq j\geq k+1$, *are all other connected components of* G *(disjoined with* C*).*

Then

1. *All components* $H_1,\dots H_t$ *and* $G_2,\dots,G_k$ *are elliptic,* $\bar{D}(G_1)<0$ *and the following inequality is equivalent to hyperbolicity of the graph* $G(C)$:

$$\bar{D}(G(C))=\bar{D}(G_1)\bar{D}(G_2)\cdots\bar{D}(G_l)(\bar{D}(G_1-\Gamma_{hyp})/(-\bar{D}(G_1))-$$

$$-\bar{D}(G_2-\Gamma_2)/\bar{D}(G_2)-\ldots\ldots-\bar{D}(G_k-\Gamma_k)/\bar{D}(G_k)) \le 0,$$

or

$$0 \le \bar{D}(G_1-\Gamma_{hyp})/(-\bar{D}(G_1))-\bar{D}(G_2-\Gamma_2)/\bar{D}(G_2)-\ldots\ldots-\bar{D}(G_k-\Gamma_k)/\bar{D}(G_k)).$$

2. rk $G(C)=\#G(C)$ *if the inequality above is strong, and* rk $G(C)=\#G(C)-1$ *if the inequality above is an equality.*

3. *If* $\mathcal{P}$ *is a parabolic subtree of the tree* $G_1=G_{hyp}$, *then*

$$m(\mathcal{P},\Gamma_{hyp}) \ge \min m(G_i,\Gamma_i),\ 2\le i\le k.$$

Proof. Use the formula (2.2) and the Propositions 2.1.3 and 2.1.4. ∎

Remark 2.3.3. If $G_{hyp}=\emptyset$, then all restrictions for the tree $G(C)$, we can give here, follow from the Propositions 2.1.1 - 2.1.5. ∎

2.4. The case $C=0$. In this case $G(C)=G$. The problem is to classify hyperbolic trees G_{hyp}. In [M], G.Maxwell investigated the case rk $G_{hyp} = \#G_{hyp}$. Fortunately, it is necessary only to reformulate his results to consider the general case which we need.

Theorem 2.4.1. *Let* G *be a connected tree of nonsingular -2 curves on a* K3 *surface. Then one of the two cases* (a) *or* (b) *holds:*

(a) *There exists a vertex* Γ *of* G *such that all connected components* $H_1,\ldots,H_t$ *of* $G_{hyp}-\Gamma$ *are parabolic or elliptic. If one of these components* $H_1,\ldots,H_t$ *is parabolic, then* G *is hyperbolic. If all the components* $H_1,\ldots,H_t$ *are elliptic then* G *is hyperbolic iff*

$$\bar{D}(H_1-\Gamma_1)/\bar{D}(H_1)+\ldots+\bar{D}(H_t-\Gamma_t)/\bar{D}(H_t)>2$$

where Γ_i *is the vertex of* H_i *joined to* Γ. *If* G *is hyperbolic, then*

$$\text{rk } G=\#G-\max\{\alpha-1,0\}$$

where α *is the number of parabolic components from* $H_1,\ldots,H_t$.

(b) *There exists an edge* $\Gamma_1\Gamma_2$ *of* G *such that all connected components of* $G-\{\Gamma_1\Gamma_2\}$ *are elliptic and, if* G_1 *is the connected component of* $G-\Gamma_2$ *containing* Γ_1 *and* G_2 *is the connected components of* $G-\Gamma_1$ *containing* Γ_2, *then both* G_1 *and* G_2 *are hyperbolic. In this situation the matrices of* G_1,G_2 and G are *hyperbolic iff* $\bar{D}(G_1)<0$, $\bar{D}(G_2)<0$ *and* $\bar{D}(G)\le 0$.

Let $H_{j1},\ldots,H_{jk_j}$ *be all connected components of* $G_j-\Gamma_j$, $j=1,2$, *and* Γ_{ji} *be a vertex of the* H_{ji} *joined to the* Γ_j. *Then the last inequalities are equivalent to*

$$A_1=\bar{D}(H_{11}-\Gamma_{11})/\bar{D}(H_{11})+\ldots+\bar{D}(H_{1k_1}-\Gamma_{1k_1})/\bar{D}(H_{1k_1})>2,$$

$$A_2=\bar{D}(H_{21}-\Gamma_{21})/\bar{D}(H_{21})+\ldots+\bar{D}(H_{2k_2}-\Gamma_{2k_2})/\bar{D}(H_{2k_2})>2,$$

and

$$(A_1-2)(A_2-2)\leq 1.$$

The rk G=#G *if the last inequality is strong, and* rk G=#G-1 *if the last inequality is equality.*

Proof. This is similar to [M]. We leave details to the reader. ∎

The Theorem 2.4.1 is sufficient to draw (in principle) all possible trees G of -2 curves on the K3 surfaces. An additional restrictions for these trees give the Propositions 2.1.1 - 2.1.5. For example, all connected trees with ≤10 vertices are either elliptic or parabolic, or hyperbolic (it is mentioned in [M]), and the full list of these trees one can find in [H]. Only the following tree with ≤10 vertices contradicts to these propositions and, hence, is impossible on K3 surfaces:

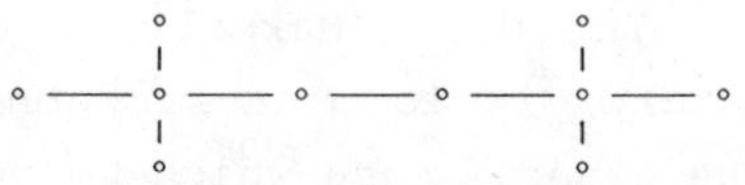

2.5. Remark. Here we only mentioned the most important and rude conditions for trees $G(C)$. We hope to give other more delicate necessary and sufficient conditions in further publications. This problem is a little similar to the problem of a description of all possible singularities of quartic singular K3 surfaces. You can see the series of Urabe's articles devoted to this subject; see [U], for example. But our problem is much more complicated. It is arithmetic and is connected with the existence of an embedding of the corresponding to $G(C)$ lattice into K3 cohomology lattice (it is an even unimodular lattice of the signature (3,19)). One can use here the discriminant form technique (see [N]).

§ 3. Fixed part of Weil linear systems on singular K3 surfaces.

3.1. General case. Let Y be a singular K3 surface and $\sigma:X\longrightarrow Y$ the minimal resolution of singularities of Y. Let $\Delta_S=\Sigma b_jF_j$, where $b_j\geq 0$ are integers and F_j are components of the exceptional divisor of σ. Let D be an effective divisor on X. A complete Weil linear system $\bar{D}$ on Y is the image $|\bar{D}|=\sigma_*(|D+\Delta_S|)$, where $|D+\Delta_S|$ is the complete linear system on X and we consider all possible $b_j\geq 0$. It is very easy to see that this image is stabilized if b_j are increased. Like in the § 1, we want to describe the moving part and the fixed part of the linear system $\bar{D}$. Evidently, the fixed part is the image of the fixed part Δ of the linear system $|D+\Delta_S|$. And the fixed components part of $|\bar{D}|$ is the image $\sigma_*(\Delta_r)$ of the part $\Delta_r=\Delta-\Delta_s$. It is not difficult to prove that when $b_j \gg 0$ then all components F_j of Δ_s belong to the fixed part of the linear system $|D+\Delta_s|$. We suppose that $b_j \gg 0$, or it is more convenient to suppose that the all $b_j=+\infty$.

Let $|D+\Delta_s|=|C|+\Delta$, where $|C|$ is the moving part and Δ is the fixed part. Then $\Delta=\Delta_r+\Delta_s$ where Δ_s is the part defined by the all components F_j of the multiplicity $+\infty$ and $\Delta_r=\Delta-\Delta_s$. Then the multiplicity a_i of an irreducible component Γ_i of Δ_r is defined and is a finite natural number. It defines the multiplicities of the corresponding irreducible components $\sigma_*(\Gamma_i)$ of the fixed part $\sigma_*(\Delta)$ of the complete linear system $|\bar{D}|$. As in § 1, we define the graph $G(C,\Delta)$. Its difference from the situation of the § 1 is that vertices of its subgraph $G(\Delta)$ are of the two kinds:

Black vertices of the multiplicity $+\infty$ corresponding to the components F_j of the exceptional divisor of σ;

White vertices of the finite multiplicity $a_i \in \mathbb{N}$, corresponding to the irreducible components $\sigma_*(\Gamma_i)$ of the fixed part of $|\bar{D}|$.

Thus, the problem we should solve, is the same as in the § 1: To describe all possible graphs $G(C,\Delta)$ of this kind such that $|C+\Delta|=|C|+\Delta$. It is a particular case of the problem we have solved in the § 1, and it is necessary to reformulate the results of § 1 in this situation only.

The analog of the condition (*) is the condition

(**) $|C|$ satisfies the condition (i), (ii), or (iii) of the Proposition 0.1, $\Delta=\Delta_r+\Delta_s$, where $\Delta_r=\sum a_i\Gamma_i$, $a_i \in \mathbb{N}$, and $\Delta_s=\sum b_j F_j$, $b_j=+\infty$ (or $b_j \gg 0$), and all Γ_i and F_j are irreducible -2 curve. (For the graph $G(C,\Delta)$, the vertices Γ_i are called white and the vertices F_j black.) If $|C|=m|E|$ where E is an elliptic curve and $m \ge 2$ then there does not exist more than one irreducible component R of Δ such that $E\cdot R \ge 1$; if here $m \ge 4$, then the vertex R is white and has the multiplicity $a=1$.

Our question is: If (**) holds, when

$$|C+\Delta|=|C|+\Delta \ ? \qquad (3.1)$$

Theorem 3.1.1. *Let $C+\Delta$ be a divisor on a nonsingular K3 surface X which satisfy the condition (**) above.*

Then $|C+\Delta|=|C|+\Delta$ (equivalently, $|\sigma_*(C+\Delta)|=\sigma_*(|C|)+\sigma_*(\Delta)$ for the contraction σ of the all black curves F_j), if and only if $G(C,\Delta)$ is a tree and $G(C,\Delta)$ has not a subtree $T=\tilde{D}_m$, $\tilde{E}_6$, $\tilde{E}_7$, $\tilde{E}_8$, $\tilde{D}_m(C)$, $\tilde{E}_6(C)$, $\tilde{E}_7(C)$, $\tilde{E}_8(C)$ $\tilde{B}_m(C)$ or $\tilde{G}_2(C)$ of the Theorem 1.1. It means that if the *tree* $G(C,\Delta_{red})$ *contains the subtree* T_{red} (red *means the reduction*), *then there exists a vertex v of* T *which is a white vertex of the tree* $G(C,\Delta)$ *and its multiplicity in* $G(C,\Delta)$ *is strongly less than the multiplicity of the vertex v in the subtree* T.

Proof. This follows from the Theorem 1.1. ■

3.2. Nef case. We use notations of **3.1.** Here we want to consider the case when a linear system $|\bar{D}|$ on a singular K3 surface Y is nef or numerically ample (in the sense of Mumford intersection pairing on a normal surface [Mu]). This case is the most interesting for applications (for Fano threefolds, for example). We use the following trivial

Lemma 3.2.1. $\bar{D}$ *is* nef *iff* $\sigma^{*}(\bar{D})$ *is* nef. *In other words, if we normalize weights* b_j *of black vertices* F_j *of the* Δ *by the condition* $F_j\cdot(C+\Delta)=0$ *(here* b_j *are rational numbers) and not change weights* a_i *of the white vertices* Γ_i *(here* a_i *are natural numbers), then for any white curve* Γ_i *we have the inequality:* $\Gamma_i\cdot(C+\Delta)\ge 0$. *If* $|\bar{D}|$ *is ample, the last inequalities are strong:* $\Gamma_i\cdot(C+\Delta)>0$. ■

Thus, it is natural to give the

Definition 3.2.2. The graph $G(C,\Delta)$ is called convex below if for the weights $\{b_j\}$ of the black vertices F_j satisfying to the condition $F_j\cdot(C+\Delta)=0$, the condition $\Gamma_i\cdot(C+\Delta)\ge 0$ holds for the white vertices Γ_i. In other words, for any component U of the Δ the inequality $U\cdot(C+\Delta)\ge 0$ holds, and, if U is black, this inequality is the equality. A diagram $G(C,\Delta)$ is called strongly convex below if it is convex below and for any white vertex Γ_i a strong inequality $\Gamma_i\cdot(C+\Delta)>0$ holds. It is sufficient to prove this conditions for connected components Δ_i of Δ only. ■

From the Theorem 3.1.1 and the Lemma 3.2.1, we get

Theorem 3.2.3. *Under the conditions of the Theorem* 3.1.1,

$$|\sigma_*(C+\Delta)|=\sigma_*(|C|)+\sigma_*(\Delta) \text{ and } \sigma_*(C+\Delta) \text{ is nef}$$

if and only if $G(C,\Delta)$ *satisfies the Theorem* 3.1.1 *for the weights* $b_j=+\infty$ *of the black vertices* F_j *and the tree* $G(C,\Delta)$ *is convex below for the weights* b_j *of the black vertices* F_j *satisfying to the condition* $F_j\cdot(C+\Delta)=0$ *of the Definition* 3.2.2. *If* $\sigma_*(C+\Delta)$ *is ample, then this tree* $G(C,\Delta)$ *should be additionally strongly convex.* ■

As an example, let us consider the case when $C^2>0$ and $G(\Delta)$ has the form $\mathbf{A}_m$ or $\mathbf{D}_m$. We denote ∘ a white vertex, · a black vertex, and ⊙ a vertex which may be either white or black. Then we get the following possible trees $G(C,\Delta)$ on the K3 surfaces over a basic field of characteristic 0, where c_i is the weight of the vertex (white or black) satisfying to the conditions of the Definition 3.2.2:

(A_m, i):

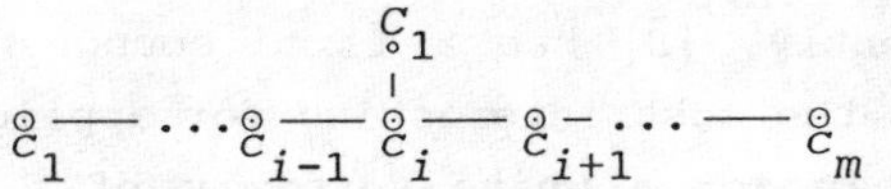

where the chains of weights $0, c_1, \ldots c_i$, and $c_i, \ldots, c_m, 0$ are convex below and $(c_i - c_{i-1}) + (c_i - c_{i+1}) \le 1$. Since $m \le 19$, the

$$\max\{c_1, \ldots, c_i, \ldots c_m\} = c_i \le i(m+1-i)/(m+1) \le (m+1)/4 \le 5 .$$

$(D_m, 1)$:

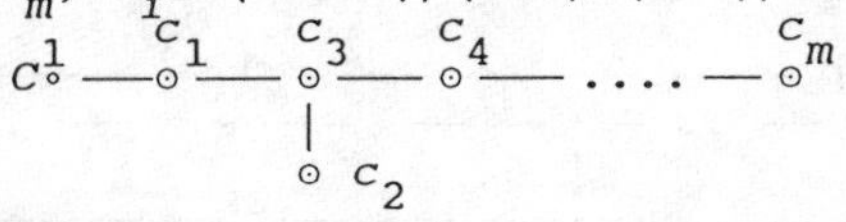

where the chains $1, c_1, c_3$, and $0, c_2, c_3$, and $c_3, c_4, \ldots, c_m, 0$ are convex below and $c_3 - c_1 + c_3 - c_2 \le c_4$. Since $m \le 19$, the

$$\max\{1, c_1, c_2, \ldots c_m\} = \max\{1, c_3\} \le (m-2)/2 \le 17/2 < 9.$$

(D_m, i), $3 < i < m$:

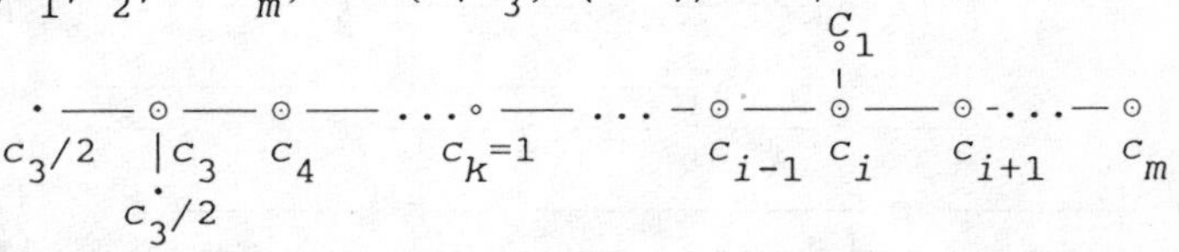

The chains of weights $c_3, c_4, \ldots, c_k = 1, \ldots c_{i-1}, c_i$ and $c_i, \ldots c_m, 0$ are convex below, and $c_4 \ge c_3$, and $(c_i - c_{i-1}) + (c_i - c_{i+1}) \le 1$. The $\max\{c_1, \ldots, c_m\} = c_i \le \dfrac{(i+1-k)(m+1-i)}{(m+1-k)} \le (m+2-k)^2/(4(m+1-k)) \le 81/17 < 5$.

(D_m, m) :

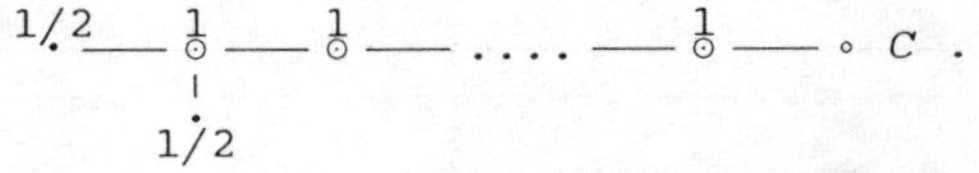

We should emphasize that this diagrams are possible for an arbitrary even $C^2 > 0$ and an arbitrary $\dim|\sigma_*(C+\Delta))| = C^2/2+1$, and here the moving part $\sigma_*|C|$ of $|\sigma_*(C+\Delta))|$ is not a pencil. A multiplicity of the fixed part components of $|\sigma_*(C+\Delta))|$ may be >1, but ≤8 (and the case of the maximum multiplicity 8 is possible). This shows the difference of the nef and ample linear systems on singular K3 surfaces comparing with the nonsingular case. But here the hyperbolic component $G_{\text{hyp}}(\Delta) = \varnothing$. We consider an opposite example below.

3.3. The case rk Pic $Y=1$. More generally, we consider the case when $\sigma_*(D_i)^2 > 0$ for any irreducible component D_i of a general member $D = \sum D_i \in |C + \Delta_r|$. This case is characterized by the following conditions: an every elliptic component of $G(\Delta)$ contains black vertices only; the tree $G(\Delta)$ does not contain parabolic components; if v is a white vertex of $G_{\text{hyp}}(\Delta)$ and $N(v)$ is the maximum connected subtree of $G_{\text{hyp}}(\Delta)$

which contains only the one white vertex v, then $N(v)$ is hyperbolic.

Using this conditions and the theory above, we get the following description of $G_{hyp}(C,\Delta)$ if it has a white vertex v of a multiplicity ≥ 2 (equivalently, $|\bar{D}|$ has a fixed component of a multiplicity ≥ 2): This case is the most interesting for applications.

We get that for a white vertex v of a multiplicity >1 the tree $N(v)$ is one of the following trees:

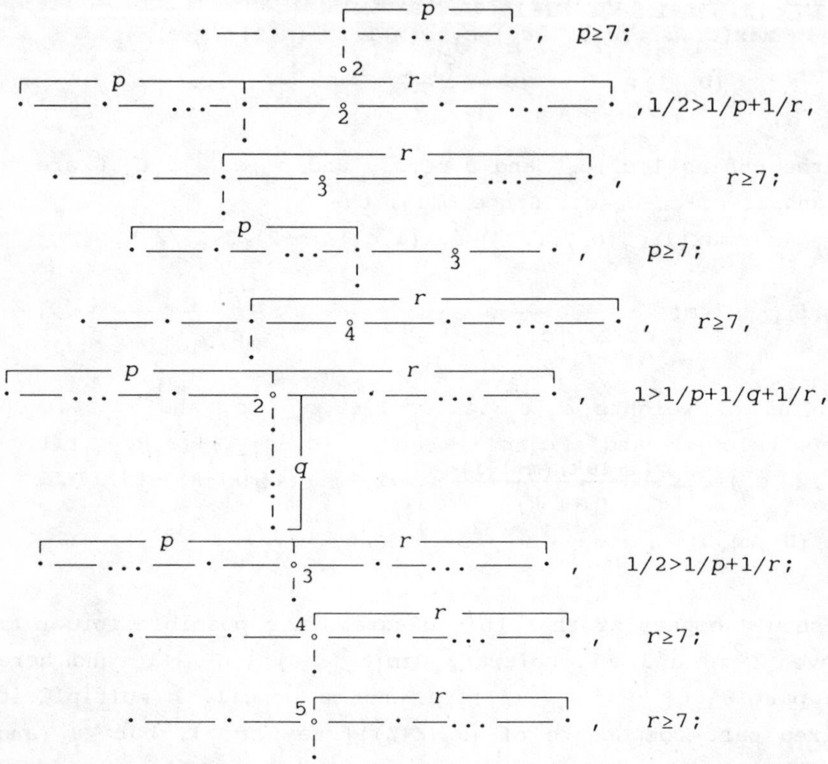

It follows very easy that this multiplicity >1 white vertex v of the $G(C,\Delta)$ is unique, and $G(C,\Delta)$ has not more then one other white vertex. We shall denote this vertex C (thus, we permit that $C^2=-2$). If there exists this additional white vertex C of the multiplicity one, then the tree $CUN(v)$ is one of the following:

$$\bullet \!-\! \cdot \!-\! \underset{\substack{|\\ \circ 2}}{\bullet} \!-\! \overset{\boxed{\quad p \quad}}{\cdots} \!-\! \cdot \!-\! \overset{1}{\underset{\circ}{}}C \; , \quad p\geq 7, \quad -2\leq C^2\leq(7-p)/(p-6)\leq 0.$$

$$C \overset{1}{\underset{\circ}{}}\!-\! \cdot \!-\! \overset{\boxed{\quad p \quad}}{\cdots} \!-\! \underset{\substack{|\\ \circ 2}}{\bullet} \!-\! \overset{\boxed{\quad r \quad}}{\cdot \!-\! \cdot \!-\!\cdots\!-\!}\bullet ,$$

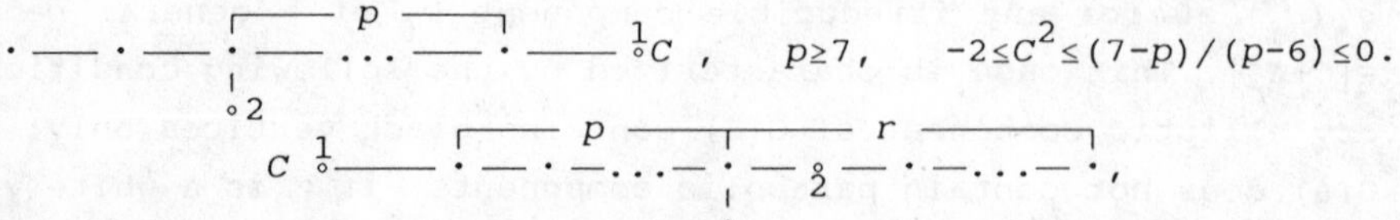

$1/2>1/p+1/r$, $-2\leq C^2\leq(2p+3r-pr-2)/(pr-2p-2r)\leq 4$.

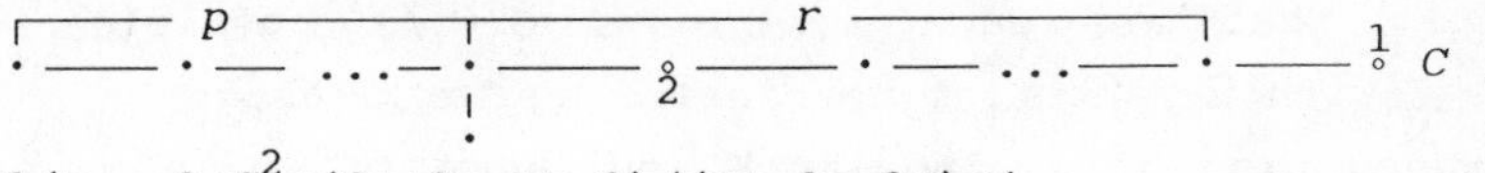

$1/2>1/p+1/r$, $-2\le C^2\le(2r+3p-pr-2)/(pr-2p-2r)\le 4$.

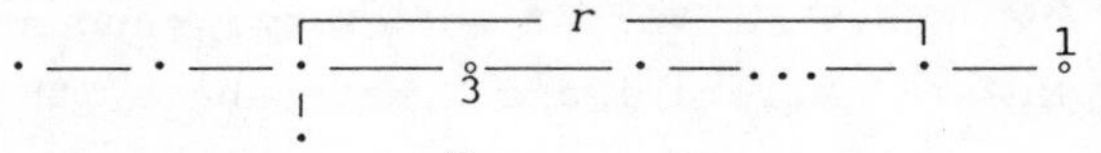 $r\ge 7$, $-2\le C^2\le(7-r)/(r-6)\le 0$.

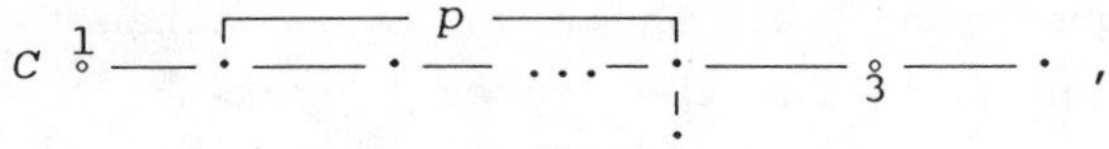 $p\ge 7$, $-2\le C^2\le(7-p)/(p-6)\le 0$.

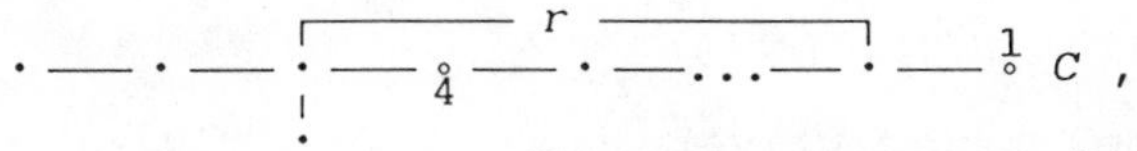 $r\ge 7$, $-2\le C^2\le(7-r)/(r-6)\le 0$.

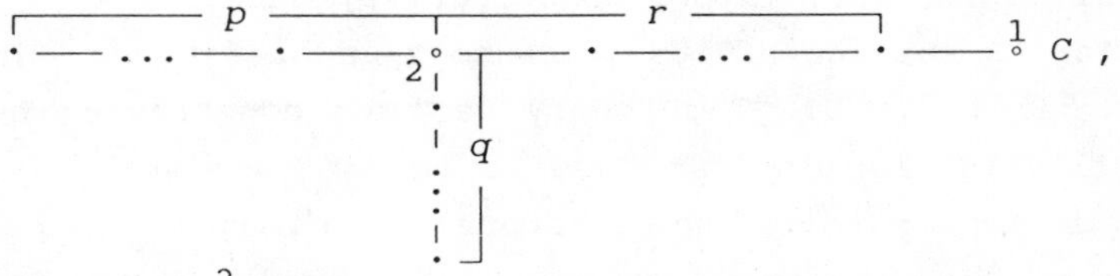

$1>1/p+1/q+1/r$, $-2\le C^2\le(2pq+pr+qr-pqr-p-q))/(pqr-pq-qr-rp)\le 10$.

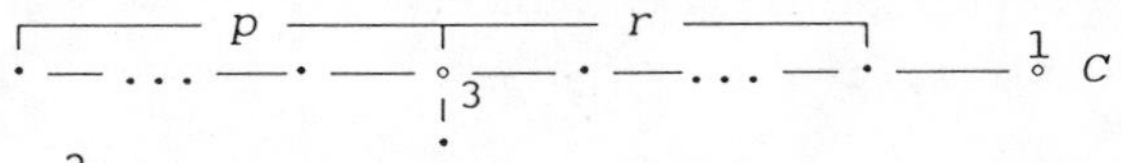

$1/2>1/p+1/r$, $-2\le C^2\le(2r+3p-pr-2)/(pr-2p-2r)\le 4$.

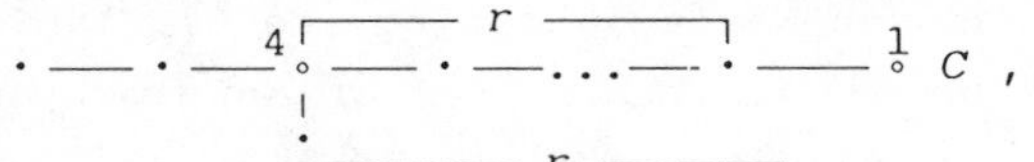 $r\ge 7$, $-2\le C^2\le(7-r)/(r-6)\le 0$.

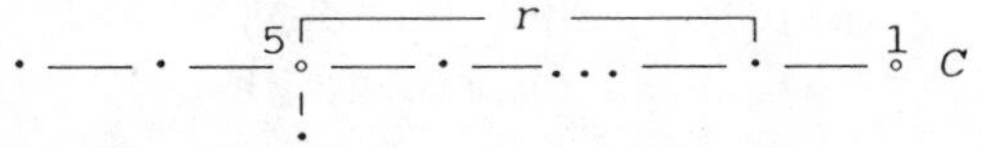 $r\ge 7$, $-2\le C^2\le(7-r)/(r-6)\le 0$.

It follows very easy the estimate $\bar{D}^2<20$ if $|\bar{D}|$ has a component of the multiplicity >1. We hope giving more precise description of the case rk Pic Y=1 in further publications.

We should mention that almost at the same time V.A.Alekseev got the same results for rk Pic Y=1 by other method (using Riemann-Roch theorem for singular K3 surfaces) and the more strong estimate: $\bar{D}^2<13$ if $|\bar{D}|$ has a fixed component of a multiplicity >1. Of course, the same estimate follows from the calculations above.

§ 4. Some open questions.

4.1. Fano threefolds. Let F be Fano threefold with $\mathbb{Q}$-factorial terminal singularities, and a good member $Y\in|-K_F|$, which is a singular K3 surface, exists. Let $\dim|-K_F|>0$. Then $|-K_F||_Y$ is an nonempty complete ample linear system on the singular K3 surface Y. Thus, some tree $G(C,\Delta)$, we have described above, corresponds to this linear system. We can consider this tree $G(C,\Delta)$ as an invariant of the Fano threefold F. What are such invariants $G(C,\Delta)$ possible for Fano threefolds F with $\mathbb{Q}$-factorial terminal singularities?

4.2. Graded ring of a singular K3 surface. I due to participants of the conference "Algebraic and Analytic Varieties" Tokyo, August 1990, the Professors Sh.Ishii, M.Reid, M.Tomari and K.Watanabe by the following very interesting question (see their articles connected with this subject): What one can say about the graded ring

$$R(Y) = \bigoplus_{m\geq 0} H^0(Y,\mathcal{O}(m\bar{D}))$$

for a nef effective (or, maybe, noneffective) integral Weil divisor $\bar{D}$ on a singular K3 surface Y, its generators and relations. The nonsingular case see in [S-D]. The theory we have constructed here gives all possibilities when it is needed to investigate this ring. Moreover, this theory permits to interpret a homogeneous constituent $H^0(Y,\mathcal{O}(m\bar{D}))$ of the ring as a some precisely described complete linear system on the nonsingular K3 surface X which is the minimal resolution of singularities of Y.

References

[H] F.Harary, "Graph theory," Addison-Wesley, Reading, Mass., 1969.

[Mu] D.Mumford, The topology of normal singularities of an algebraic surface and a criterion for simplicity, IHES 9 (1961) 5-22.

[M] G.Maxwell, Hyperbolic trees, J. Algebra. 54 (1978) 46-49.

[N] V.V.Nikulin, Integral symmetric bilinear forms and some of their geometrical applications. Izv AN SSSR, ser. matem. 43 (1979) 111-177 (English transl: Math. USSR Izv. 14 (1980) 103-167.

[S-D] B.Saint-Donat, Projective models of K3 surfaces, Amer. J. Math. 96 (1974) 602-639.

[U] T.Urabe, Tie transformations of Dynkin graphs and singularities on quartic surfaces, Invent. math. 100 (1990), 207-230.

The Euler numbers of ℓ-adic sheaves of rank 1 in positive characteristic.

TAKESHI SAITO

Department of Mathematics, University of Tokyo, Tokyo 113 Japan

One of the most important themes in ramification theory is the formula for the Euler characteristic of ℓ-adic sheaves. Although we have the Grothendieck-Ogg-Shafarevich formula [G] in one dimensional case, we don't have a general formula in higher dimension even in the form of a conjecture. However for sheaves of rank 1, K.Kato formulated a conjecture in arbitrary dimension and actually proved it in dimension 2 in [K2]. In this paper, we will prove it in arbitrary dimension under a certain hypothesis, which is hoped to hold when the variety is sufficiently blowed up.

We consider a smooth ℓ-adic sheaf $\mathcal{F}$ of rank 1 on a dense open subscheme U of a proper smooth variety X over an algebraically closed field k of characteristic $p \neq \ell$ such that the complement $X - U$ is a divisor with normal crossing. At each generic point of the complement divisor, we have the ramification theory of $\mathcal{F}$ by Kato's theory on the abelianized absolute Galois group of complete discrete valuation field in [K3]. When $\mathcal{F}$ is clean, which means the ramification of $\mathcal{F}$ is understood by those at codimension 1 points, the characteristic variety $Ch(\mathcal{F})$ in the cotangent bundle with logarithmic poles and the characteristic 0-cycle

$$c_{\mathcal{F}} = (-1)^{\dim(X)-1} \cdot (Ch(\mathcal{F}),\ \text{0-section}) \in CH_0(X)$$

are defined in [K1]. The definition will be reviewed in the text. The main result of this paper is

THEOREM (VAGUE). *Assume $\mathcal{F}$ as above is clean and something arising from $\mathcal{F}$ are also clean. Then we have*

$$\chi_c(U, \mathcal{F}) - \chi_c(U) = -\deg\ c_{\mathcal{F}}.$$

We didn't give the exact form of the theorem here because to write down the precise assumption will need some preparation. The formula is conjectured by Kato in [K2] without redundant assumption. If dim $X = 1$, the assumption is automatically satisfied and the formula is the rank 1 case of the Grothendieck-Ogg-Shafarevich formula. In general, if X, U and $\mathcal{F}$ are given, it is hoped that there is a resolution $\pi : X' \to X$

such that $\pi^*U \simeq U$ and that $\pi^*\mathcal{F}$ satisfies the assumption of the theorem and also hoped that $\pi_*(c_{\pi^*\mathcal{F}})$ does not depend on the choice of X'. If this is true, we can eliminate the redundant assumption. Actually, if $\dim X = 2$, it is proved to be true in [K1] and the formula is shown in [K2] by Kato. Although the theorem is not a final definitive one, the author believes that the argument used in the proof will be a critical step in a full proof of the conjecture.

Recently, P.Berthelot has made a great progress in p-adic ramification theory by constructing a theory of p-adic $\mathcal{D}$-modules. The author hopes that this will shed some light on ℓ-adic theory.

The author thanks Professor K.Kato greatly who kindly taught him a proof of the main result of his unpublished paper [K2]. The discussion with him was indespensible to complete this work. The author also appriciate the hospitality of the Department of Mathematics of the Johns Hopkins University and JAMI where this work was done.

First we briefly review the ramification theory of sheaves of rank 1 by Kato ([K1] and [K3]) and fix the notations. Let K be a complete discrete valuation field with arbitrary residue field $\overline{K}$ and let A be the integer ring of K. We define a $\overline{K}$-vector space ω_K to be $\Omega^1_A(\log \overline{K}) \otimes_A \overline{K}$ where $\Omega^1_A(\log \overline{K})$ is the A-module of differential forms with logarithmic poles $(\Omega^1_A \oplus (A \otimes K^\times))/(da - a \otimes a, a \in A, \neq 0)$ (cf. [K1] (2.2)). For $a \in A, \neq 0$, the class of $1 \otimes a$ is denoted by $d\log a$. Then ω_K is generated by $\Omega^1_{\overline{K}}$ and $d\log \pi$, where π is a uniformizer of K, and fits in an exact sequence

$$0 \to \Omega^1_{\overline{K}} \to \omega_K \xrightarrow{\text{residue}} \overline{K} \to 0 \quad (\text{res} \cdot d\log \pi = 1).$$

Let $\overline{G}_K$ be the abelianized absolute Galois group $Gal(K^{ab}/K)$ and $X_K = Hom(\overline{G}_K, \mathbf{Q}/\mathbf{Z})$. Define the pairing $\{\ ,\ \}_K : X_K \times K^\times \to Br(K)$ by the cup product $H^2(K, \mathbf{Z}) \times H^0(K, \mathbf{G}_m) \to H^2(K, \mathbf{G}_m)$. Let K' be the completion of the field $K(T)$ with respect to the discrete valuation corresponding to the prime ideal $m_A \cdot A[T]$ of $A[T]$. For $n \in \mathbf{N}, > 0$, let $U^n_K = 1 + m^n_K$. Then the filtration by ramification on X_K is defined by

$$X^n_K = \{\chi \in X_K;\ \{\chi_{K'}, U^{n+1}_{K'}\}_{K'} = 0\}$$

for $n \in \mathbf{N}$, where $\chi_{K'}$ is the image of χ in $X_{K'}$. The set of at most tamely ramified characters is X^0_K and $X_K = \bigcup X^n_K$. There is a canonical morphism $\omega_K \to Br(K)$ extending the Artin-Schreier theory $\overline{K} \to X_{\overline{K}}$

and $\Omega^1_{\overline{K}} \to Br(\overline{K})$. We will give its definition in the proof of Lemma 5. For $n > 0$, we have a commutative diagram

$$\begin{array}{ccc} X_K^n \times U_K^n & \longrightarrow & \omega_K \\ \downarrow & & \downarrow \\ X_K \times K^\times & \longrightarrow & Br(K). \end{array}$$

The upper horizontal map induces a pairing $\{\ ,\ \}_K^n : gr^n(X_K) \times N_K^n \to \omega_K$ where $N_K^n = m_K^n/m_K^{n+1} \simeq U_K^n/U_K^{n+1}$. This pairing is $\overline{K}$-linear with respect to N_K^n and characterized by the property that the pairing $X_K^n \times U_{K'}^n \to Br(K')$ is given by the composite

$$\begin{aligned} X_K^n \times U_{K'}^n \to & gr^n(X_K) \times N_{K'}^n \simeq \\ & gr^n(X_K) \times N_K^n \otimes_{\overline{K}} \overline{K}' \to \omega_K \otimes_{\overline{K}} \overline{K}' \hookrightarrow \omega_{K'} \to Br(K') \end{aligned}$$

since $\omega_K \xrightarrow{\times T} \omega_{K'} \to Br(K')$ is injective. For a character $\chi \in X_K$, we define the Swan conductor $sw(\chi)$ to be the minimum integer $n \in \mathbb{N}$ such that $\chi \in X_K^n$. We define the refined Swan conductor $rsw(\chi)$ of a character χ with $sw(\chi) = n > 0$ to be the $\overline{K}$-homomorphism

$$rsw(\chi) = \{\chi,\ \}_K^n : N_K^n \to \omega_K.$$

Let k be an algebraically closed field of characteristic p. Let X be a smooth variety over k of dimension d and U be an open subscheme of X whose complement is a divisor with (Zariski locally) normal crossings. We call such (X, U) an RN-pair over k. Let $\ell \neq p$ be a prime number and $\mathcal{F}$ be the smooth ℓ-adic sheaf on U of rank 1 corresponding to a character $\chi : \pi_1(U)^{ab} \to \overline{\mathbb{Q}}_\ell^\times$ of finite order. We fix an isomorphism $\mathbb{Q}/\mathbb{Z}(1)_{\overline{\mathbb{Q}}_\ell} \simeq \mathbb{Q}/\mathbb{Z}$. For each irreducible component C_i of $X - U$, let K_i be the completion of the function field of X with respect to the discrete valuation corresponding to C_i. Then by applying the theory reviewed above to the pull-back of χ to each K_i, we get $sw_i(\chi) \in \mathbb{N}$ and $rsw_i(\chi) : N_{K_i}^{sw_i(\chi)} \to \omega_{K_i}$ if $sw_i(\chi) > 0$. We define the Swan divisor D_χ of $\mathcal{F}$ by

$$D_\chi = \sum sw_i(\chi) \cdot C_i.$$

Let ω_X denote the sheaf $\Omega^1_{X/k}(\log(X - U))$ of differential 1-forms with logarithmic poles at $X - U$. Beware of the unusual notation. Generally, for a point x of a scheme X (resp. a closed immersion $i : Z \to X$) and a quasi-coherent $\mathcal{O}_X$-module $\mathcal{E}$, we put $\mathcal{E}(x) = \mathcal{E}_x \otimes_{\mathcal{O}_{X,x}} \kappa(x)$ (resp. $\mathcal{E}|_Z = i^*\mathcal{E}$). Then at each generic point ξ_i of C_i, we have $\omega_X(\xi_i) = \omega_{K_i}$ and hence $rsw_i(\chi)$ is a map $\mathcal{O}_X(-D_\chi)(\xi_i) \to \omega_X(\xi_i)$ for i with $sw_i(\chi) > 0$. By [K3] Theorem (7.1), this extends to a morphism $rsw_i(\chi) : \mathcal{O}_X(-D_\chi)|_{C_i} \to \omega_X|_{C_i}$. Further it is shown there that they extends to a morphism $rsw(\chi) : \mathcal{O}_X(-D_\chi)|_D \to \omega_X|_D$ where D is the support of D_χ with the reduced structure. The sheaf $\mathcal{F}$ is said to be clean if $rsw_i(\chi)$ is locally an isomorphism onto a direct summand for every i with $sw_i(\chi) > 0$. Assume $\mathcal{F}$ is clean. Then the characteristic variety $Ch(\mathcal{F})$ of $\mathcal{F}$ is a dimension d cycle of the cotangent bundle $\mathbf{V}(\omega_X) = \mathbf{V}(\Omega^1_{X/k}(\log(X-U)))$ with logarithmic poles

$$Ch(\mathcal{F}) = \sum sw_i(\mathcal{F}) \cdot [\mathbf{V}(\mathcal{O}_X(-D_\chi)|_{C_i})].$$

Here $\mathbf{V}$ denotes the covariant vector bundle associated to a locally free sheaf and $\mathbf{V}(\mathcal{O}_X(-D_\chi)|_{C_i})$ is regarded as a subvariety of $\mathbf{V}(\omega_X)$ by $rsw_i(\chi)$. The characteristic 0-cycle $c_\mathcal{F}$ is

$$c_\mathcal{F} = (-1)^{d-1} s^* Ch(\mathcal{F}) \in CH_0(X)$$

where $s : X \to \mathbf{V}(\omega_X)$ is the 0-section. By an elementary calculation, we have

$$c_\mathcal{F} = \{c^*(\omega_X) \cdot (1 + D_\chi)^{-1} \cdot D_\chi\}_{\dim 0}$$

where $c(\omega_X)$ is the total chern class of $\omega_X = \Omega^1_{X/k}(\log(X-U))$ and $*$ denotes the operator multiplying $(-1)^i$ on codimension i part.

CONJECTURE (KATO). *Let (X, U) be an RN-pair over k and $\mathcal{F}$ be the smooth ℓ-adic sheaf of rank 1 corresponding to a character of $\pi_1(U)^{ab}$ of finite order as above. Assume X is proper and $\mathcal{F}$ is clean. Then*

$$\chi_c(U, \mathcal{F}) - \chi_c(U) = -\ \deg c_\mathcal{F}.$$

Here $\chi_c(U, \mathcal{F}) = \sum (-1)^i \dim H^i_c(U, \mathcal{F})$ and $\chi_c(U) = \chi_c(U, \mathbb{Q}_\ell)$. If dim $X = 2$, this conjecture is proved by Kato in [K2]. The Euler number $\chi_c(U)$ itself is the degree of a 0-cycle $c_{X,U} = c^*(\omega_X)_{\dim 0}$ (cf. Lemma 0 below).

We give the precise statement of the assumption of our main theorem. Let (X, U) be an RN-pair over k as above. Let θ be a character of $\pi_1(U)^{ab}$ of order p. It is called s-clean if it is clean and for each C_i with $sw_i(\theta) > 0$, the composite $\mathcal{O}(-D_\theta)|_{C_i} \xrightarrow{rsw_i(\theta)} \omega_X|_{C_i} \xrightarrow{res} \mathcal{O}_{C_i}$ is either an isomorphism or a zero map (depending on C_i). Let $\pi : Y \to X$ be the integral closure of X in the etale covering V of U of degree p trivializing θ and $j : V \to Y$ be the inclusion. Then as we will show in Lemma 1, the logarithmic structure $M = \mathcal{O}_Y \cap j_*\mathcal{O}_V^\times$ on Y is regular. We will briefly review the theory of logarithmic structure later (or see [K4]). Hence we have the resolution $\tilde{Y} \to Y$ associated to a suitable proper subdivision of the fan associated to (Y, M) by the procedure of [K4] (10.4). We refer to [K4] Sections 5, 9 and 10 for the theory of fan and associated resolution (cf. Proof of Lemma 2). This $\tilde{Y} \to Y$ is an isomorphism on V and $(\tilde{Y}, V)$ is an RN-pair over k.

Let (X, U) be an RN-pair over k as above and χ be a character of $\pi_1(U)^{ab}$ of order n. We consider the following condition $*$ on (X, U, χ) which says the above construction works inductively.

$*$ There is a sequence $(X_i,\ U_i,\ \chi_i)$ for $0 \leqslant i \leqslant e = \mathrm{ord}_p n$, satisfying the following conditions

1). $(X_0, U_0, \chi_0) = (X,\ U,\ \chi)$.

2). For $0 \leqslant i < e$, the character $\theta_i = \frac{n}{p^{i+1}} \cdot \chi_i$ is s-clean and of order p, the pair (X_{i+1}, U_{i+1}) is $(\tilde{Y}, V)$ constructed by the above procedure from (X_i, U_i, θ_i) and χ_{i+1} is the pull-back of χ_i.

Then our main result is

THEOREM. *Let (X, U) be an RN-pair over k and χ be a character of $\pi_1(U)^{ab}$ of order n. Let $\mathcal{F}$ be the smooth ℓ-adic sheaf of rank 1 on U corresponding to χ. Assume that X is proper, the condition $*$ is satisfied and that (X_i, U_i, χ_i) is clean for every $0 \leqslant i \leqslant e = \mathrm{ord}_p n$. Then we have*

$$\chi_c(U, \mathcal{F}) - \chi_c(U) = -\deg c_{\mathcal{F}}.$$

REMARK: If dim $X = 2$, by [K1] Theorem (4.1), there is an RN-pair (X', U') with a proper morphism $\pi : X' \to X$ such that $\pi : U' \simeq U$ and that $(X', U', \pi^*\mathcal{F})$ satisfies the assumption of Theorem and $\pi_*(c_{\pi^*\mathcal{F}}) = c_{\mathcal{F}}$ by Theorem (5.2) loc. cit. Therefore if dim $X = 2$, Theorem implies Conjecture above as is shown in [K2].

PROOF: First we reduce it to the case where n is a power of p. Let χ_0 be the character of order p^e such that $\chi \cdot \chi_0^{-1}$ is of order prime to p. Then it is clear that χ_0 satisfies the assumption of Theorem and $c_\chi = c_{\chi_0}$. Let V be the etale covering of U of degree n trivializing χ and $G = Aut_U(V) \cong \mathbb{Z}/n$. For $\sigma \in G$, we put $Tr_V(\sigma) = \sum (-1)^i Tr(\sigma; H_c^i(V,\ \mathbb{Q}_\ell))$. Then

THEOREM DL ([D-L] THEOREM 3.2 AND PROPOSITION 3.3). *For all $\sigma \in G$, $Tr_V(\sigma)$ is a rational integer independent of ℓ and, if the order of σ is not a power of p, it is zero.*

Since $\chi_c(U,\mathcal{F}_\chi) = (\chi, Tr_V)_G$, we have $\chi_c(U,\mathcal{F}_\chi) = \chi_c(U,\mathcal{F}_{\chi_0})$. Therefore it is sufficient to prove Theorem in the case $n = p^e$ and $p \neq 0$.

Before continuing the proof, we need to review the theory of logarithmic structure and that of ramification of an automorphism by Kato [K4] and [K2]. A logarithmic structure on a scheme X is a morphism of sheaves of commutative monoids $\alpha : M \to \mathcal{O}_X$ with respect to the multiplication of $\mathcal{O}_X$ such that $\alpha^{-1}(\mathcal{O}_X^\times) \simeq \mathcal{O}_X^\times$. A log structure is called trivial if $M = \mathcal{O}_X^\times$. A scheme X with log structure M is called a log scheme (X, M). In this paper, we only consider such a log structure M that, locally on X, there is a finitely generated integral monoid P and $\alpha : P \to \mathcal{O}_X$ such that M is induced by P. This condition is slightly weaker than (S) in [K4] (1.5). Here M is induced by P means that M is the amalgamated sum of P and $\mathcal{O}_X^\times$ over $\alpha^{-1}(\mathcal{O}_X^\times)$. A commutative monoid P is said to be integral if the canonical morphism $P \to P^{gr}$ is injective and it is said to be saturated further if every $a \in P^{gr}$ such that $a^n \in P$ for some $n > 0$ is contained in P. A log structure M on a noetherian scheme X is called regular if it is locally induced by a finitely generated saturated monoid and if for every $x \in X$, the following condition is satisfied. If I_x denotes the ideal generated by the image of M_x in $\mathcal{O}_{X,x} - \mathcal{O}_{X,x}^\times$, then $\mathcal{O}_{X,x}/I_x$ is a regular local ring and $\dim\, \mathcal{O}_{X,x} = \dim(\mathcal{O}_{X,x}/I_x) + \mathrm{rank}(M_x^{gp}/\mathcal{O}_{X,x}^\times)$. A scheme X with a regular log structure M is said to be log regular and is normal ([K4] Theorem (4.1)). Further, for the largest open subscheme $j : U \hookrightarrow X$ such that M is trivial on U, we have $M = j_*\mathcal{O}_U^\times \cap \mathcal{O}_X$ and $M^{gp} = j_*\mathcal{O}_U^\times$ (loc. cit. Theorem (11.6)). We call this M the log structure associated to U. If (X, U) is an RN-pair, the log structure associated to U is regular and U is the largest open. For a log structure M on X, we define the sheaf $\omega_{X,M}$ of the differential forms with logarithmic poles in M to be the quasi-coherent $\mathcal{O}_X$-module

$$\omega_{X,M} = (\Omega^1_X \oplus (\mathcal{O}_X \otimes M^{gp}))/(d\alpha(a) - \alpha(a) \otimes a;\ a \in M).$$

If X is of finite type over k and M is regular, then $\omega_{X,M}$ is locally free of rank $\dim X$. In fact, for every $x \in X$, there is an exact sequence

$$0 \to \Omega^1_{\kappa(x)/k} \to \omega_{X,M}(x) \to \kappa(x) \otimes (M_x^{gp}/\mathcal{O}_{X,x}^\times) \to 0.$$

We put $c_{X,M} = c^*(\omega_{X,M})_{\dim 0}$. When (X, U) is an RN-pair over k and M is associated to U, the sheaf $\omega_{X,M}$ is equal to ω_X hence $c_{X,U} = c_{X,M}$.

LEMMA 0. *Let X be a proper k-scheme with a regular logarithmic structure M. Let U be the largest open subscheme of X where M is trivial. Then*

$$\chi_c(U) = \deg\ c_{X,M}.$$

PROOF: Take the resolution $\pi : X' \to X$ associated to a regular proper subdivision of the fan $F(X)$ associated to (X, M) ([K4] (10.4)). Then $U' = \pi^* U$ is isomorphic to U, (X', U') is an RN-pair and $\omega_{X'} \simeq \pi^* \omega_{X,M}$. Hence Lemma is reduced to the case where (X, U) is an RN-pair. The proof is easy in this case and left to the reader.

Let (X, M) be a logarithmic scheme and σ be an automorphism of (X, M). We define the fixed part $X^\sigma \subset X$ by the cartesian diagram

$$\begin{array}{ccc} X^\sigma & \longrightarrow & X \\ \downarrow & & \downarrow{\scriptstyle \text{graph of } \sigma} \\ X & \xrightarrow[\text{diagonal}]{} & X \times X \end{array}$$

and I_σ to be the ideal sheaf of $\mathcal{O}_X$ corresponding to X^σ. We say that σ is admissible if the action of σ on $M_x / \mathcal{O}_{X,x}^\times$ is trivial for all $x \in X^\sigma$. If σ is admissible, we define an ideal sheaf J_σ to be that generated by I_σ and $1 - (\sigma(a)/a)$ for $a \in M_x$ at $x \in X^\sigma$. The action of σ is called clean if J_σ is an invertible ideal. Then we let D_σ denote the Cartier divisor of J_σ. Assume X is of finite type over k, the log structure M is regular and that k-automorphism σ of (X, M) is clean. Then we define the 0-cycle c_σ by

$$c_\sigma = \{c^*(\omega_{X,M}) \cdot (1 + D_\sigma)^{-1} \cdot D_\sigma\}_{\dim 0} \in CH_0(X).$$

This 0-cycle has the following property which will not be used in the sequel.

PROPOSITION. *Let X be a proper k-scheme, M be a regular log structure on X and σ be a clean k-automorphism of (X, M). Let U be the largest open subscheme of X where M is trivial. Assume one of the followings*

1). There is an open covering of X by σ-stable affine subschemes.

2). (X, U) is an RN-pair.

Then we have

$$Tr_U(\sigma) = \deg c_\sigma.$$

This will be proved in [K2] at least in the case 2). In the present version of [K2], only a k-automorphism σ of an RN-pair (X, U) is treated and the definitions given there coincide with those here. We can reduce the case 1) to case 2) but we omit the detail (cf. Proof of Lemmas 0 and 2).

We return to the proof of Theorem.

LEMMA 1. *Let (X, U) be an RN-pair over k and θ be an s-clean character of $\pi_1(U)^{ab}$ of order p. Let $\pi : Y \to X$ be the integral closure of X in the etale covering V of U of degree p trivializing θ and $G = Aut_X(Y)$. Then*

1). The log structure M associated to $V \hookrightarrow Y$ is regular. Let $\sigma \in G$ be a non-trivial element. Then

2). The action of σ on Y is clean with respect to M and we have $D_\theta = \pi_* D_\sigma$, $\pi^* D_\theta = pD_\sigma$ and $\pi_* c_\sigma = c_\theta$.

3). The sheaf $\omega_{Y/X} = Coker(\pi^*\omega_X \to \omega_{Y,M})$ is an invertible $\mathcal{O}_{D_{Y/X}}$-module where $D_{Y/X}$ is the divisor $\pi^*D_\theta - D_\sigma$. The map $\varphi_\sigma : \omega_{Y/X}|_{D_\sigma} \to \mathcal{O}(-D_\sigma)|_{D_\sigma}$ defined by $a \cdot d\log b \mapsto a(1 - \sigma(b)/b)$ is an isomorphism. The total chern class $c(\omega_{Y/X})$ is equal to $(1 - D_\sigma) \cdot (1 - \pi^* D_\theta)^{-1}$ and $c_{Y/X} = c_{Y,M} - \pi^* c_{X,U}$ is

$$c_{Y/X} = -\{c^*(\omega_{Y,M}) \cdot (1 + D_\sigma)^{-1} \cdot D_{Y/X}\}_{\dim 0} = -(p-1) \cdot c_\sigma.$$

4). Let D be the support of D_σ with reduced structure. Then the sequence

$$0 \to \pi^*\mathcal{O}_X(-D_\theta)|_D \xrightarrow{\pi^*(rsw(\theta))|_D} \pi^*\omega_X|_D \to$$

$$\omega_{Y,M}|_D \xrightarrow{\varphi_\sigma|_D} \mathcal{O}(-D_\sigma)|_D \to 0$$

is exact hence locally homotope to 0.

We show that Lemma 1 implies Theorem in the case $n = p$. By Lemma 0, we have $\chi_c(V) - p \cdot \chi_c(U) = \deg\ c_{Y/X}$. Since $Tr_V(\sigma) \in \mathbb{Z}$ for $\sigma \in G$ by Theorem DL, we have $(p-1) \cdot \chi_c(U, \mathcal{F}_\theta) = Tr_{\mathbb{Q}(\zeta_p)/\mathbb{Q}}(\theta, Tr_V)_G = \chi_c(V) - \chi_c(U)$. Hence $(p-1) \cdot (\chi_c(U, \mathcal{F}_\theta) - \chi_c(U)) = \deg\ c_{Y/X}$. Therefore by 2) and 3) of Lemma 1, we have

$$\chi_c(U, \mathcal{F}_\theta) - \chi_c(U) = -\deg c_\sigma = -\deg c_\theta.$$

PROOF OF LEMMA 1: Let $D_\theta = \sum n_i C_i$ and E be the union of the divisors with $p \nmid n_i$. Let $x \in X$ be a point of X, I be the set of indices i such that $x \in C_i$ and for every $i \in I$, π_i be a section of $\mathcal{O}_X$ defining C_i at x. Then at a neighborhood of x, the character θ corresponds to an Artin-Schreier extension

$$t^p - t = \frac{u}{\prod \pi_i^{n_i}}$$

such that

1). If $x \notin E$, every n_i is divisible by p and du is everywhere non-vanishing in $\Omega^1_{C_J}$ for $C_J = \cap_{i\in J} C_i$ where $J \subset I$ runs every subset $J \subset I$ containing at least one element i with $n_i \neq 0$.

2). If $x \in E$, one of n_i is not divisible by p and u is a unit.

Outside E, the integral closure (Y, V) is an RN-pair since Y is given by $Y = X[s]/(s^p - \prod \pi_i^{(p-1)n_i/p} \cdot s - u)$ where $s = (\prod \pi_i^{n_i/p}) \cdot t$. We consider at $x \in E$. Let Y_1 be the scheme

$$Y_1 = X[s, w]/(s^p w - \Pi, s^{p-1} w - w + u)$$

where Π denotes $\prod \pi_i^{n_i}$. It is easy to check that $Y_1 \times_X U = V$ by putting $s = t^{-1}$ and $w = u + \Pi \cdot t$. The scheme Y_1 is finite over X since w satisfies $w^p - \Pi^{p-1} \cdot w - u^p = 0$ and is a unit.

Let Q_1 be the integral submonoid of $\mathbf{Q}^I \times \mathbf{Q} \cdot e_w$ generated by $\mathbf{N}^I \times \mathbf{Z} \cdot e_w$ and $e_s = \frac{1}{p}(\sum n_i e_i - e_w)$. Let M_1 be the log structure of Y_1 induced by $Q_1 \to \mathcal{O}_{Y_1}$ defined by $e_i \mapsto \pi_i, e_s \mapsto s$ and $e_w \mapsto w$. Let $Q = \{a \in Q^{gp}; \exists n > 0, n \cdot a \in Q_1\}$ be the saturation of Q_1.

Claim. *The scheme $Y_2 = Y_1 \otimes_{k[Q_1]} k[Q]$ with the logarithmic structure M_2 induced by $Q \to \mathcal{O}_{Y_2}$ is log regular at the inverse image of x.*

Proof of Claim: Let y be a unique point of Y_1 lying on x. Let I_y be the ideal of $\mathcal{O}_{Y_1,y}$ generated by the image of $Q_1 - \mathcal{O}^\times_{Y_1,y}$. By [K4] Proposition (12.2), it is sufficient to check that

0). $Q_1^{gp}/Q_1^\times$ is torsion free

1). $\mathcal{O}_{Y_1,y}/I_y$ is regular

2). dim $\mathcal{O}_{Y_1,y} =$ dim $(\mathcal{O}_{Y_1,y}/I_y)+$ rank$(Q_1^{gp}/Q_1^\times)$.

The condition 0) is clear. By definition of Y_1, it is clear that $\mathcal{O}_{Y_1,y}/I_y \simeq \mathcal{O}_{X,x}/(\pi_i, i \in I)$ and is regular. Since Y_1 is finite over X, we see that Y is locally of complete intersection of dimension $d =$ dim X. Thus Claim is proved.

Since a log regular scheme is normal, $Y_2 = Y$ at y. Further it is easy to see that V is the largest open set where M_2 is trivial. Hence the log structure M associated to V coincides with M_2 and is regular.

Next we prove 2). Outside E, we have $\sigma(s) = s + \prod_i \pi_i^{n_i/p}$. Using this, it is easy to check that σ is admissible and clean and that J_σ is generated by $\prod_i \pi_i^{n_i/p}$. We consider at E. The log scheme (Y_1, M_1) above has an admissible action of G. In fact, $\sigma(s)/s = (s+1)^{-1}$ is invertible since $s^{p-1} - 1 = -u \cdot w^{-1}$ and $\sigma(w) = w + \Pi$ is also invertible at E. Using this and the fact that s divides Π, we can also easily check that σ is clean and J_σ is generated by s. By the definition of Y_2, it is easy to see that the action of σ is admissible and clean on Y_2 and that $J_{\sigma,Y_2} \simeq \varphi_1^* J_{\sigma,Y_1}$, where $\varphi_1 : Y_2 \to Y_1$. Thus we have the cleanness of σ and the equalities $\pi^* D_\theta = pD_\sigma$ and $\pi_* D_\sigma = D_\theta$.

We prove 3) and the equality $\pi_* c_\sigma = c_\theta$. Outside E, $\omega_{Y/X}$ is generated by ds and the relation is $(\prod \pi_i^{n_i/p})^{p-1} ds = 0$. On the other hand at E, $\omega_{Y_1/X}$ is generated by $d\log s$ and $d\log w$ and the relations are $s^{p-1} \cdot d\log s = d\log w = 0$. By the definition of Y_2, we see that $\omega_{Y_2/X} \simeq \varphi_1^* \omega_{Y_1/X}$. Thus we have the assertion on the structure of $\omega_{Y/X}$. Since φ_σ is a surjection of invertible sheaves, it is an isomorphism. From these fact, it is easy to check the formula for $c(\omega_{Y/X})$ and that for $c_{Y/X}$. The equality $\pi_* c_\sigma = c_\theta$ also immediately follows from the definition, the formula for $c(\omega_{Y/X})$ and from $\pi_* D_\sigma = D_\theta$.

We prove 4). In 3) we have already shown the exactness at $\omega_{Y,M}|_D$ and at $\mathcal{O}(-D_\sigma)|_D$. By ranks and the cleanness of θ, it is sufficient to show that the composite $\pi^* \mathcal{O}_X(-D_\theta)|_D \to \omega_{Y,M}|_D$ is zero. Outside E, $rsw\ \theta$ is given by $\prod \pi_i^{n_i} \mapsto du$. Since $u \equiv s^p \bmod \prod \pi_i^{n_i}$ on Y, the composite is zero. At E, $rsw\ \theta$ is given by $\prod \pi_i^{n_i} \mapsto du - u \sum n_i d\log \pi_i$. Since $u \equiv w \bmod s$ and $d\log w = \sum n_i d\log \pi_i$ on Y_1, the composite is also zero. Thus we have completed the proof of Lemma 1 and therefore that of Theorem in the case $n = p$.

We prove Theorem in the case $n = p^e$ by induction on e. Assume $e > 1$. Let $(\tilde{Y}, V, \chi')$ be (X_1, U_1, χ_1) in Theorem and let $\pi : \tilde{Y} \to X$ and $\varphi : \tilde{Y} \to Y$. Let W be the etale covering of U of degree $n = p^e$ trivializing χ and $G = Aut_U(W) \cong \mathbb{Z}/p^e$. Since $Tr_W(\sigma) \in \mathbb{Z}$ for $\sigma \in G$ by Theorem DL, we have $p \cdot \chi_c(U, \mathcal{F}_\chi) = Tr_{\mathbb{Q}(\zeta_{p^e})/\mathbb{Q}(\zeta_{p^{e-1}})}(\chi, Tr_W)_G = \chi_c(V, \mathcal{F}_{\chi'})$. Therefore by the assumption of induction and the equality $\deg c_{Y/X} = \chi_c(V) - p \cdot \chi_c(U)$ of Lemma 0, it is sufficient to show that

$$- \pi^* c_\chi = -\, c_{\chi'} + \varphi^* c_{Y/X}.$$

LEMMA 2. *Let (X, U) be an RN-pair over k and let χ and θ be clean characters of $\pi_1(U)^{ab}$ of finite order. Assume that θ is of order p and s-clean and that $D_\theta = 0$ (resp. $D_\theta < D_\chi$) where $D_\chi = 0$ (resp. $D_\chi \neq 0$). Let (Y, M) be as in Lemma 1, $\varphi : \tilde{Y} \to Y$ be the resolution associated to a regular proper subdivision $F' \to F(Y)$ and π denote the map $\tilde{Y} \to X$. If $\chi' = \pi^*(\chi)$ is also clean, we have*

$$\pi^* c_\chi = c_{\chi'} - \varphi^* c_{Y/X}.$$

We see Lemma 2 implies the induction step. In fact, it is clear that $\theta = \frac{n}{p} \cdot \chi$ satisfies $D_\theta = 0$ where $D_\chi = 0$ and $D_\theta < D_\chi$ where $D_\chi \neq 0$. Therefore we will complete the proof of Theorem by showing Lemma 2.

PROOF OF LEMMA 2: We show that everything appeared in Lemma 1 has its counterpart on $\tilde{Y}$ and that it is a pull-back to $\tilde{Y}$ of that on Y. By the definition [K4] (10.4), locally on $\tilde{Y}, \varphi : \tilde{Y} \to Y$ is described as follows. Let $Y = \mathrm{Spec} A$ and $P \to A$ be a morphism of monoids such that $P \times A^{\times} \simeq M$. Then locally on $\tilde{Y}$, there is a submonoid $\tilde{P}$ of P^{gp} containing P and isomorphic to $\mathbf{N}^r \times \mathbf{Z}^s$ for some r and $s \in \mathbf{N}$ such that $\tilde{Y} \to Y$ is given by $A \to A \otimes_{k[P]} k[\tilde{P}] = B$. By loc. cit., we see that φ induces an isomorphism on V, that $(\tilde{Y}, V)$ is an RN-pair, and that the log structure induced by $\tilde{P} \to B$ coincides with that associated to V. Here we identified V and $\varphi^* V$. From this description, it is easy to show that $\varphi^* \omega_{Y,M} = \omega_{\tilde{Y}}$. We show that the action of G extends to $\tilde{Y}$. Since Y is affine over X, we may assume G acts on A above. For all $a \in M, \neq 0$, we have $\sigma(a)/a \in A^{\times}$, since the action of σ on (Y, M) is admissible. It is straight forward to check that the action of G extends to $B = A \otimes_{k[P]} k[\tilde{P}]$ hence to $\tilde{Y}$. (More canonically, $B = A \otimes_{k[M]} k[M \times_{P^{gp}} \tilde{P}]$ and hence G acts on it.) Once we have the action of G, it is quite easy to check that the action of each $\sigma \in G$ is admissible ane clean and that $J_{\sigma,\tilde{Y}} = \varphi^* J_{\sigma,Y}$. Hence every statement in Lemma 1 applies to $\tilde{Y}$.

Now by definition and 3) of Lemma 2, $\pi^* c_X - (c_{X'} - \varphi^* c_{Y/X})$ is equal to the dimension 0 part of

$$c^*(\omega_{\tilde{Y}}) \left(\frac{(1 + \pi^* D_\theta) \cdot \pi^* D_X}{(1 + D_\sigma)(1 + \pi^* D_X)} - \left(\frac{D_{X'}}{1 + D_{X'}} + \frac{D_{\tilde{Y}/X}}{1 + D_\sigma} \right) \right).$$

By an elementary calculation, this is equal to

$$c^*(\omega_{\tilde{Y}}) \left(\frac{\pi^* D_X - (D_{X'} + D_{\tilde{Y}/X})}{(1 + D_{X'})(1 + D_\sigma)} - \frac{(\pi^* D_X - \pi^* D_\theta) \cdot (\pi^* D_X - D_{X'})}{(1 + \pi^* D_X)(1 + D_{X'})(1 + D_\sigma)} \right).$$

LEMMA 3. *Let the notation be as in Lemma 2 except that we do not assume the relations between D_θ and D_X. Then*

1). We have $D_{X'} \leqslant \pi^* D_X$. If C is a component of $\tilde{Y} - V$ such that $D_{X'} < \pi^* D_X$ at C, then $\pi^* D_\theta > 0$ at C and the sequence

$$0 \to \pi^* \mathcal{O}(-D_X)|_C \xrightarrow{\pi^*(rsw_X)} \pi^* \omega_X|_C \to \omega_{\tilde{Y}}|_C$$

is exact.

2). Assume $\pi^* D_\theta < \pi^* D_X$ at a component C of $\tilde{Y} - V$. Then we have $\pi^* D_X \leqslant D_{X'} + D_{\tilde{Y}/X}$ and $D_{X'} > 0$ at C. Further if $\pi^* D_X < D_{X'} + D_{\tilde{Y}/X}$ at C, then $\pi^* D_\theta > 0$ at C and the sequence

$$0 \to \mathcal{O}(-D_{X'})|_C \xrightarrow{rsw_{X'}} \omega_{\tilde{Y}}|_C \xrightarrow{\varphi_\sigma|_C} \mathcal{O}(-D_\sigma)|_C \to 0$$

is a complex. The cohomology sheaves are zero except at $\omega_{\tilde{Y}}|_C$ and it is locally free of rank dim $X-2$ there.

We show Lemma 3 implies Lemma 2. It is sufficient to show the following equalities.

$$1).\quad (\pi^*D_\chi - \pi^*D_\theta)(\pi^*D_\chi - D_{\chi'}) = 0.$$

$$2).\quad \left(\frac{c^*(\omega_{\tilde{Y}})}{(1+D_{\chi'})(1+D_\sigma)}(\pi^*D_\chi - (D_{\chi'} + D_{\tilde{Y}/X}))\right)_{\dim 0} = 0.$$

To prove 1), it is sufficient to show that at each component C where $D_{\chi'} \neq \pi^*D_\chi$, there is an isomorphism $\pi^*\mathcal{O}_X(-D_\chi)|_C \simeq \pi^*\mathcal{O}_\chi(-D_\theta)|_C$. By 1) of Lemma 3 and 4) of Lemma 1, they are both equal to the kernel of $\pi^*\omega_X|_C \to \omega_Y|_C$. To prove 2), it is sufficient to show

$$\left(\frac{c(\omega_{\tilde{Y}})}{(1-D_{\chi'})(1-D_\sigma)}\cdot C\right)_{\dim 0} = 0$$

for each component C where $\pi^*D_\chi \neq D_{\chi'} + D_{Y/X}$. By the assumption, we have $\pi^*D_\theta < \pi^*D_\chi$ at C and 2) of Lemma 3 applies. Therefore it is equal to the (dim $X-1$)-th chern class of the cohomology sheaf there, which is of rank dim $X-2$, and is zero.

PROOF OF LEMMA 3: The assertions are reduced to those at the generic point of each irreducible component of $\tilde{Y} - V$. In fact, this is clear for the inequalities and, for the rest, it follows from the cleaness of χ and χ' and 4) of Lemma 1. Furthur by 4) of Lemma 1, for the assertions concerning on the sequences and the cohomology sheaves, it is sufficient to show that the sequences are complexes i.e. the composites of the maps are zero at each generic point. We will show that we may assume $\tilde{Y} = Y$ i.e. Y is finite over X.

We need a lemma on fans ([K4] Section 9) as below. Let F and G be fans satisfying (S^{fan}) (loc. cit (9.4)). We call a morphism $f : G \to F$ an isogeny if the following conditions are satisfied.

1). f is a homeomorphism of underlying spaces.

2). $M^{gp}_{F,t} \to M^{gp}_{G,f^{-1}(t)}$ is injective for all $t \in F$.

3). There is an integer $n > 0$ such that $n(M_{G,f^{-1}(t)}) \subset$ *Image* $M_{F,t}$ for all $t \in F$.

LEMMA 4. *Let $f : G \to F$ be an isogeny and $G' \to G$ be a subdivision.*

Then there is a cartesian diagram of fans

$$\begin{array}{ccc} G' & \longrightarrow & F' \\ \downarrow & & \downarrow \\ G & \longrightarrow & F. \end{array}$$

Here $F' \to F$ is a subdivision and $G' \to F'$ is an isogeny. When $G' \to G$ is proper, $F' \to F$ is also proper.

The proof of Lemma 4 is easy by using $F \to G$ such that the composite $F \to F$ is induced by the multiplication by n and is left to the reader.

Let F and G be the fans associated to X and Y respectively. By the construction of Y given in the proof of Lemma 1, there is a natural map $G \to F$ induced by $Y \to X$. Let $G' \to G$ be the proper subdivision to which $\tilde{Y} \to Y$ is associated. We apply Lemma 4 to $G' \to G \to F$. Then by [K4] Proposition (9.9) and (9.10), we have $\varphi : X_1 \to X$ associated to the proper subdivision F' of $F = F(X)$. It is clear that $\tilde{Y}$ is the integral closure of X_1 in V. Let $U_1 = \varphi^* U$ and $\chi_1 = \varphi^* \chi$. To reduce Lemma 3 to the case where $\tilde{Y} = Y$, it is sufficient to show the following. At the generic point ξ of every component C of $X_1 - U_1$, if χ_1 is not unramified, then χ_1 is clean and we have $D_{\chi_1} = \varphi^* D_\chi$ and $rsw\ \chi_1 = \varphi^* rsw\ \chi$. By the definition of X_1, there is an open neighborhood $X^{(m)}$ of ξ and a sequence

$$X^{(m)} \to \cdots \to X^{(i+1)} \to X^{(i)} \to \cdots \to X^{(0)} \subset X.$$

Here $X^{(0)}$ is an open subscheme of X and $X^{(i+1)}$ is an open subscheme of the blowing-up of $X^{(i)}$ at the closure $C^{(i)}$ of the image of ξ. Further for each i, $C^{(i)}$ is the intersection of some irreducible components of the divisor $X^{(i)} - U^{(i)}$ with normal crossing, where $U^{(i)}$ is the inverse image of U. If χ is unramified at $C^{(0)}$, there is nothing to prove. Assume χ is ramified at $C^{(0)}$. Then χ is strongly clean at the generic point of $C^{(0)}$ ([K3] Definition (7.4)). In fact, for the generic point of the intersection of some components of the divisor, the cleaness is equivalent to the strong cleaness. Hence by applying inductively [K3] Theorem (8.1), we see that χ_1 is clean, $D_{\chi_1} = \varphi^* D_\chi$ and $rsw\ \chi_1 = \varphi^* rsw\ \chi$ at ξ. Thus we have reduced Lemma 3 to the case $\tilde{Y} = Y$. Namely Lemma 3 has been reduced to

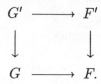

LEMMA 3'. *Let K be a complete discrete valuation field with residue field $\overline{K}$ of $ch = p$. Let χ and θ be characters of $G_K = Gal\ (K^{ab}/K)$ of*

finite order. Assume that the order of θ is p. Let L be the extension of K of degree p trivializing θ and χ' be the restriction of χ to L. Then

1). We have $sw(\chi') \leqslant e_{L/K} \cdot sw(\chi)$. If $sw(\chi') < e_{L/K} \cdot sw(\chi)$, then L is ramified over K and the sequence

$$0 \to N_K^{sw(\chi)} \otimes_{\overline{K}} \overline{L} \xrightarrow{rsw\chi \otimes 1_{\overline{L}}} \omega_K \otimes_{\overline{K}} \overline{L} \to \omega_L$$

is exact.

2). Assume $sw(\theta) < sw(\chi)$. Then we have $e_{L/K} \cdot sw(\chi) \leqslant sw(\chi') + d_{L/K}$ and $sw(\chi') > 0$. Here $d_{L/K}$ denotes $e_{L/K} \cdot sw(\theta) - s_\sigma$ and s_σ is the integer n such that m_L^n is generated by $1 - \sigma(a)/a, a \in \mathcal{O}_L, \neq 0$. Further assume $e_{L/K} \cdot sw(\chi) < sw(\chi') + d_{L/K}$. Then L is ramified over K and the composite

$$N_L^{sw\chi'} \xrightarrow{rsw\ \chi'} \omega_L \xrightarrow{\varphi_\sigma} N_L^{s_\sigma}$$

is zero, where $\varphi_\sigma : \omega_L \to N_L^{s_\sigma}$ is defined by $a \cdot d\log b \mapsto a \cdot (1 - \sigma(b)/b)$.

REMARK: The integer $d_{L/K}$ is equal to the length of $\Omega^1_{\mathcal{O}_L}(\log \overline{L}) / \mathcal{O}_L \otimes_{\mathcal{O}_K} \Omega^1_{\mathcal{O}_K}(\log \overline{K})$ (cf. 3) of Lemma 1). It is also equal to $\delta_{L/K} - (e_{L/K} - 1)$ where $\delta_{L/K}$ is the valuation of the different of L over K.

PROOF: Let K' be the completion of $K(t)$ appeared in the definition of the Swan conductor reviewed before and $L' = L \otimes_K K'$. We put $s = sw(\chi), s' = sw(\chi'), s_0 = sw(\theta), e = e_{L/K}$ and $d = d_{L/K}$ for short.

First we show 1). By [K3] Proposition (6.3), we see $\chi_{L'}$ annihilates $U_{L'}^{es+1}$ which means $sw(\chi') \leqslant e_{L/K} \cdot sw(\chi)$. Assume $s' < es$. Then L is ramified over K by [K3] Lemma (6.2). For the exactness, it is sufficient to show that the composite $N_K^s \to \omega_L$ is zero since the kernel of $\omega_K \otimes \overline{L} \to \omega_L$ is of dimension 1. By commutativity of the diagram

$$\begin{array}{ccc} K'^{\times} & \xrightarrow{\{\chi_{K'},\ \}_{K'}} & Br(K') \\ \downarrow & & \downarrow \\ L'^{\times} & \xrightarrow{\{\chi_{L'},\ \}_{L'}} & Br(L'), \end{array}$$

we see the composite $U_{K'}^s \xrightarrow{\{\chi,\ \}_{K'}} Br(K') \to Br(L')$ is zero. This implies the composite $N_K^s \xrightarrow{rsw\ \chi} \omega_K \to \omega_L \xrightarrow{\times T} \omega_{L'} \to Br(L')$ is zero. Since $\omega_L \xrightarrow{\times T} \omega_{L'} \to Br(L')$ is injective, we have $N_K^s \to \omega_L$ is zero.

We prove 2). By [K3] Proposition (6.8), we have $d = (p-1)es_0/p$. For $n > s_0$, we have $en - d > es_0$ and, by a similar computation as in [S] Chap.V § 3, we have $U^n_{K'} = N_{L'/K'}(U^{en-d}_{L'})$. The diagram

$$\begin{array}{ccc} L'^{\times} & \xrightarrow{\{\chi_{L'},\ \}_{L'}} & Br(L') \\ {\scriptstyle N_{L'/K'}}\downarrow & & \downarrow{\scriptstyle Cor_{L'/K'}} \\ K'^{\times} & \xrightarrow{\{\chi_{K'},\ \}_{K'}} & Br(K'), \end{array}$$

is commutative. Hence by taking $s = sw\chi > s_0$ as n above, we see that $\chi_{L'}$ does not annihilate $U^{es-d}_{L'}$. Therefore we have $sw(\chi') \geqslant es - d > es_0/p \geqslant 0$.

To complete the proof, we need the trace map $Tr_{L/K} : \omega_L \to \omega_K$. If L is unramified over K, it is simply $Tr_{\overline{L}/\overline{K}} : \omega_L \simeq \omega_K \otimes_{\overline{K}} \overline{L} \to \omega_K$. We assume L is ramified over K. Then it is defined as follows. It is easily seen that the exterier differential $d : \omega_K \to \Lambda^2\omega_K$ and the Cartier operator $C : \omega_{K,d=0} \to \omega_K$ are defined in the same way as in the usual case. Namely, d is defined by $d(a \cdot d\log b) = da \wedge d\log b$, the kernel $\omega_{K,d=0}$ of d is generated by da and $a^p \cdot d\log b$ as an abelian group and C is defined by $C(da) = 0$ and $C(a^p \cdot d\log b) = a \cdot d\log b$. We define the trace map $Tr_{L/K} : \omega_L \to \omega_K$ by $Tr_{L/K}(a \cdot d\log b) = C(a^p \cdot d\log N_{L/K}b)$. By an elementary computation, we check it is well-defined. It is also easily checked that it is non zero, $\overline{K}$-linear and annihilates the image of $\omega_K \otimes_{\overline{K}} \overline{L}$.

LEMMA 5. *Let K, θ and L be as in Lemma 3'. Then*

1). The diagram below is commutative.

$$\begin{array}{ccc} \omega_L & \longrightarrow & Br(L) \\ {\scriptstyle Tr_{L/K}}\downarrow & & \downarrow{\scriptstyle Cor_{L/K}} \\ \omega_K & \longrightarrow & Br(K). \end{array}$$

Here the horizontal arrows are the canonical maps, whose definition is reviewed below.

2). Let χ and χ' be as in Lemma 3' and assume $sw(\theta) < sw(\chi)$. Then

$rsw(\chi')$ is defined and there is a commutative diagram

$$\begin{array}{ccc} N_L^{s'} & \xrightarrow{rsw\ \chi'} & \omega_L \\ \downarrow & & \downarrow Tr_{L/K} \\ N_K^s & \xrightarrow{rsw\ \chi} & \omega_K \end{array}$$

where the left vertical map is the one induced by $Tr_{L/K}$ if $s' = es - d$ and 0 if $s' > es - d$.

PROOF: First we give the definition of the canonical map $\omega_K \to Br(K)$ used in the definition of the refined Swan conductor. The kernel $Br(K_{nr}/K)$ of $Br(K) \to Br(K_{nr})$ is isomorphic to $H^2(\overline{K}, K_{nr}^\times/U^1)$. It is easy to see that we have an exact sequence of $Gal(\overline{K}_{sep}/\overline{K})$-module

$$0 \to (K_{nr}^\times/U^1)/p \xrightarrow{d\log} \omega_{K_{nr},d=0} \xrightarrow{1-C} \omega_{K_{nr}} \to 0$$

extending the Artin-Schreier sequence of $\overline{K}$. Hence by taking cohomology, we have an isomorphism

$$\omega_K/(1-C)\omega_{K,d=0} \simeq {}_pBr(K_{nr}/K)$$

and the canonical homomorphism $\omega_K \to Br(K)$.

We show 1). If L is unramified over K, it immediately follows from the definition of the canonical map. If L is ramified over K, it also follows from the definition and the commutative diagram of exact sequences

$$\begin{array}{ccccccccc} 0 & \longrightarrow & (L_{nr}^\times/U^1)/p & \xrightarrow{d\log} & \omega_{L_{nr},d=0} & \xrightarrow{1-C} & \omega_{L_{nr}} & \longrightarrow & 0 \\ & & \downarrow Norm & & \downarrow Tr & & \downarrow Tr & & \\ 0 & \longrightarrow & (K_{nr}^\times/U^1)/p & \xrightarrow{d\log} & \omega_{K_{nr},d=0} & \xrightarrow{1-C} & \omega_{K_{nr}} & \longrightarrow & 0. \end{array}$$

We prove 2). Since we have already shown $sw(\chi') > 0$, the refined swan conductor $rsw(\chi')$ is defined. By definition of rsw and by 1), we have a commutative diagram

$$\begin{array}{ccccccc} U_L^{s'} & \longrightarrow & N_L^{s'} & \xrightarrow{rsw\ \chi'} & \omega_L & \longrightarrow & Br(L) \\ \downarrow N_{L/K} & & & & \downarrow Tr_{L/K} & & \downarrow Cor_{L/K} \\ U_K^s & \longrightarrow & N_K^s & \xrightarrow{rsw\ \chi} & \omega_K & \longrightarrow & Br(K). \end{array}$$

By a similar computation as in [S] Chap. V §3, if $n > s_0$, we have $N_{L/K}(U_L^{en-d+1}) \subset U_K^{n+1}, Tr_{L/K}(m_L^{en-d+1}) \subset m_K^{n+1}$ and a commutative diagram

$$\begin{array}{ccc} U_L^{en-d}/U_L^{en-d+1} & \xrightarrow{\sim} & N_L^{en-d} \\ N_{L/K}\big\downarrow & & \big\downarrow Tr_{L/K} \\ U_K^n/U_K^{n+1} & \xrightarrow{\sim} & N_K^n. \end{array}$$

Now it is easy to see the diagram of 2) is commutative. Thus Lemma 5 is proved.

We complete the proof of 2) of Lemma 3'. Assume $s' > es - d$. Then by 2) of Lemma 5, the composite $N_L^{s'} \xrightarrow{rsw\ \chi'} \omega_L \xrightarrow{Tr_{L/K}} \omega_K$ is zero. By [K3] Lemma (6.2), L is ramified over K. Hence $\varphi_\sigma : \omega_L \to N_L^{s_\sigma}$ is well-defined. It is clear that the composite $\omega_K \otimes \overline{L} \to \omega_L \xrightarrow{\varphi_\sigma} N_L^{s_\sigma}$ is zero and the cokernel of $\omega_K \otimes \overline{L} \to \omega_L$ is one dimensional. Therefore, if the composite $N_L^{s'} \xrightarrow{rsw\ \chi'} \omega_L \xrightarrow{\varphi_\sigma} N_L^{s_\sigma}$ was not zero, ω_L would be the sum of the image of $rsw\ \chi'$ and that of $\omega_K \otimes_{\overline{K}} \overline{L}$. But this is a contradiction since $Tr_{L/K}(\omega_L) \neq 0$. Thus we have completed the proof of Lemma 3' and hence of Theorem. Q.E.D.

References

[DL] P.Deligne - G.Lustig, *Representation of reductive groups over finite fields*, Annals of Math. **103** (1976), 103-161.

[G] A.Grothendieck, *Formule d'Euler-Poincare en cohomologie etale*, in "SGA 5," LNM 589, Springer-Verlag, Berlin-Heidelberg-New York, 1977, pp. 372-406.

[K1] K.Kato, *Class field theory, $\mathcal{D}$-modules, and ramification on higher dimensional schemes I*, American Journal of Math. (to appear).

[K2] ———, *Class field theory, $\mathcal{D}$-modules, and ramification on higher dimensional schemes II*. in preparation

[K3] ———, *Swan conductors for characters of degree one in the imperfect residue field case*, Contemporary Math **83** (1989), 101-131.

[K4] ———, *Toric singularities*, submitted to American Journal of Math..

[S] J.-P.Serre, "Corps Locaux," 3-ieme edition, Hermann, Paris, 1968.

QUANTUM GROUPS AND HOMOLOGY OF LOCAL SYSTEMS

Vadim V. Schechtman,[1] Alexander N. Varchenko[1]
School of Mathematics,
The Institute for Advanced Study, Princeton, NJ 08540

In this paper we "quantize" results of [SV2], Part II. We establish and study the connection between (co)homology of local systems introduced in [SV1], [SV2] and homology of nilpotent subalgebras of certain Hopf algebras very close to Drinfeld-Jimbo q-analogues of Kac-Moody algebras.

The main result is an explicit version of Kohno type theorem "half-monodromy = R-matrix" (see §4). Details of proofs will be published elsewhere.

Notations.

$\mathbb{N}$ the set of nonnegative integers;

Σ_N symmetric group on N letters; for $\sigma \in \Sigma_N$ $|\sigma| \in \mathbb{Z}/2\mathbb{Z}$ is the sign. For $r \in \mathbb{N}$ $[r] = \{1, \ldots, r\}$; $[a, b] = [b] \setminus [a-1]$.

$\# I$ cardinality of a set I.

For a vector space V V^* is the dual space.

1. Quantum groups.

Let us fix the following data:

(a) a finite dimensional complex vector space $\mathfrak{h}$;

(b) a non-degenerate symmetric bilinear form $(\, , \,)$ on $\mathfrak{h}$;

(c) linearly independent covectors (" simple roots") $\alpha_1, \ldots, \alpha_r \in \mathfrak{h}^*$;

(d) a non-zero complex number $\varkappa$.

We will denote by $b : \mathfrak{h} \xrightarrow{\sim} \mathfrak{h}^*$ the isomorphism induced by $(\, , \,)$. We will transfer the form $(\, , \,)$ to $\mathfrak{h}^*$ using b. Put $b_{ij} = (\alpha_i, \alpha_j)$; $B = (b_{ij}) \in \mathrm{Mat}_r(\mathbb{C})$, $h_i = b^{-1}(\alpha_i) \in \mathfrak{h}$. Put $q = \exp(2\pi i/\varkappa)$; for $a \in \mathbb{C}$ put $q^a = \exp(2\pi i a/\varkappa)$.

We shall denote by $U_q\mathfrak{g} = U_q\mathfrak{g}(B)$ the $\mathbb{C}$-algebra generated by elements e_i, f_i, $i = 1, \ldots, r$, and the space $\mathfrak{h}$, subject to relations

$$[h, e_i] = \langle \alpha_i, h \rangle e_i; \ [h, f_i] = -\langle \alpha_i, h \rangle f_i; \tag{1.1}$$

$$[e_i, f_j] = \left(q^{h_i/2} - q^{-h_i/2} \right) \cdot \delta_{ij}; \tag{1.2}$$

$$[h, h'] = 0 \tag{1.3}$$

for all i, j, $h, h' \in \mathfrak{h}$. By definition, $q^{ah} = \exp\left(ah \cdot \frac{2\pi i}{\varkappa}\right)$. (We include in our algebra convergent power series in $h \in \mathfrak{h}$, i.e. such power series that become convergent in any $\mathfrak{h}$-diagonalizable representation.)

[1] This work was supported in part by NSF Grant DMS-8610730.

Define comultiplication $\Delta : U_q\mathfrak{g} \to U_q\mathfrak{g}\hat{\otimes}U_q\mathfrak{g}$ by the rule

$$\Delta(h) = h \otimes 1 + 1 \otimes h \tag{1.4}$$

$$\Delta(f_i) = f_i \otimes q^{h_i/4} + q^{-h_i/4} \otimes f_i$$

$$\Delta(e_i) = e_i \otimes q^{h_i/4} + q^{-h_i/4} \otimes e_i \tag{1.5}$$

This makes $U_q\mathfrak{g}$ a Hopf algebra.

Define the counit $\varepsilon : U_q\mathfrak{g} \to \mathbb{C}$ and the antipode $A : U_q\mathfrak{g} \to U_q\mathfrak{g}$ by

$$\varepsilon(f_i) = \varepsilon(e_i) = \varepsilon(h) = 0 \tag{1.6}$$

$$A(h) = -h,\ A(e_i) = -q^{b_{ii}/4}e_i,\ A(f_i) = -q^{-b_{ii}/4}f_i \tag{1.7}$$

We denote by $U_q\mathfrak{n}_-$ (resp. $U_q\mathfrak{n}_+, U_q\mathfrak{h}$) subalgebras generated by f_i (resp. $e_i, h \in \mathfrak{h}$), $i = 1, \dots, r$. $U_q\mathfrak{n}_\pm$ are free. We have $U_q\mathfrak{g} = U_q\mathfrak{n}_- \cdot U_q\mathfrak{h} \cdot U_q\mathfrak{n}_+$. We put $U_q\mathfrak{b}_\pm = U_q\mathfrak{n}_\pm \cdot U_q\mathfrak{h}$. These are Hopf subalgebras of $U_q\mathfrak{g}$.

For $\Lambda \in \mathfrak{h}^*$ we denote by $M(\Lambda)$ the Verma module over $U_q\mathfrak{g}$ generated by a vector v subject to relations $hv = \langle\Lambda, h\rangle v$; $U_q\mathfrak{n}_+v = 0$.

For $\lambda = (k_1, \dots, k_r) \in \mathbb{N}^r$ put

$$(U_q\mathfrak{n}_-)_\lambda = \{x \in U_q\mathfrak{n}_- \mid [h, x] = \langle -\sum k_i\alpha_i; h\rangle x \text{ for all } h \in \mathfrak{h}\};$$

$$M(\Lambda)_\lambda = \{x \in M(\Lambda) \mid hx = \langle \Lambda - \sum k_i\alpha_i, h\rangle x \text{ for all } h \in \mathfrak{h}\}$$

We have $U_q\mathfrak{n}_- = \bigoplus_\lambda (U_q\mathfrak{n}_-)_\lambda$, $M(\Lambda) = \bigoplus_\lambda M(\Lambda)_\lambda$.

Let

$$\tau : U_q\mathfrak{g} \to U_q\mathfrak{g}$$

be the algebra antihomomorphism such that $\tau(e_i) = f_i$, $\tau(f_i) = e_i$, $\tau(h) = h$, $h \in \mathfrak{h}$.

Put $M(\Lambda)^* = \bigoplus_\lambda M(\Lambda)^*_\lambda$. Define the structure of a $U_q\mathfrak{g}$-module on $M(\Lambda)^*$ by the rule $\langle g\varphi, x\rangle = \langle \varphi, \tau(g)x\rangle$, $\varphi \in M^*$, $g \in U_q\mathfrak{g}$, $x \in M$.

Contravariant forms. There exists a unique symmetric bilinear form S on $M(\Lambda)$ such that

$$S(v, v) = 1 \tag{1.8}$$

$$S(f_ix, y) = S(x, e_iy) \tag{1.9}$$

for all $i, x, y \in M(\Lambda)$.

Different subspaces $M(\Lambda)_\lambda$ are pairwise orthogonal with respect to S.

Let us define symmetric bilinear forms on spaces $(U_q\mathfrak{n}_-)_\lambda$. First suppose that $k_1 = k_2 = \dots = k_r = 1$. If $f_I = f_{i_1} \cdot f_{i_2} \cdot \dots \cdot f_{i_r}$, $f_J = f_{j_1} \cdot f_{j_2} \cdot \dots \cdot f_{j_r} \in (U_q\mathfrak{n}_-)_\lambda$, then there exists a unique $\sigma = \sigma(I, J) \in \Sigma_r$ such that $j_p = i_{\sigma(p)}$ for all p. We put

$$S(f_I, f_J) = q^{\left(\sum_{p<q} \pm b_{i_pi_q}\right)/4} \tag{1.10}$$

where in the sum we take the sign + before $b_{i_p i_q}$ if $\sigma(p) > \sigma(q)$ and − otherwise.

To define S for an arbitrary $\lambda = (k_1, \dots, k_r)$, let us introduce operators $g_i : (U_q\mathfrak{n}_-)_\lambda \to (U_q\mathfrak{n}_-)_{\lambda'_i}$, $i = 1, \dots, r$. Here $\lambda'_i = (k_1, \dots, k_i - 1, \dots, k_r)$. Namely, for $f_I = f_{i_1} \cdot \dots \cdot f_{i_N} \in (U_q\mathfrak{n}_-)_\lambda$ put

$$g_i(f_I) = \sum_{p:i_p=i} q^{\sum_{\ell<p} b_{ii_\ell}/4 - \sum_{\ell>p} b_{ii_\ell}/4} \cdot f_{i_1} \cdot \dots \hat{f}_{i_p} \dots f_{i_N} \quad .$$

Now S is defined as the unique form satisfying

$$S(1,1) = 1, \; S(f_i x, y) = S(x, g_i y) \tag{1.11}$$

for all , $x, y \in U_q\mathfrak{n}_-$.

One easily verifies that S is symmetric.

Serre relations. Drinfeld-Jimbo quantized Kac-Moody algebras.

Suppose that $\forall i \; b_{ii} \neq 0$. Put $a_{ij} = \frac{2b_{ij}}{b_{ii}}$. Suppose that the condition

$$\text{for } i \neq j \quad a_{ij} \in \mathbb{Z}, \quad a_{ij} \leq 0 \tag{GCM}$$

holds. So, $A = (a_{ij})$ is a Generalized Cartan Matrix. Suppose that (a)-(c) is its realization [K, §1.1]. Introduce the following notations.
For $a \in \mathbb{C}$

$$(a)_q = q^{a/2} - q^{-a/2} \tag{1.12}$$

For $b, a \in \mathbb{N}$

$$(a)_q! = \prod_{i=1}^{a} (i)_q \tag{1.13}$$

$$\binom{a}{b}_q = \frac{(a)_q!}{(b)_q!(a-b)_q!} \tag{1.14}$$

Put $n_{ij} = -a_{ij} + 1$, $q_i = q^{\frac{b_{ii}}{2}}$. Introduce *Chevalley-Serre elements*

$$\begin{aligned} R_{ij}(f) &= \sum_{v=0}^{n_{ij}} (-1)^v f_i^v \cdot f_j \cdot f_i^{n_{ij}-v} \cdot \binom{n_{ij}}{v}_{q_i} \in U_q\mathfrak{n}_- \\ R_{ij}(e) &= (R_{ij}(f) \text{ with } f_{i,j} \text{ replaced by } e_{i,j}) \in U_q\mathfrak{n}_+ \end{aligned} \tag{1.15}$$

1.16. Lemma. *All $R_{ij}(f)$ lie in the kernel of S: $U_q\mathfrak{n}_- \to (U_q\mathfrak{n}_-)^*$* □

Let $U_q\mathfrak{g}^{\mathrm{DJ}}$ be the quotient algebra of $U_q\mathfrak{g}$ by the two-sided ideal generated by all $R_{ij}(f)$,$R_{ij}(e)$, cf. [D1], [D2]. Let $\overline{U_q\mathfrak{g}}$ be the quotient algebra of $U_q\mathfrak{g}$ by the two-sided ideal generated by ker S and by ker $S_+ : U_q\mathfrak{n}_+ \to (U_q\mathfrak{n}_+)^*$ (S_+ is defined in the same way as S).

By 1.16 we have a natural epimorphism

$$U_q\mathfrak{g}^{\mathrm{DJ}} \longrightarrow \overline{U_q\mathfrak{g}} \tag{1.17}$$

1.18. Conjecture. *If $x \notin \mathbb{Q}$ then (1.17) is an isomorphism.*

It is a q-analogue of Gabber-Kac theorem [K, Th. 9.11].)

It seems also plausible that when q is a root of unity, (1.17) defines an isomorhism of $\overline{U_q\mathfrak{g}}$ with a finite dimensional quotient of $U_q\mathfrak{g}^{\mathrm{DJ}}$ defined by Lusztig, [L].

Coalgebra structure on $U_q\mathfrak{n}_-$ and a comodule structure on $M(\Lambda)$. For general matrices B and weights Λ maps $S : U_q\mathfrak{n}_- \to (U_q\mathfrak{n}_-)^*$ and $S : M(\Lambda) \to M(\Lambda)^*$ are isomorhisms. Define maps

$$\mu : U_q\mathfrak{n}_- \longrightarrow U_q\mathfrak{n}_- \otimes U_q\mathfrak{n}_- \tag{1.19}$$

$$\nu : M(\Lambda) \longrightarrow U_q\mathfrak{n}_- \otimes M(\Lambda) \tag{1.20}$$

for such B, Λ by the condition of the commutativity of squares

$$\begin{array}{ccc} (U_q\mathfrak{n}_-)^* \otimes (U_q\mathfrak{n}_-)^* & \xrightarrow{\mu^*} & (U_q\mathfrak{n}_-)^* \\ \uparrow S & & \uparrow S \\ U_q\mathfrak{n}_- \otimes U_q\mathfrak{n}_- & \xrightarrow{\text{mult.}} & U_q\mathfrak{n}_- \end{array}$$

$$\begin{array}{ccc} (U_q\mathfrak{n}_-)^* \otimes M(\Lambda)^* & \xrightarrow{\nu^*} & M(\Lambda)^* \\ S\uparrow & & \uparrow S \\ U_q\mathfrak{n}_- \otimes M(\lambda) & \xrightarrow{\text{mult.}} & M(\Lambda) \end{array}$$

One can show that the maps (1.19), (1.20) which are some universal functions on b_{ij}, (Λ, α_i), do not have singularities, so they are defined by continuity for *all* B, Λ. They make $(U_q\mathfrak{n}_-)^*$ a coalgebra and $M(\Lambda)^*$ a $(U_q\mathfrak{n}_-)^*$-comodule. Analogous construction works with $M(\Lambda)$ replaced by $M(\Lambda_1) \otimes \ldots \otimes M(\Lambda_n)$ (cf. [SV2, 6.15-6.17]).

Warning. $U_q\mathfrak{n}_-$ with the usual multiplication and μ is *not* a Hopf algebra: μ is not an algebra map.

***R*-matrix.** Introduce the action

$$U_q\mathfrak{n}_+^* \otimes M(\Lambda)^* \to M(\Lambda)^*$$

as the composition

$$U_q\mathfrak{n}_+^* \otimes M(\Lambda)^* \xrightarrow{\tau^*\otimes id} U_q\mathfrak{n}_-^* \otimes M(\Lambda)^* \xrightarrow{\mu^*} M(\Lambda)^*$$

Consider the expression

$$(1.21) \qquad R = \sum_{\lambda \in \mathbf{N}^r} q^{\Omega_0/2 + \frac{1}{4}(h_\lambda \otimes 1 - 1 \otimes h_\lambda) + d(\lambda)} \cdot \Omega_\lambda$$

Here $\Omega_0 \in \mathfrak{h} \otimes \mathfrak{h}$ is the element corresponding to the form $(\ ,\ )$; for $\lambda = (k_1, \dots, k_r)$ $h_\lambda := b^{-1}\left(\sum_{i=1}^{r} k_i \alpha_i\right) \in \mathfrak{h}$; $d(\lambda) \in \mathbf{C}$ is a constant defined as follows: represent $\alpha(\lambda) = \sum k_i \alpha_i$ as a sum of simple roots $\alpha(\lambda) = \alpha_{i_1} + \dots + \alpha_{i_N}$. Then

$$d(\lambda) := -\sum_{p \leq q} b_{i_p, i_q}/4.$$

$\Omega_\lambda \in (U_q\mathfrak{n}_+ \otimes U_q\mathfrak{n}_+^*)_\lambda$ is the canonical element, cf. [D1], [D2].

For any pair of weights Λ, Λ' (1.21) defines an operator

$$(1.22) \qquad R(\Lambda, \Lambda')^* : M(\Lambda)^* \otimes M(\Lambda')^* \longrightarrow M(\Lambda)^* \otimes M(\Lambda')^*$$

Example 1. Suppose that $n = 2$, $\lambda = (\delta_{i1}, \dots, \delta_{ir})$ for some i, $1 \leq i \leq r$. M_λ has the base $\{f_i v \otimes v, v \otimes f_i v\}$. Denote by $\{\delta^{(1)}, \delta^{(2)}\}$ the dual base of M_λ^*. $\mathfrak{n}_{+\lambda}$ has the base $\{e_i\}$. Let $\{e_i^*\}$ be the dual base of $\mathfrak{n}_{+\lambda}^*$. The part of R acting non-trivially on M_λ^* is

$$q^{\Omega_0/2} + q^{\Omega_0/2 + \frac{1}{4}(h_i \otimes 1 - 1 \otimes h_i)} q^{-\frac{b_{ii}}{4}} e_i \otimes e_i^*$$

One has

$$R(\delta^{(1)}) = q^{(\Lambda_1 - \alpha_i, \Lambda_2)/2} \cdot \delta^{(1)} + (\Lambda_2, \alpha_i)_q \cdot q^{(\Lambda_1, \Lambda_2)/2 - 1/4(\alpha_i, \Lambda_1) - 1/4(\alpha_i, \Lambda_2)} \cdot \delta^{(2)}$$
$$R(\delta^{(2)}) = q^{(\Lambda_1, \Lambda_2 - \alpha_i)/2} \cdot \delta^{(2)}$$

Remark. $M(\Lambda)^*$ has a stucture of a module over the quantum double $\mathcal{D}(\mathfrak{b}_+)$ such that the action of R is induced by the canonical element in $\mathcal{D}(\mathfrak{b}_+) \otimes \mathcal{D}(\mathfrak{b}_+)$ (cf. [D1], n°13). The "quasiclassical" analogue of this structure is introduced in [SV2, 6.17].

Hochschild homology. For an algebra A and a left A-module M we will denote by $C_.(A; M)$ the standard Hochschild complex of A with coefficients in M: $C_i(A; M) = A^{\otimes i} \otimes M$; $d_i : C_i(A, M) \to C_{i-1}(A; M)$ maps $a_i \otimes \dots \otimes a_1 \otimes m$ to

$$a_i \otimes \dots \otimes a_2 \otimes a_1 m + \sum_{p=1}^{i-1} (-1)^p a_i \otimes \dots \otimes a_{p+1} a_p \otimes \dots \otimes a_1 \otimes m.$$

We will also use the notation $a_1 | a_2 | \dots a_i \mid m$ for $a_1 \otimes \dots \otimes a_i \otimes m$.

We put $H_.(A; M) = H_.(C_.(A; M))$

§2. Local systems.

Consider an affine complex space $\mathbf{C}^M$ with coordinates $t_1, \dots, t_M$. Denote by H_{ij} hyperplane $t_i - t_j = 0$ $(i \neq j)$. We put $U_M = \mathbf{C}^M \setminus \bigcup_{i \pm j} H_{ij} \subset \mathbf{C}^M$. Fix n

weights $\Lambda_1, \ldots, \Lambda_N \in \mathfrak{h}^*$. Consider $\lambda = (k_1, \ldots, k_r) \in \mathbf{N}^r$, $|\lambda| = N := \sum_{i=1}^{r} k_i$. Fix an epimorphism $\pi : [N] \to [r]$ with $\#\pi^{-1}(i) = k_i$ for all i. Consider the space $\mathbf{C}^{n+N}$; assign to every coordinate t_j a covector $\alpha(t_j) \in \mathfrak{h}^*$ by the rule

$$\alpha(t_j) = \begin{cases} \Lambda_j & \text{if } 1 \le j \le n \\ -\alpha_{\pi(j-n)} & \text{if } n < j \le n+N \end{cases}$$

Assign to every hyperplane $H_{ij} \subset \mathbf{C}^{n+N}$ a number

$$a(H_{ij}) = (\alpha(t_i), \alpha(t_j))/\kappa.$$

Denote by $\mathcal{L}_\lambda$ the trivial 1-dimensional vector bundle over U_{n+N} with the integrable connection ∇_λ with the connection form $\omega = \sum_{i<j} a(H_{ij}) \frac{d(t_j - t_i)}{t_j - t_i}$. Denote by $\mathcal{S}_\lambda$ the local system of horizontal sections of ∇_λ over U_{n+N}.

Consider the projection on the first n coordinates $p : \mathbf{C}^{n+N} \to \mathbf{C}^n$. For $z = (z_1, \ldots, z_n) \in U_n$ put $V(z) = V_{n;N}(z) = p^{-1}(z) \subset \mathbf{C}^{n+N}$; $U(z) = U_{n;N}(z) = U_{n+N} \cap V(z)$.

Denote by $\mathcal{L}_\lambda(z)$, $\mathcal{S}_\lambda(z)$ restrictions of $\mathcal{L}_\lambda$, $\mathcal{S}_\lambda$ to $U(z)$.

§3. Homology.

Let X be a space; $Y \subset X$ a closed subspace; $U = X \setminus Y \overset{j}{\hookrightarrow} X$; $\mathcal{S}$ a local system (a locally constant sheaf) over U. We will denote by $\mathcal{S}^*$ the dual local system. We will consider two sorts of (co)homology.

(a) **Extension by star:** $H^\cdot(U, \mathcal{S}) = H^\cdot(X, j_*\mathcal{S})$.

We will denote by $H_\cdot(U, \mathcal{S}^*)$ the dual homology space. It may be calculated using the following complex. Let us call an "n-cell" e a closed convex polytope in $\mathbf{R}^N$. A singular n-cell is a continuous map $f : e \to X$. A complex $C_\cdot(U, \mathcal{S}^*)$ has as n-chains finite linear combinations of pairs (a singular cell $f : e \to U$, a section $\in \Gamma(e, f^*\mathcal{S}^*)$). We have $H_\cdot C_\cdot(U, \mathcal{S}^*) = H_\cdot(U, \mathcal{S}^*)$.

(b) **Extension by !:** $H^\cdot_!(U, \mathcal{S}) := H^\cdot(X, j_!\mathcal{S})$.

We will denote the dual space by $H_{!\cdot}(U, \mathcal{S}^*)$. Let us denote by $C_{!\cdot}(U, \mathcal{S})$ the complex whose n-chains are finite linear combination of pairs (a singular n-cell $f : e \to X$ such that $f^{-1}(f(e) \cap Y)$ is a union of subcells $e' \subset \partial e$; a section $\in \Gamma(f^{-1}(U \cap f(e)), f^*\mathcal{S}^*)$. We have $H_\cdot C_{!\cdot}(U, \mathcal{S}^*) = H_{!\cdot}(U, \mathcal{S}^*)$.

One has an evident map $C_\cdot(U, \mathcal{S}^*) \to C_{!\cdot}(U, \mathcal{S}^*)$ inducing the canonical map in homology.

Let us return to the situation of n°2. Fix $\lambda = (k_1, \ldots, k_r) \in \mathbf{N}^r$, $|\lambda| = N$. Put $\Sigma_\lambda = \Sigma_{k_1} \times \ldots \times \Sigma_{k_r}$. For each i fix a faithfull action of Σ_{k_i} on $\pi^{-1}(i)$. This gives the action of Σ_λ on $[N] = \bigcup_i \pi^{-1}(i)$. It induces an action on U_{n+N} respecting fibers of the projection $U_{n+N} \to U_n$. Local system $\mathcal{S}_\lambda$ is Σ_λ-equivariant. So, we get an action of Σ_λ on complexes $C_.(U_{n;N}(z), \mathcal{S}^*(z))$; $C_{!.}(U_{n;N}(z), \mathcal{S}^*(z))$, $z \in U_n$. We put

$$C_{(!).}(\dots)^- = \{x \in C_{(!).}(\dots) \mid \sigma x = (-1)^{|\sigma|}x \text{ for all } \sigma \in \Sigma_\lambda\}.$$

Fix $z = (z_1, \dots, z_n) \in U_n$ such that all $z_i \in \mathbf{R}$, $z_1 < z_2 < \dots < z_n$. The aim of this n° is to construct a certain inclusion of complexes

$$\psi_. = \psi_{\lambda.} : C_.(U_q\mathfrak{n}_-, M)^*_\lambda \longrightarrow C_{!N-.}(U_{n;N}(z); \mathcal{S}^*_\lambda(z))^- \tag{3.1}$$

where $M = M(\Lambda_1) \otimes \dots \otimes M(\Lambda_n)$; $C_.(U_q\mathfrak{n}_-, M)_\lambda := \{x \in C_.(U_q\mathfrak{n}_-, M) \mid h \cdot x = \left\langle \sum_{i=1}^{n} \Lambda_i - \sum_{j=1}^{r} k_j\alpha_j, h \right\rangle x \ \forall h \in \mathfrak{h}\}$, the action of $\mathfrak{h}$ on $C_.(U_q\mathfrak{n}_-, M)$ being induced by the adjoint action on $U_q\mathfrak{n}_-$ and the given action on M.

In other words, we will construct for all p, $0 \le p \le N$, maps

$$\psi_{N-p} : (U_q\mathfrak{n}_-^{*\otimes p} \otimes M^*)_\lambda \longrightarrow C_{!N-p}(U(z), \mathcal{S}_\lambda(z))^- \tag{3.2}$$

such that the ladder

$$\begin{array}{ccccc}
\cdots \longleftarrow & (U_q\mathfrak{n}^* \otimes M^*)_\lambda & \longleftarrow & M^*_\lambda \\
& \big\downarrow \psi_{N-1} & & \big\downarrow \psi_N \\
\cdots \longleftarrow & C^-_{!N-1} & \longleftarrow & C^-_{!N}
\end{array}$$

commutes. Cf. [SV2], §5.

Moreover, (3.1) will be a quasiisomorphism.

Put $\boldsymbol{\Delta}[p] = \{(u_1, \dots, u_p) \in \mathbf{R}^p \mid 0 \le u_1 \le u_2 \le \dots \le u_p \le 1\}$: $\boldsymbol{\Delta}[0] = \{*\}$. We begin with the simplest.

Example 2. $N = r = 1$; $\lambda = (1)$. So, we have one simple root $\alpha = \alpha_1 \in \mathfrak{h}^*$; and one generator $f = f_1 \in U_q\mathfrak{n}_-$. Denote by v_i generators of $M(\Lambda_i)$; put $v = v_1 \otimes \dots \otimes v_n \in M$; $f^{(i)} = v_1 \otimes \dots \otimes v_{i-1} \otimes fv_i \otimes v_{i+1} \otimes \dots \otimes v_n \in M$. Elements $\{f^{(i)}\}$ $i = 1, \dots, n$, form a base of M_λ. Denote by $\{\delta^{(i)}\}$ the dual base of M^*_λ. $(U_q\mathfrak{n}_- \otimes M)_\lambda$ admits as a base one element $f \mid v$. Let $\delta \in (U_q\mathfrak{n}_- \otimes M)^*_\lambda$ be the dual functional. The complex $C_.(U_q\mathfrak{n}_-, M)_\lambda$ has length 1:

$$0 \to (U_q\mathfrak{n}_- \otimes M)_\lambda \xrightarrow{d} M_\lambda \to 0$$

Here

$$d(f \mid v) = \sum_{i=1}^{n} q^{\left[-\sum_{j<i}(\alpha,\Lambda_j) + \sum_{j>i}(\alpha,\Lambda_j)\right]/4} \cdot f^{(i)} \tag{3.3}$$

(cf. (1.5)).

Let us consider a one-dimensional complex affine space $V \cong \mathbf{C}$ with a fixed coordinate t, and n marked points $t = z_i$, $i = 1, \dots, n$. We will identify $V_{1;n}(z)$ with V, t_{n+1}

corresponding to t. Suppose that all z_i are real, and $z_1 < z_2 < \ldots < z_n$. Fix a point $t = T$ in the upper half-plane; draw a "fork" in V:

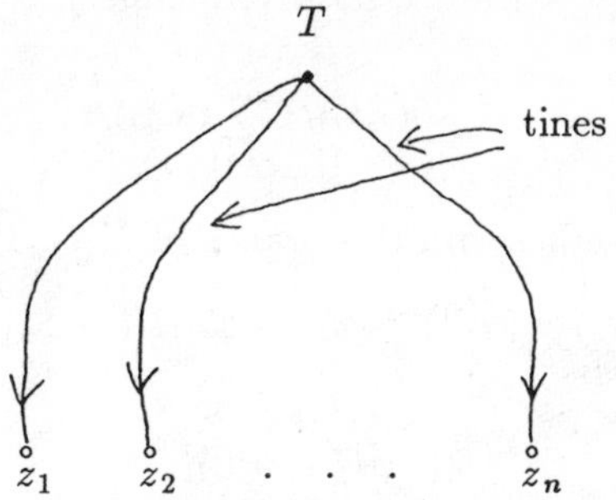

Fig. 3.1 – A Fork

It has n "tines" going from T to z_i, $i = 1, \ldots, n$. Each tine goes vertically near its end at z_i. Choose continuous maps $c^{(i)} : \mathbf{\Delta}[1] \to V$ such that $c^{(i)}(u) \in i$-th tine for all $u \in \mathbf{\Delta}[1]$; $c^{(i)}(0) = T$; $c^{(i)}(1) = z_i$; and $u < u'$ implies $\mathbf{Im} c^{(i)}(u) > \mathbf{Im} c^{(i)}(u')$. In other words, a point $t = c^{(i)}(u)$ moves along the i-th tine from T to z_i.

Now let us define 1-simplices $\tilde{c}^{(i)} \in C_{!1}(U(z), \mathcal{S}_\lambda^*(z))$ as follows. As a singular simplex $\tilde{c}^{(i)}$ is equal to $c^{(i)}$. Let us choose a section of $\mathcal{S}_\lambda(z)^*$ over $c^{(i)}$. Consider a multivalued function

$$\ell_\lambda = \ell_\lambda(t) = \prod_{i<j} (z_j - z_i)^{(\Lambda_i, \Lambda_j)/\varkappa} \prod_i (t - z_i)^{-(\alpha, \Lambda_i)/\varkappa}$$

To give a section of $\mathcal{S}(z)^*$ is the same as to give a branch of ℓ_λ multiplied by a number. Let us define such a section over the i-th tine by the rule: when a point $t = c^{(i)}(u)$ is close to z_i:

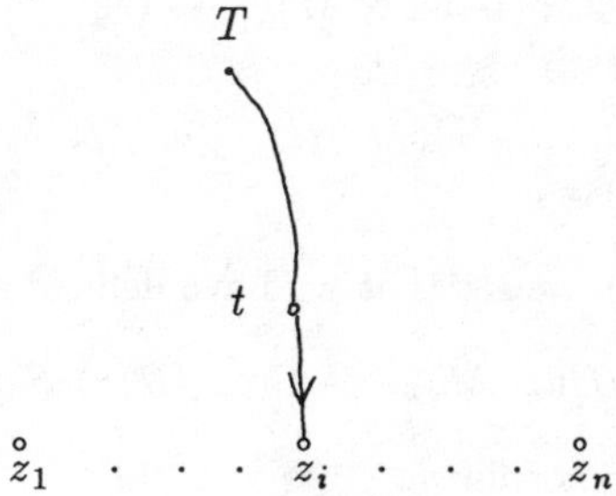

Fig. 3.2

then

$$\arg(t - z_j) \approx \begin{cases} 0 & \text{if } j < i \\ \frac{\pi}{2} & \text{if } j = i \\ \pi & \text{if } j > i \end{cases}$$

$$\arg(z_j - z_i) = 0 \text{ if } j > i$$

This defines a branch of $\ell_\lambda(t)$ over $c^{(i)}$. If all $(\Lambda_i, \Lambda_j)/\varkappa$ and $(\alpha_i, \Lambda)/\varkappa$ are real, then for t close to z_i

$$\ell_\lambda(t) \approx (\text{ a real positive number}) \cdot \gamma^{(i)}$$

on this branch, where

$$\gamma^{(i)} = q^{-(\alpha,\Lambda_i)/4 - \sum\limits_{j>i}(\alpha,\Lambda_j)/2}$$

By definition, $\tilde{c}^{(i)}$ is $c^{(i)}$ together with this branch of $\ell_\lambda(t)$. Put

$$\Psi_1(\delta^{(i)}) = (\gamma^{(i)})^{-1} \cdot \tilde{c}^{(i)} \in C_{!1}(U(z); \mathcal{S}_\lambda(z)^*)$$

This defines a map

$$\Psi_1 : M_\lambda^* \longrightarrow C_{!1}.$$

Define a singular 0-simplex $c : \mathbf{\Delta}[0] \to V : c(*) = T$. Define a branch of ℓ_λ at c by the convention $0 < \arg(T - z_i) < \pi$ for all i. In other words, if all $(\Lambda_i, \Lambda_j)/\varkappa$, $(\alpha, \Lambda_i)/\kappa$ are real, and $|\mathrm{Im}\, T| \gg |z_n - z_1|$, $\mathrm{Re}\, T \approx \mathrm{Re}\, z_i$, then $\arg(T - z_i) \approx +\frac{\pi}{2}$, and

$$\ell_\lambda(T) \approx (\text{real positive}) \cdot \gamma$$

where

$$\gamma = q^{-\sum\limits_i (\alpha,\Lambda_i)/4}$$

c together with this branch of ℓ_λ defines a simplex $\tilde{c} \in C_{!0}$. We put

$$\Psi_0(\delta) = \gamma^{-1} \cdot \tilde{c} \in C_{!0}(U(z); \mathcal{S}_\lambda^*(z))$$

This defines

$$\Psi_0 : (U_q\mathfrak{n}_- \otimes M)_\lambda^* \longrightarrow C_{!0}$$

By definition,

$$\partial\Psi_1(\delta^{(i)}) = q^{-\sum\limits_{j<i}(\alpha,\Lambda_j)/4 + \sum\limits_{i>i}(\alpha,\Lambda_j)/4} \cdot \Psi_0(\delta)$$

Comparing this with (3.3), we see that we have defined a map of complexes

$$\Psi_\lambda : C_.(U_q\mathfrak{n}_-, M)_\lambda^* \longrightarrow C_{!1-.}(U(z), \mathcal{S}_\lambda(z))$$

It is clear that Ψ_λ is a quasiisomorphism.

Now we will define Ψ_λ for the case

$$\lambda = (1, 1, \ldots, 1);\ N = r;\ \pi = id : [N] \longrightarrow [r].$$

For a subset $I = \{i_1, \ldots, i_k\} \subset [N]$, with a linear order $i_1 < \ldots < i_k$ (which may be not induced from $[N]$) put $f_I = f_{i_k} \cdot f_{i_{k-1}} \cdot \ldots \cdot f_{i_1} \in U_q\mathfrak{n}_-$; $f_\phi = 1$. The space $C_p(U_q\mathfrak{n}_-, M)_\lambda$ is generated by elements of the form

$$f_{I(p)}|f_{I(p-1)}|\ldots|f_{I(1)}|f_{J(1)}v_1 \otimes \ldots \otimes f_{J(n)}v_n \tag{3.4}$$

where $[N]$ is the disjoint union $\coprod_k I(k) \coprod \coprod_s J(s)$.

For a non-decreasing map $\rho : [N] \to [n+p]$ put $J(s) = \rho^{-1}(s), s \in [n]$, $I(k) = \rho^{-1}(n+k)$, $k \in [p]$; the order on $J(\cdot)$, $I(\cdot)$ being induced from $[N]$. Suppose tht all $I(k)$ are non-empty. Denote by $f(\rho)$ the element (3.4). The group Σ_N acts naturally on $C_p(U_q\mathfrak{n}_-, M)_\lambda$ by permuting f_i's. For $\sigma \in \Sigma_N$ put $f(\sigma; p) = \sigma f(\rho)$. All elements $f(\sigma; \rho)$ form a base of $C_p(U_q\mathfrak{n}_-, M)_\lambda$. Denote by $\{\delta(\sigma; \rho)\}$ the dual base of $C_p(U_q\mathfrak{n}_-, M)^*_\lambda$. Let $m(k)$ be the maximal element of $I(k)$. Put

$$s(\rho) = \sum_{k=1}^{p} m(k)$$

Let us consider the space V as in Example 1. We identify $V_{n;N}(z)$ with $\mathrm{Map}([N], V) =$ the space of N-tuples of points $(t_1, \ldots, t_N)$ in V. For $I \subset [N]$ put $V_I = \mathrm{Map}(I, V)$. If $[N] = I(1) \coprod \ldots \coprod I(k)$ then we have an evident identification $V_{n;N}(z) = \prod_s V_{I(s)}$.

Fix $T_1, \ldots, T_N \in V$ with $\mathrm{Re}\, T_1 = \ldots = \mathrm{Re}\, T_N \approx z_i$; $\mathrm{Im}\, T_i > 0$, $(z_n - z_1) \ll \mathrm{Im}\, T_1 < \mathrm{Im}\, T_2 < \ldots < \mathrm{Im}\, T_N$; draw a fork with the handle in V:

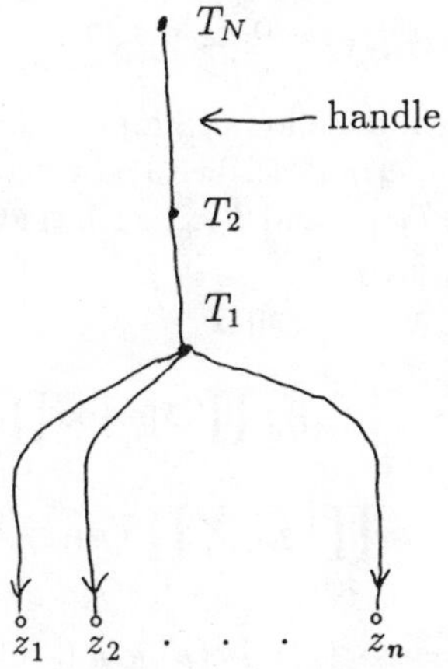

Fig. 3.3 – A fork with the handle

Let ρ, $I(\cdot)$, $J(\cdot)$ be as above. Put $p_i = \#J(i)$; $q_i = \#I(i)$, $[M] = \rho^{-1}([n])$. Define singular symplices

$$c_{J(i)} : \Delta[p_i] \longrightarrow V_{J(i)}$$

as follows. If $J(i) = [a+1, a+p_i]$, $c_{J(i)}$ defines the movement of points $t_{a+1}, \dots, t_{a+p_i}$ along the path shown in Fig. 3.4:

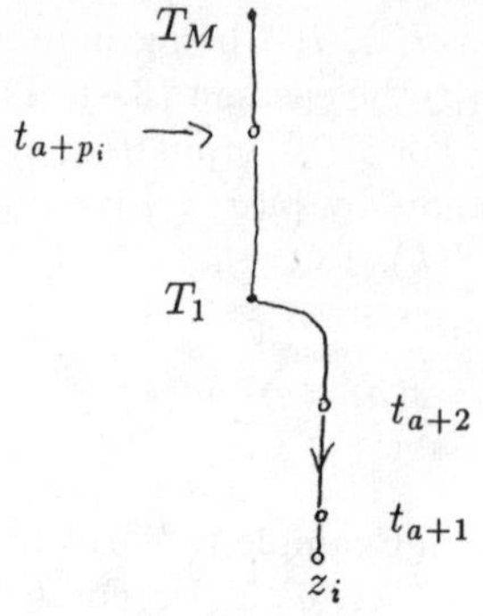

Fig. 3.4 $c_{J(i)}$

with $\operatorname{Im} t_j \le \operatorname{Im} t_{j'}$ for $j < j'$.

Define singular simplices

$$c_{I(i)} : e^{q_i - 1} \to V_{I(i)}$$

where e^{q_i-1} is a certain $(q_i - 1)$-cell, as follows. Suppose that $\rho^{-1}([n+i-1]) = [K]$; so $I(i) = [K+1, K+q_i]$. By definition, $c_{I(i)}$ describes a movement of points $t_{K+1}, \dots, t_{K+q_i}$ along the handle of a fork between T_{K+q_i} and T_{K+1} such that t_{K+j} is moving from T_{K+q_i} to T_{K+j}, and $\operatorname{Im} t_{K+j} \le \operatorname{Im} t_{K+j'}$ for $j < j'$.

Now let us define a singular $(N-p)$-cell

$$c(\rho) = \prod_j c_{J(j)} \times \prod_i c_{I(i)} : \prod_j \Delta[p_j] \times \prod_i e^{q_i - 1} \longrightarrow$$
$$\longrightarrow \prod_j V_{J(i)} \times \prod_i V_{I(i)} = V_{n;N}(z)$$

For example, if $p = N$, then $\rho(i) = n + i$, and $c(\rho)$ is a 0-cell $\{t_i = T_i\}_{i=1,\dots,N}$.

Consider a multivalued function

$$\ell_\lambda(t; z) = \prod_{1 \le i < j \le n} (z_j - z_i)^{(\Lambda_i, \Lambda_j)/\varkappa} \cdot \prod_{\substack{i \in [N] \\ j \in [n]}} (t_i - z_j)^{-(\alpha_i, \Lambda_j)/\varkappa} \cdot$$
$$\cdot \prod_{1 \le i < j \le N} (t_j - t_i)^{(\alpha_i, \alpha_j)/\varkappa}$$

Choose some branch of ℓ_λ on $c(\rho)$; the cell $c(\rho)$ together with this branch is denoted by $\widetilde{c(\rho)}$. Let us consider the position of t_i's when all points from groups $J(i)$ are close to z_i; and all points from the handle are in their lowest position (Fig. 3.5).

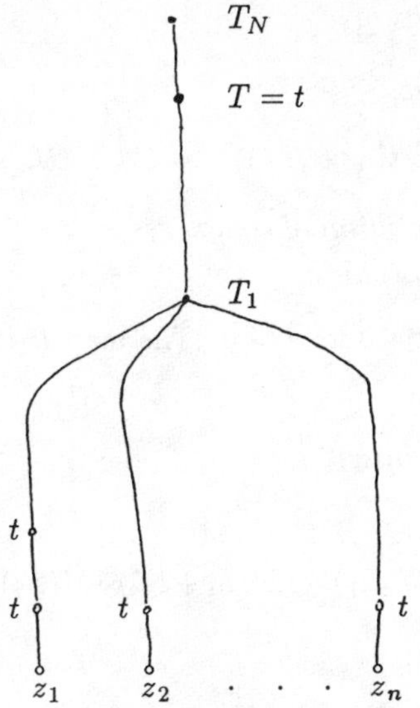

Fig. 3.5

For such t, if all $(\Lambda_i, \Lambda_j)/\varkappa$, $(\Lambda_i, \alpha_j)/\varkappa$, $(\alpha_i, \alpha_j)/\varkappa$ are real then

$$\ell_\lambda(t) \approx \gamma(\rho) \times (\text{positive real}) \tag{3.5}$$

for some constant $\gamma(\rho) = q^{\cdots}$ (cf. Ex. 1). We put

$$\Psi_p(\delta(\mathrm{id}; \rho)) = (-1)^{s(\rho)} \gamma(\rho)^{-1} \cdot \tilde{c}(\rho) \in C_{!N-p}$$

The group Σ_N acts on the set of singular cells of $V_{n;N}(z)$ by permutation of t_i's. Put $c(\sigma; \rho) = \sigma c(\rho)$, $\sigma \in \Sigma_N$. Choose some branch of ℓ_λ on $c(\sigma; \rho)$; let $\tilde{c}(\sigma; \rho) \in C_{!N-p}$ be the corresponding chain. Define the constant $\gamma(\sigma; \rho) = q^{\cdots}$ by the same positivity condition as (3.5) for "upper position" of $t \in c(\sigma; \rho)$. Put

$$\Psi_p(\delta(\sigma; \rho)) = (-1)^{|\sigma| + s(\rho)} \gamma(\sigma; \rho)^{-1} \cdot \tilde{c}(\sigma; \rho) \tag{3.6}$$

This defines Ψ_λ for $\lambda = (1, \ldots, 1)$, $\pi = \mathrm{id}$.

For arbitrary π Ψ_λ is obtained simply by permuting t_i's. If $\lambda = (k_1, \ldots, k_r)$ and $i_i \neq 0 \Rightarrow k_i = 1$ or all i then $\Psi_\lambda = \Psi_{\bar{\lambda}}$ where $\bar{\lambda} = (1, \ldots, 1)$ consists of all the ones contained in λ.

Now let $\lambda = (k_1, \ldots, k_r)$ be arbitrary, $N = |\lambda| := \Sigma k_i$. Fix $\pi : [N] \to [r]$ with $\#\pi^{-1}(i) = k_i$ for all i. Define an action of Σ_λ on $[N]$ as at the beginning of this Section. Define the $N \times N$-matrix $\widetilde{B} = (\tilde{b}_{ij}) := (b_{\pi(i)\pi(j)})$. Mark by tilde objects connected with $\widetilde{B} : \widetilde{U_q\mathfrak{n}_-}$ etc. Put $\widetilde{M} = \widetilde{M}(\Lambda_1) \otimes \ldots \otimes \widetilde{M}(\Lambda_n)$, $\tilde{\lambda} = (1, 1, \ldots, 1) \in \mathbf{N}^N$. We have a natural inclusion

$$C_.(U_q\mathfrak{n}_-, M)^*_\lambda \longrightarrow C_.(\widetilde{U_q\mathfrak{n}_-}, \widetilde{M})^*_{\tilde{\lambda}}, \tag{3.7}$$

$x \mapsto \sum\limits_{\sigma\in\Sigma_\lambda} \sigma x$. ($\Sigma_\lambda$ acts naturally on the r.h.s:) By definition, ψ_λ equals the composition of (3.7) with $\psi_{\bar\lambda}$ (cf. [SV1], [SV2]).

3.8. Theorem. (i) *The map*

$$\Psi_\lambda : C.(U_q\mathfrak{n}_-, M)^*_\lambda \longrightarrow C_{!N-}.(U_{n;N}(z), \mathcal{S}^*_\lambda)^- \tag{3.8}$$

defined above, is the morphism of complexes.

(ii) *This map is a quasiisomorphism.* □

Note that $H_i(U_q\mathfrak{n}_-; M) = 0$ for $i > 0$. The map $H.(\psi_\lambda)$ does not depend on the choice of a fork.

§4. *R* matrix and monodromy.

Put

$$M^*_{(i,i+1)} = M(\Lambda_1)^*\otimes\ldots\otimes M(\Lambda_{i-1})^*\otimes M(\Lambda_i+1)^*\otimes M(\Lambda_i)^*\otimes M(\Lambda_{i+2})^*\otimes\ldots\otimes M(\Lambda_n)$$

Define operators

$$R_{i,i+1} : M^* \longrightarrow M^*_{(i,i+1)} \tag{4.1}$$

to be induced by

$$M(\Lambda_i)^*\otimes M(\Lambda_{i+1})^* \xrightarrow{R(\Lambda_i,\Lambda_{i+1})^*} M(\Lambda_i)^*\otimes M(\Lambda_{i+1})^* \xrightarrow{p} M(\Lambda_{i+1})^*\otimes M(\Lambda_i)$$

where the first arrow is (1.22), $p(x\otimes y) = y\otimes x$. On other factors $M(\Lambda_p)$, $p\neq i, i+1$, $R_{i,i+1}$ is the identity. (4.1) induces morphisms of complexes

$$R_{i,i+1} : C.(U\mathfrak{n}_-, M^*) \longrightarrow C.(U\mathfrak{n}_-, M^*_{(i,i+1)}) \tag{4.2}$$

On the other hand, in the situation of n°3, consider a path in U_n which interchanges z_i with z_j:

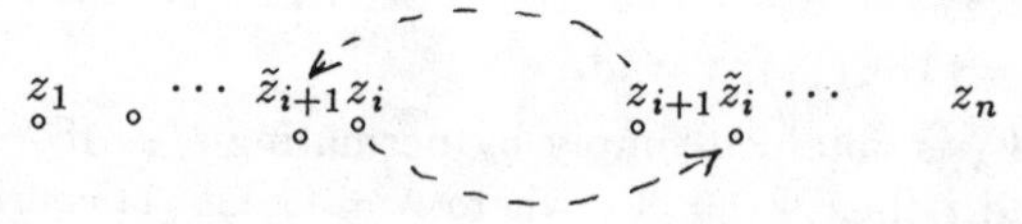

Fig. 4.1

It induces maps of complexes

$$T_{i,i+1} : C_!.(U_(z), \mathcal{S}^*_\lambda)_- \longrightarrow C_!.(U(z), \mathcal{S}^*_{\lambda(i,i+1)})^- \tag{4.3}$$

where $\mathcal{S}_{\lambda(ij)}$ is the local system corresponding to the sequence of weights $(\Lambda_1,\ldots,\Lambda_{i+1}, \Lambda_i,\ldots,\Lambda_n)$, i.e. to $M^*_{(i,i+1)}$.

Here is the main result of the paper:

4.4. Theorem. *The diagrams*

$$\begin{array}{ccc} C_{\cdot}(U\mathfrak{n}_-, M^*)_\lambda & \xrightarrow{\psi_\lambda} & C_{!N-\cdot}(U(z), \mathcal{S}^*_\lambda)^- \\ {\scriptstyle R_{i,i+1}}\downarrow & & \downarrow{\scriptstyle T_{i,i+1}} \\ C_{\cdot}(U\mathfrak{n}_-, M^*_{(i,i+1)})_\lambda & \xrightarrow{\psi_\lambda} & C_{!N-\cdot}(U(z), \mathcal{S}^*_{\lambda(i,i+1)}) \end{array}$$

commute up to homotopies. □

So, maps induced by $R_{i,i+1}$ and $T_{i,i+1}$ in cohomology coincide (cf. [Ko]).

Theorem is proved by direct verification.

Example 3. In the situation of n°3, Example 2, suppose that $n = 2$. Put $m_i = (\alpha, \Lambda_i)$, $b = (\alpha, \alpha)$.

We have:

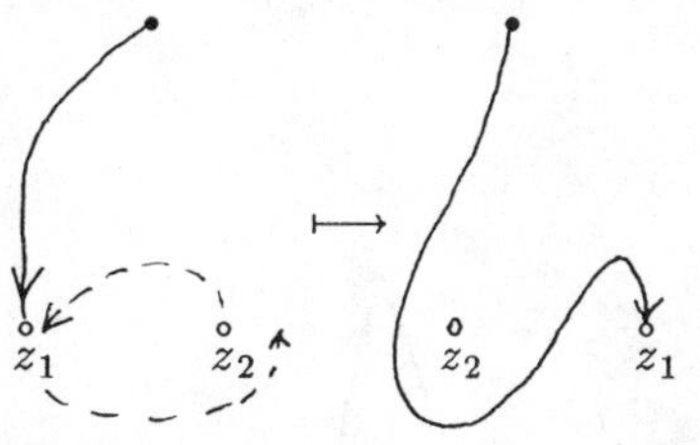

$$T_{12}(\psi(\delta^{(1)})) \simeq q^{\frac{m_1 m_2}{2b}} \left[(m_2)_q \cdot q^{-\frac{m_1 - m_2}{4}} \psi(\delta^{(1)}) + q^{-\frac{m_2}{2}} \psi(\delta^{(2)}) \right]$$

(a)

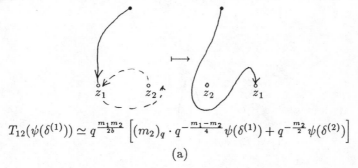

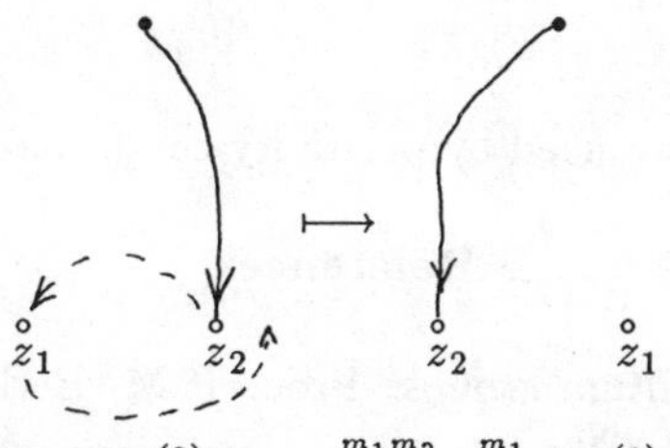

$$T_{12}(\psi(\delta^{(2)})) \simeq q^{\frac{m_1 m_2}{2b} - \frac{m_1}{2}} \psi(\delta^{(1)})$$

(b)

Fig. 4.2

Compare Example 1.

§5. Cohomology of the complement.

In the situation of n°3 one can also construct maps of complexes

$$\eta_\lambda : C_{\cdot}(U\mathfrak{n}^*_-, M^*)^*_\lambda \longrightarrow C_{N-\cdot}(U(z), \mathcal{S}^*_\lambda)^- \tag{5.1}$$

where the structure of $U\mathfrak{n}^*$-module on M^* is descried in n°1.

The squares

$$(5.2)\qquad \begin{array}{ccc} C_{\cdot}(U\mathfrak{n}_-, M)^*_\lambda & \xrightarrow{\ \psi\ } & C_{!N-\cdot}(U(z), \mathcal{S}^*_\lambda)^- \\ {\scriptstyle S^\bullet}\uparrow & & \uparrow \\ C_{\cdot}(U\mathfrak{n}^*_-, M^*)^*_\lambda & \xrightarrow{\ \eta\ } & C_{N-\cdot}(U(z), \mathcal{S}^*_\lambda)^- \end{array}$$

commute up to a homotopy. Here the left vertical map is induced by contravariant forms, and the right vertical one is the canonical one.

(5.2) clarifies the geometric meaning of forms S (cf. [SV2], §6). Moreover, $\eta_{\cdot}$ are quasiisomorphisms. We shall not give here the full construction of maps η, and restrict ourselves to the simplest

Example 4. In the situation of Example 2 $\eta_0(f|v) = \psi_0(\delta)\cdot\eta_1(f^{(i)})$ are drawn in the picture:

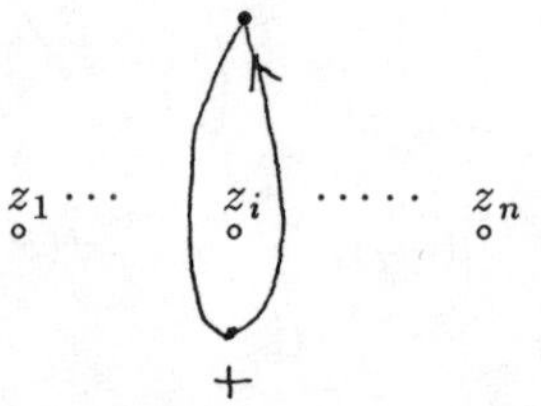

Fig. 5.1 $\eta(f^{(i)})$.

A coefficient and a branch are fixed by positivity condition at the point marked by +.

References

[D1] Drinfeld, V. G., *Quantum groups.* Proc. ICM (Berkeley, 1986), vol. 1. Amer. Math. Soc. 1987, 198-820.

[D2] Drinfeld, V. G., *On almost cocommutative Hopf algebras.* Algebra & Analysis, v. 1, No. 2, 1987, 30–46 (Russian).

[K] Kac, V. G., *Infinite dimensional Lie algebras.* Cambridge Univ. Press, 1985.

[Ko] Kohno, T., *Quantized universal enveloping algebras and monodromy of braid groups.* Preprint, 1988.

[L] Lusztig, G., *Finite dimensional Hopf algebras arising from quantum groups.* J. Amer. Math. Soc. v. 3, 1990, 257-296.

[SV1] Schechtman, V. V. and Varchenko, A. N., *Integral representations of N-point conformal correlators in the WZW model.* Preprint MPI/89-51, Bonn 1989. *Hypergeometric solutions of Knizhnik-Zamolodchikov equations.* Lett. Math. Phys. 20, 279-283, 1990.

[SV2] Schechtman, V. V. and Varchenko, A. N., *Arrangements of hyperplanes and Lie algebra homology.* Preprint, Moscow 1990.

A lower bound for the size of monodromy of systems of ordinary differential equations

Carlos T. Simpson[1]
Princeton University, Princeton NJ 08544, USA

Introduction

The material discussed in my talk in Tokyo was mostly contained in the manuscript submitted for the proceedings of ICM-90. The present paper concerns a subject which was not mentioned in Tokyo—I would have discussed it briefly, had there been more time at the end of my talk. Then, the discussion would have been purely conjectural, but in the months after the conference I have been able to give the proof in the rank two case. I apologize for this somewhat tenuous connection between my talk and the material presented here.

Let X be a compact Riemann surface. The estimate in this paper concerns the complex analytic isomorphism between the moduli spaces of rank two determinant one systems of ordinary differential equations on X, and representations of $\pi_1(X)$ in $S\ell(2,\mathbf{C})$. We will state the relevant facts about the moduli spaces, without proof. Then we will state the main theorem and a corollary.

First, the easy part. $\pi_1(X)$ has standard generators $\gamma_1,\ldots,\gamma_{2g}$, subject to one relation $rel(\gamma_1,\ldots,\gamma_{2g})=1$. The space R_B of representations of π_1 into $S\ell(2,\mathbf{C})$ is the set of $2g$-tuples of 2×2 matrices of determinant one $(A_1,\ldots,A_{2g})$ which satisfy the equation $rel(A_1,\ldots,A_{2g})=1$. This is a closed subvariety of the affine variety $S\ell(2)^{2g}$. The group $S\ell(2)$ acts on R_B by simultaneous conjugation of all the matrices. It is well known that there exists an affine variety M_B which is the universal categorical quotient of R_B by the action of $S\ell(2)$. The points of M_B are in one to one correspondence with the isomorphism classes of semisimple representations. A point in $\rho\in R_B$ maps to the point in M_B corresponding to the semisimple representation obtained by adding the quotients in a Jordan-Hölder series for ρ.

There is another way to look at a representation of $\pi_1(X)$. A *system of ODE's* on X is a pair (N,∇) consisting of a locally free sheaf N of $\mathcal{O}_X$-modules, and a morphism of sheaves

$$\nabla: N\to N\otimes_{\mathcal{O}_X}\Omega^1_X$$

such that $\nabla(av)=a\nabla(v)+(da)v$ for sections a of $\mathcal{O}_X$ and v of N. (If X had dimension 2 or more, one would add the condition $\nabla^2=0$.) If (N,∇) is a system of ODE's, then the equation $\nabla(v)=0$ looks in local coordinates like an analytic system of ordinary differential equations in the usual sense,

$$\frac{dv_i}{dz}=\sum_j A_{ij}(z)v_j.$$

[1]Partially supported by NSF grant DMS-9007857.

Thus, for given initial conditions $v(p)$, there is a unique local solution v which is a section of the sheaf of analytic sections N^{an}. The sheaf of analytic solutions $\mathcal{V} = (N^{an})^{\nabla}$ is a locally constant sheaf of $\mathbf{C}$-vector spaces. As such, it has a monodromy representation $\rho : \pi_1(X,p) \to G\ell(\mathcal{V}_p)$ representing the effect of continuation of solutions along paths beginning and ending at p.

This map taking a system of ODE's to its monodromy representation provides an isomorphism between the set of isomorphism classes of systems of ODE's and the set of isomorphism classes of representations. In particular, the set of representations in $S\ell(2)$ is equal to the set of systems of ODE's (N,∇) of rank two such that $(\bigwedge^2 N, \nabla) \cong (\mathcal{O}_X, d)$. A good reference is Deligne's book [5]. If one starts with a representation ρ, then one can make a locally constant sheaf $\mathcal{V}$ with that monodromy. Then $(N^{an}, \nabla) = \mathcal{V} \otimes_{\mathbf{C}} (\mathcal{O}_X^{an}, d)$. This provides an analytic system of ODE's, and Serre's GAGA result implies that this analytic system on a compact variety X must come from an algebraic system.

The notion of system of ODE's on X is an algebraic one, so we can formulate an algebraic moduli problem. To fix ideas, we mention the notion of an algebraic family of systems of ODE's. An algebraic family of systems of ODE's, indexed by a variety S, is a pair (N,∇) where N is a locally free sheaf on $X \times S$ and

$$\nabla : N \to N \otimes \Omega^1_{X\times S/S}$$

is an operator satisfying the same rule as before, using the exterior derivative in the X direction. The equations $\nabla_{X\times\{s\}}(v) = 0$ form a family of systems of ODE's which depend algebraically on the parameter $s \in S$.

There exists a line bundle L such that for any rank two system of ODE's (N,∇), $H^1(N \otimes L) = 0$ and $N \otimes L$ is generated by global sections. This was proved by Atiyah [2]. A proof is given in [14], using the boundedness results of Maruyama [11]. In the course of our arguments, we will obtain an alternative analytic proof—see Corollary 27.

There exists a parameter space R_{DR} parametrizing quadruples $(N,\nabla,\delta,\varphi)$ where (N,∇) is a system of ODE's of rank two on X, $\delta : \bigwedge^2(N,\nabla) \cong (\mathcal{O}_X, d)$, and φ is a frame for N_x. There exists a universal categorical quotient $M_{DR} = R_{DR}/G\ell(2)$ by the action of $G\ell(2)$ on the choice of φ. The points of M_{DR} are in one to one correspondence with the isomorphism classes of semisimple systems of ODE's of rank two and determinant one. Given an algebraic family of rank two systems of ODE's of determinant one, indexed by a space S, there is a unique algebraic map from S to M_{DR} giving the required map on the set of points.

The map of sets discussed above gives a complex analytic isomorphism $R_{DR} \cong R_B$ which leads to $M_{DR}^{an} \cong M_B^{an}$. This is proved in [14]. (The argument for proving the second isomorphism was incorrect in the original version of [14]—since then, I have fixed it.)

A function on a variety is called an *algebraic exhaustion function* if it is comparable to the inverse of the distance to infinity in some projective compactification (see the section after next for a more precise statement). Let $\psi_B(\rho)$ and $\psi_{DR}(N,\nabla)$ denote algebraic exhaustion functions for the spaces M_B and M_{DR} respectively. We can now state our main theorem.

Theorem 1 *There are constants C and k such that if ρ is the point in M_B corresponding to a point (N, ∇) in M_{DR}, then*

$$\psi_{DR}(N, \nabla) \leq C(\log \psi_B(\rho))^k.$$

This theorem should be interpreted as giving a lower bound on the size of the monodromy representation ρ attached to a system of ODE's (N, ∇)—it says that the size of the monodromy representation is at least the exponential of the k-th root of the size of the system of ordinary differential equations.

Corollary *There are no algebraic invariants of rank two determinant one systems of ODE's on X: $H^0(M_{DR}, \mathcal{O}^{alg}_{M_{DR}}) = \mathbf{C}$.*

Proof: Suppose f is a regular algebraic function on M_{DR}. Then, in particular, f is an analytic function, and there exists κ with $|f(x)| \leq \psi_{M_{DR}}(x)^\kappa$. Using the isomorphism $M^{an}_{DR} \cong M^{an}_B$, f may be considered as an analytic function on M_B. By Theorem 1, it satisfies the estimate $|f(x)| \leq C_1 + C_2(\log \psi_{M_B}(x))^{k\kappa}$. But on an algebraic variety, there are no analytic functions with such logarithmic growth at infinity, other than the constants. (These moduli varieties are connected [14].) Thus, f is a constant.

The proof of the theorem will take up the remainder of the paper. Here are some notations used throughout. X is a fixed compact Riemann surface of genus g, with a metric. Denote by $\sqrt{-1}\omega$ the volume form of X (or any other metrized Riemann surface in context), so ω is proportional to $dz \wedge d\overline{z}$ in local coordinates. If V is a bundle with holomorphic structure $\overline{\partial}$ and metric K, there is a unique connection compatible with $\overline{\partial}$ and K. We will denote its curvature by $R(V, \overline{\partial}, K)$. We may divide by the two-form ω to get a self-adjoint section $\omega^{-1}R$ of $End(V)$. The condition of "positive curvature" is that the eigenvalues of $\omega^{-1}R$ are positive (and vice-versa for negative curvature). If L is a holomorphic line bundle, we usually denote by S a standard metric on L, such that $\omega^{-1}R(L, S)$ is a constant. We use the notation U for the unit disc in $\mathbf{C}$, and the notation $B(y, \epsilon)$ for a geodesic disc of radius ϵ around $y \in X$. Finally, a word about constants. Constants denoted C or k are assumed to be indexed by order of occurrence unless otherwise indicated. Later values of C or k are assumed to be larger (but they will not depend on the rank two harmonic bundle V being considered). Some constants will be numbered for cross-referencing. There may be factors of 2 missing in formulas which depend on the normalization of the metric.

I would like to thank J. Bland and J. Kohn for helpful discussions.

A Nicole

Parameter spaces

In order to be able to analyze an algebraic exhaustion function for the space M_{DR}, we need to describe in detail a parameter space for systems of ODE's. The Hilbert schemes, which seem to be most convenient for the abstract construction of M_{DR}, are more difficult to understand concretely, so for the present purposes we will use a parameter space similar to the type defined by Gieseker [7].

The set of all vector bundles underlying systems of ODE's of rank two is bounded. This is proved in [14] using results of Maruyama; we will give an alternative analytic proof in the process of choosing a line bundle to prove the main theorem. Assume for now that a line bundle L has been fixed, so that $N \otimes L$ is generated by global sections, and $H^1(N \otimes L) = 0$, whenever (N, ∇) is a system of ODE's of rank two on X. Let $p = 2(deg(L) + 1 - g) = dim\, H^0(N \otimes L)$.

Proposition 1 *There exists a parameter space* $\mathbf{Z}$ *for quadruples* (N, ∇, δ, ν) *where* (N, ∇) *is a rank two system of ODE's,* $\delta : \bigwedge^2(N, \nabla) \cong (\mathcal{O}_X, d)$*, and* $\nu = (\nu_1, \ldots, \nu_p)$ *is a collection of sections in* $H^0(N \otimes L)$ *which generate* $N \otimes L$*.*

Proof: Let $G = G(p, 2)$ be the Grassmanian of quotients $\mathbf{C}^p \to V \to 0$ of rank 2. It has a natural very ample line bundle $\mathcal{O}_G(1)$, whose fiber over a point $\mathbf{C}^p \to V$ is the line $\bigwedge^2 V$. Let $\mathrm{M{\scriptsize AP}}_L(X, G)$ denote the space of pairs (f, δ) where f is a map from X to $G(p, 2)$, and $\delta : f^*\mathcal{O}_G(1) \cong L^{\otimes 2}$. $\mathrm{M{\scriptsize AP}}_L(X, G)$ is a parameter space for triples (N, δ, ν) where N is a vector bundle of rank two, $\delta : \bigwedge^2 N \cong \mathcal{O}_X$, and $\nu = (\nu_1, \ldots, \nu_p)$ is a collection of sections generating $N \otimes L$. Such a triple is associated with the surjection $\nu : \mathcal{O}_X^p \to N \otimes L \to 0$, which in turn is associated with a map f from X to G such that $f^*(\mathcal{O}_G(1)) = L^{\otimes 2} \otimes \bigwedge^2 N$. Conversely, a map from X to G gives a vector bundle and collection of sections. In either direction, the conditions $\delta : \bigwedge^2 N \cong \mathcal{O}_X$ and $\delta : f^*(\mathcal{O}_G(1)) \cong L^{\otimes 2}$ are the same. The space $\mathrm{M{\scriptsize AP}}_L(X, G)$ has a natural structure of quasiprojective variety, which we'll discuss in detail below.

There is a universal bundle N on $X \times \mathrm{M{\scriptsize AP}}_L(X, G)$ (with trivialization δ of the determinant). We will see how to construct a space $\mathbf{Z}$ over $\mathrm{M{\scriptsize AP}}_L(X, G)$ which parametrizes choices of connection ∇ on this universal bundle N. We will work informally, not always distinguishing objects on X and the universal objects on $X \times \mathrm{M{\scriptsize AP}}_L(X, G)$.

Recall some standard facts about the sheaf of rings of differential operators $\mathcal{D} = \mathcal{D}_X$. It is a union of locally free sheaves $\mathcal{D} = \bigcup_{k=0}^{\infty} \mathcal{D}_k$. Each $\mathcal{D}_k$ is the sheaf of differential operators of order $\leq k$. The associated graded ring for this filtration is the symmetric algebra on TX. If z is a local coordinate on X, then a section of $\mathcal{D}_k$ can be written locally as

$$a_k(z)\frac{\partial^k}{\partial z^k} + \ldots + a_0(z).$$

One can also multiply sections of $\mathcal{D}$ by functions on the right, but the left and right $\mathcal{O}_X$-module structures are not the same. Each of the $\mathcal{D}_k$ are locally free with respect to either structure, and the two structures on the associated graded are the same. If N is a vector bundle on X, then the sheaf $\mathcal{D}_1 \otimes N$ is locally free on $X \times \mathrm{M{\scriptsize AP}}_L(X, G)$. Some care must be taken since the tensor product contracts the right $\mathcal{O}_X$-module structure of $\mathcal{D}_1$ with the $\mathcal{O}_X$-module structure of N, leaving the left structure of $\mathcal{D}_1$.

Lemma 2 *Let N be a vector bundle on X. To give a connection ∇ on N is the same as to give a map*

$$\tilde{\nabla} : \mathcal{D}_1 \otimes N \to N$$

such that the induced map $N = \mathcal{D}_0 \otimes N \to N$ is the identity.

Proof: Given a connection ∇, one gets such a map, in local coordinates by

$$(a(z)\frac{\partial}{\partial z} + b(z)) \otimes n \mapsto a(z)\langle \nabla(n), \frac{\partial}{\partial z}\rangle + b(z)n.$$

Conversely, given such a map, one obtains a connection ∇ (which is well defined globally, independent of the choice of coordinate z).

In general, given some locally free sheaves E and F on $X \times \text{MAP}_L(X, G)$, there is a total space over $\text{MAP}_L(X, G)$ representing the functor parametrizing homomorphisms from E to F on the fibers X. Denote this total space by $\mathbf{Hom}_X(E, F)$. Maps between sheaves give rise to maps between these total spaces. We obtain spaces $\mathbf{Hom}_X(\mathcal{D}_1 \otimes N, N)$ and $\mathbf{Hom}_X(N, N)$ over $\text{MAP}_L(X, G)$, with a map

$$\mathbf{Hom}_X(\mathcal{D}_1 \otimes N, N) \to \mathbf{Hom}_X(N, N)$$

induced by the inclusion $\mathcal{O}_X = \mathcal{D}_0 \subset \mathcal{D}_1$. Define the space $\mathbf{Z}'$ to be the inverse image in $\mathbf{Hom}_X(\mathcal{D}_1 \otimes N, N)$ of the section $1 : \text{MAP}_L(X, G) \to \mathbf{Hom}_X(N, N)$ corresponding to the identity map $N \to N$. Then $\mathbf{Z}'$ is a parameter space for connections ∇ on the universal bundle. Each such connection induces a connection on $\bigwedge^2 N \cong \mathcal{O}_X$. Let $\mathbf{Z} \subset \mathbf{Z}'$ be the closed subset parametrizing connections ∇ such that $\delta \bigwedge^2(N, \nabla) = (\mathcal{O}_X, d)$. In particular, note that $\mathbf{Z}$ is a closed subscheme of $\mathbf{Hom}_X(\mathcal{D}_1 \otimes N, N)$.

We will now describe an explicit embedding of $\mathbf{Z}$ as a closed subvariety of an open subset of an affine space. First, let us describe explicitly $\text{MAP}_L(X, G)$ as a quasiprojective variety. There is a natural projective embedding $G \subset \mathbf{P}^{m-1}$ given by the line bundle $\mathcal{O}_G(1)$ with the natural choice of basis $\{e_{ij} = e_i \wedge e_j\}_{i<j}$ for

$$H^0(G, \mathcal{O}_G(1)) \cong \bigwedge^2 \mathbf{C}^p = \mathbf{C}^m, \qquad m = \binom{p}{2}.$$

An element of $\bigwedge^2 \mathbf{C}^p$ projects in each quotient $\mathbf{C}^p \to V$, to an element of the line $\bigwedge^2 V$. Thus, for a point in the Grassmanian, we obtain elements $u_{ij} \in \bigwedge^2 V$. Description of a map $f : X \to \mathbf{P}^{m-1}$ together with isomorphism $f^*\mathcal{O}(1) \cong L^{\otimes 2}$, is equivalent to providing m sections $u_{ij} \in H^0(X, L^{\otimes 2})$, such that they generate $L^{\otimes 2}$. The space of choices of $u_{ij} \in H^0(X, L^{\otimes 2})$ is $U = H^0(X, L^{\otimes 2})^m$. The subset where the sections do not generate $L^{\otimes 2}$ is the union

$$\Sigma = \bigcup_{x \in X} H^0(X, L^{\otimes 2}(-x))^m.$$

We have $\text{MAP}_L(X, \mathbf{P}^{m-1}) = U - \Sigma$, so our space $\text{MAP}_L(X, G)$ is a closed subvariety of $U - \Sigma$. A point (N, δ, ν) in $\text{MAP}_L(X, G)$ is mapped to the point of $U - \Sigma$ with coordinates $u_{ij} = \delta(\nu_i \wedge \nu_j) \in H^0(X, L^{\otimes 2})$. This gives $\mathrm{MAP}_L(X,G)$ a structure of quasiprojective variety. Furthermore, it has a projective completion where the complement is contained in the union of the projective space at infinity $\mathbf{P}(U)$, and the subset Σ.

The original closed embedding of the parameter space $\mathbf{Z}$ is not so useful for our purposes, since the scheme $\mathbf{Hom}_X(\mathcal{D}_1 \otimes N, N)$ varies with N. We will further embed that into a vector space which is independent of N, to obtain a nice embedding of $\mathbf{Z}$. Over $\mathrm{MAP}_L(X,G)$, there is a canonical surjection $(L^*)^p \to N \to 0$. This gives a surjection

$$\mathcal{D}_1 \otimes (L^*)^p \to \mathcal{D}_1 \otimes N \to 0.$$

The isomorphism $\delta : \bigwedge^2 N \cong \mathcal{O}_X$ gives an isomorphism $\delta : N \cong N^*$. Therefore the original surjection may be dualized to give a strict injection $0 \to N \to L^p$. Hence there are closed immersions

$$\mathbf{Z} \subset \mathbf{Hom}_X(\mathcal{D}_1 \otimes N, N) \subset \mathbf{Hom}_X(\mathcal{D}_1 \otimes (L^*)^p, L^p).$$

This last is a trivial vector bundle, the product of $\mathrm{MAP}_L(X,G)$ with the vector space $Hom(\mathcal{D}_1 \otimes (L^*)^p, L^p)$ (since this space no longer depends on N). Combining this with our closed immersion $\mathrm{MAP}_L(X,G) \subset U - \Sigma$, we obtain a closed immersion

$$\mathbf{Z} \subset (U - \Sigma) \times Hom(\mathcal{D}_1 \otimes (L^*)^p, L^p).$$

Let's review how to describe the point on the right which corresponds to a choice of point (N, ∇, δ, ν) in $\mathbf{Z}$. The point $u \in U - \Sigma$ is given by the coordinates $u_{ij} = \delta(\nu_i \wedge \nu_j) \in H^0(X, L^{\otimes 2})$. The point in $Hom(\mathcal{D}_1 \otimes (L^*)^p, L^p)$ is the map $f : \mathcal{D}_1 \otimes (L^*)^p \to L^p$ which is given by

$$f(\zeta_1 \otimes \lambda_1, \ldots, \zeta_p \otimes \lambda_p)_i = \delta\left(\nu_i \wedge \tilde{\nabla}(\sum_{j=1}^{p} \zeta_j \otimes (\lambda_j(\nu_j)))\right).$$

Exhaustion functions

Suppose W is a quasiprojective algebraic variety. Suppose $W \subset \mathbf{P}^n$ is an embedding as a locally closed subset of projective space, and set $W_\infty = \overline{W} - W$. A function ψ_W is called an *algebraic exhaustion function* for W if for some constants C and k,

$$C^{-1} d(z, W_\infty)^{-1/k} \le \psi_W(z) \le C d(z, W_\infty)^{-k}.$$

This condition is independent of the projective embedding. If $W_1 \subset W_2$ is a closed subvariety, then $\psi_{W_2}|_{W_1}$ is an algebraic exhaustion function for W_1. If W_1 and W_2 are varieties, then $p_1^*\psi_{W_1} + p_2^*\psi_{W_2}$ is an algebraic exhaustion function for $W_1 \times W_2$. And if W is a vector space, we can take any norm as an algebraic exhaustion function. If M is an algebraic variety and $Z \to M$ is a surjective map, and if ψ_Z is an algebraic exhaustion function for Z, then there exists an algebraic exhaustion function for M with $\psi_M(m) \le \inf_{z \mapsto m} \psi_Z(z)$.

Lemma 3 *Fix generators γ_i of $\pi_1(X)$. We can define an algebraic exhaustion function for M_B by setting*

$$\psi_{M_B}(m) = \inf_{\rho^{ss}\mapsto m} (\sup_i |\rho(\gamma_i)|),$$

where the infimum is taken over all semisimple representations corresponding to the point m.

Proof: We may define an algebraic exhaustion function on R_B by setting $\psi_{R_B}(\rho) = \sup_i |\rho(\gamma_i)|$. This is because R_B is a closed subvariety of $S\ell(2)^{2g}$, and hence a closed subvariety of $Mat(2\times 2)^{2g}$—and the function given is the restriction of a norm for the vector space $Mat(2\times 2)^{2g}$.

Next, embed R_B in a projective variety $\overline{R}_B$, with an action of $S\ell(2)$, such that M_B is an open affine subset of the categorical quotient $\overline{M}_B = \overline{R}_B^{ss}/S\ell(2)$ (see [12]). Let $\tilde{R}_B$ be a projective closure of $\overline{R}_B^{ss}$ which maps to the quotient $\overline{M}_B$. Call the map ϖ. By the results of Kempf–Ness and Kirwan [9], there is a compact subset $F \subset \overline{R}_B^{ss}$ such that $\varpi(F) = \overline{M}_B$. Let

$$D_1 = \tilde{R}_B - \overline{R}_B^{ss}, \qquad D_2 = \varpi^{-1}(\overline{M}_B - M_B).$$

Thus $R_B = \tilde{R}_B - D_1 - D_2$. Let ψ_1 be an algebraic exhaustion function for $\tilde{R}_B - D_1$. Let ψ' be an algebraic exhaustion function for M_B, so $\psi_2 = \varpi^*\psi'$ is an algebraic exhaustion function for $\tilde{R}_B - D_2$. Thus $\psi_{R_B} \le C(\psi_1 + \psi_2)^k$. If $m \in M_B$ then there is a point $f \in F$ which maps to m. Since $F \subset \tilde{R}_B - D_1$ is compact, we have a uniform bound $\psi_1(f) \le C$. Thus

$$\inf_{\rho\mapsto m} \psi_{R_B}(\rho) \le \psi_{R_B}(f) \le C\psi'(m)^k.$$

By the last remark before the lemma, there is another algebraic exhaustion function ψ'' for M_B such that $\psi''(m) \le \inf_{\rho\mapsto m} \psi_{R_B}(\rho)$. Therefore $\inf_{\rho\mapsto m} \psi_{R_B}(\rho)$ is an algebraic exhaustion function of $m \in M_B$.

If ρ is a representation, there exists a semisimple representation ρ^{ss} which is Jordan-Hölder equivalent to ρ, with $\psi_{R_B}(\rho^{ss}) \le \psi_{R_B}(\rho)$. For if ρ is not semisimple itself, then there is a fixed subspace $V_1 \subset \mathbf{C}^2$. We can choose a unitary basis u_1, u_2 such that $u_1 \in V_1$. When expressed in terms of this basis, the matrices $\rho(\gamma)$ are upper triangular. Define $\rho^{ss}(\gamma)$ to be the diagonal part of $\rho(\gamma)$. Since the basis is unitary, we may use the new matrix expressions to find the norms, so $|\rho^{ss}(\gamma)| \le |\rho(\gamma)|$. Therefore ψ_{M_B}, as defined in the lemma, is an algebraic exhaustion function for M_B.

We may apply the remark at the end of the paragraph preceding the lemma to the situation $\mathbf{Z} \to M_{DR}$. Let $\psi_{\mathbf{Z}}$ be an algebraic exhaustion function for our parameter space (it will be described in more detail below), which gives rise to an exhaustion functions $\psi_{M_{DR}}$. Let ψ_{M_B} be the exhaustion function provided by the lemma. If we use these as our exhaustion functions, then the main theorem may be restated.

Theorem 1′ *Fix generators γ_i of $\pi_1(X)$. There are constants C and k. Suppose (N, ∇, δ) is a semisimple rank two system of ODE's with determinant one. Suppose* *the monodromy matrices are bounded by $|\rho(\gamma_i)| \leq T$. Then there is a collection of generating sections $\nu_1, \ldots, \nu_p$ in $H^0(N \otimes L)$ such that*

$$\psi_{\mathbf{Z}}(N, \nabla, \delta, \nu) \leq C(\log T)^k.$$

In the remainder of this section we will describe precisely some bounds for algebraic exhaustion function for the parameter space $\mathbf{Z}$. Since $\mathbf{Z} \subset (U - \Sigma) \times Hom(\mathcal{D}_1 \otimes (L^*)^P, L^P)$ is a closed subvariety, we can take

$$\psi_{\mathbf{Z}} = \psi_{U-\Sigma} + \psi_{Hom}$$

as our exhaustion function. The ψ_{Hom} is just a norm on the vector space $Hom(\ldots)$. We can further decompose

$$\psi_{U-\Sigma} = \psi_U + \psi_{\tilde{\Sigma}}$$

where ψ_U is a norm on the vector space U, and

$$\psi_{\tilde{\Sigma}}(u) = \sup_{x \in X} d(u, H^0(L^{\otimes 2}(-x))^m)^{-1}.$$

Suppose (N, ∇, δ, ν) is a quadruple consisting of a rank two vector bundle, a connection, a trivialization of $\Lambda^2 N$, and a collection of generating sections $\nu_1, \ldots, \nu_p$ in $H^0(N \otimes L)$. Let $\mathbf{z}$ denote the corresponding point of $\mathbf{Z}$. Let S denote a standard metric on L. The point $u \in H^0(X, L^{\otimes 2})^m$ is given by the collection $u_{ij} = \delta(\nu_i \wedge \nu_j)$. Hence

$$\psi_U(\mathbf{z}) = \sup_{i<j} \|\delta(\nu_i \wedge \nu_j)\|_{L^1(S^{\otimes 2})}.$$

For each $i < j$ and each $x \in X$, let $v_{ij,x} \in H^0(L^{\otimes 2}(-x))$ be the closest vector to u_{ij}. Then

$$\psi_{\tilde{\Sigma}}(\mathbf{z}) = \sup_x (\sup_{i<j,y} |(u_{ij} - v_{ij,x})(y)|_{S^{\otimes 2}})^{-1}.$$

Note that $(u_{ij} - v_{ij,x})(x) = u_{ij}(x)$, so

$$\psi_{\tilde{\Sigma}}(\mathbf{z}) \leq \sup_x (\sup_{i<j} |\delta(\nu_i \wedge \nu_j)(x)|_{S^{\otimes 2}})^{-1}.$$

Finally, we come back to ψ_{Hom}. It is a bit complicated to describe a norm on the bundle $\mathcal{D}_1 \otimes (L^*)^P$, since this is a right $\mathcal{O}$-module which is obtained by taking a tensor product which contracts the left module structure of $\mathcal{D}_1$. So instead, we will make a norm on the space $Hom(\ldots)$ in the following way. We can choose an open subset $X_0 \subset X$, with bounded local coordinate z whose derivative is bounded. We can choose a local generating section λ for L^*. Then

$$g_i = (\ldots, 0, \frac{\partial}{\partial z} \otimes \lambda, 0, \ldots)$$

and

$$h_i = (\ldots, 0, 1 \otimes \lambda, 0, \ldots)$$

make a frame for $\mathcal{D}_1 \otimes (L^*)^P$ over X_0. Since the space $Hom(\mathcal{D}_1 \otimes (L^*)^P, L^P)$ is finite dimensional, we can obtain a norm by setting

$$|f| = \sup_{i,j}(\|\lambda(f(g_i)_j)\|_{L^1(X_0)} + \|\lambda(f(h_i)_j)\|_{L^1}(X_0)).$$

Use this norm to make the exhaustion function ψ_{Hom}. Using the concrete description given at the end of the last section, we have

$$\psi_{Hom}(\mathbf{z}) = \sup_{i,j}\left(\|\lambda\delta(\nu_j \wedge \nabla_{\frac{\partial}{\partial z}}(\lambda(\nu_i)))\|_{L^1(X_0)} + \|\lambda\delta(\nu_j \wedge \lambda(\nu_i))\|_{L^1(X_0)}\right).$$

Harmonic metrics

The main analytic tool we will use for obtaining the estimate is the notion of harmonic metric for a local system. The basic idea is that if the monodromy representation has matrices of size less than T, then there is an equivariant map from the universal cover of X to $U(2)\backslash G\ell(2)$, with derivative bounded by $\log T$. This can be replaced by a harmonic map, so the resulting harmonic metric on the local system will have Higgs field bounded by $|\theta| \leq \log T$. Then we can use the harmonic metric to select a point representing our system of differential equations in the parameter space, with the desired estimate for $\psi_{M_{DR}}$.

Suppose (N, ∇) is a rank two bundle with flat connection. Fix a base point $q \in X$. A smoothly varying metric K on N may be described as a family of metrics $K(x, \tau)$ on N_q, depending on points (x, τ) in the universal cover of X. Here x denotes a point in X and τ a path from q to x. The metric $K(x, \tau)$ is the metric on N_q obtained by pulling back the metric K_x by parallel translation along the path τ. These metrics satisfy an equivariance condition with respect to the monodromy representation ρ of the system (N, ∇). Namely, if $\gamma \in \pi_1(X, q)$ then

$$K(x, \tau\gamma)(u, v) = K(x, \tau)(\rho(\gamma)u, \rho(\gamma)v).$$

Fix a frame $\mathbf{C}^2 \cong N_q$. The space of metrics on N_q is then identified with the symmetric space $SU(2)\backslash S\ell(2)$ in the following way. Let (u, v) denote the standard metric on $\mathbf{C}^2$. If $g \in S\ell(2)$, then we obtain a new metric $K_g(u, v) = (gu, gv)$. If $w \in SU(2)$ then $(wgu, wgv) = (gu, gv)$, so $K_{wg} = K_g$. The correspondence between $U(2)g$ and the metric K_g provides an isomorphism between $SU(2)\backslash S\ell(2)$, and the space of metrics on $\mathbf{C}^2$.

There is (up to a scalar) a unique riemannian metric on $SU(2)\backslash S\ell(2)$ which is invariant by right translation by elements of $S\ell(2)$. This corresponds to a metric on the space of metrics on $\mathbf{C}^2$. The tangent space to the space of metrics, at a given metric K, is equal to the space of K-selfadjoint matrices s, in other words matrices with $K(su, v) = K(u, sv)$. The geodesic path starting at K with initial velocity s is the path of metrics $K(e^{ts}u, v)$. The length of the tangent vector s is given by $|s| = \sqrt{Tr(s^2)}$. The length of the geodesic $\{Ke^{ts} : 0 \leq t \leq 1\}$ is equal to $|s|$.

Given an equivariant map $K : \tilde{X} \to SU(2)\backslash S\ell(2)$, we can define its *energy* by integrating the square norm of the derivative, over the fundamental region:

$$\mathcal{E}(K) = \int_X |dK|^2.$$

By equivariance of K and of the metric on the target, the integrand descends to a function on X, so the integral over the fundamental region has been written as an integral over X. A metric is *harmonic* if the energy is a minimum (over all maps equivariant for the given representation).

Proposition 4 (Corlette [4], Donaldson [6]) *If (N, ∇) has semisimple monodromy representation, then there exists a smooth harmonic metric. The energy of this new metric is smaller than the energy of any smooth initial metric.*

We will apply this theorem by giving a geometric construction of a map K, with a bound on the energy. Then the harmonic map exists, with energy satisfying the same bound.

Here are some linear algebra constructions which can be made for any metric K; and the equations satisfied in the case of harmonic metrics. These are principally due to Hitchin [8]. Discussions may be found in [4] and (in the present notation) [13]. Let V denote the C^∞ bundle underlying N. The sum of ∇ and the operator d'' defining the holomorphic structure of N is a flat connection D on V. Using the metric, this decomposes as $D = \partial + \overline{\partial} + \theta + \overline{\theta}$, such that $\partial + \overline{\partial}$ is a connection compatible with the unitary structure, and $K(\theta u, v) = K(u, \overline{\theta} v)$. Thus $d'' = \overline{\partial} + \overline{\theta}$, and $\nabla = \partial + \theta$ are the operators defining the structure of the system of ODE's. On the other hand, the equation that K is harmonic is $\overline{\partial}(\theta) = 0$. The holomorphic bundle $E = (V, \overline{\partial})$ and its holomorphic endomorphism-valued one-form θ form the *Higgs bundle* (E, θ) corresponding to the harmonic bundle V and system of ODE's (N, ∇). Finally, one point to notice is that $dK = -2(\theta + \overline{\theta})$ (this is the operator which measures the failure of D to be unitary). Thus $|dK|^2 = 4|\theta|^2 + 4|\overline{\theta}|^2 = 8|\theta|^2$.

Finally, we describe one of the main constructions:

Lemma 5 *Suppose (N, ∇) is a rank two determinant one system of ODE's with a semisimple monodromy representation. Suppose that $\gamma_1, \ldots, \gamma_{2g}$ are the standard generators for $\pi_1(X, q)$. Suppose that there exists a frame $N_q \cong \mathbf{C}^2$ so that, regarding the resulting expression of $\rho(\gamma_i)$ as 2×2 matrices, we have the bounds $|\rho(\gamma_i)| \leq T$. Then there exists a smooth metric K on N, with energy bounded by $\mathcal{E}(K) \leq C(\log T)^2$.*

Proof: Draw a fundamental region for the universal cover of X, as a convex polygon in $\mathbf{R}^2$. Make the cuts so that the corners correspond to the base point q, and so that the edges correspond to the elements $\gamma_i \in \pi_1(X, q)$. Our references to a metric on the fundamental region will be to the euclidean metric of the polygon in $\mathbf{R}^2$; this will be comparable to a metric induced from the Riemann surface. We may assume that the metrics along the edges of the polygon are constant multiples of the metrics on the paths γ_i in X.

Fix a frame for N_q as hypothesized in the lemma, then define $K(q,1)$ to be the standard metric corresponding to that frame. The points $K(q,\gamma_i)$ are now determined by the equivariance condition. The equivariance condition fixes also the images of the corners of the fundamental region.

The first claim is that the distance from $K(q,1)$ to any of the $K(q,\gamma_i)$ is bounded by $C\log T$. Note that

$$K(q,\gamma_i)(u,v) = K(q,1)(\rho(\gamma_i)u, \rho(\gamma_i)v) = K(q,1)((\rho(\gamma_i))^\dagger \rho(\gamma_i)u, v).$$

The matrix $(\rho(\gamma_i))^\dagger\rho(\gamma_i)$ is selfadjoint with respect to $K(p,1)$. Therefore there is a selfadjoint endomorphism s with

$$e^s = (\rho(\gamma_i))^+\rho(\gamma_i),$$

with the bound

$$|s| = \sqrt{Tr(s^2)} \leq C\log Tr(e^s).$$

This bound comes from the fact that $|s|$ is essentially the same as the biggest eigenvalue of s (note that $det(\rho) = 1$ so $Tr(s) = 0$), and the biggest eigenvalue of s is the logarithm of the biggest eigenvalue of e^s. Note that

$$Tr(e^s) = Tr((\rho(\gamma_i))^\dagger\rho(\gamma_i)) = |\rho(\gamma_i)|^2.$$

Thus we get

$$|s| \leq C\log T.$$

The geodesic $\{K(p,1)e^{ts} : 0 \leq t \leq 1\}$ begins at $K(p,1)$ and ends at $K(p,\gamma_i)$. It has length $|s|$. Therefore the points $K(p,\gamma_i)$ are at distance less than $C\log T$ from $K(p,1)$. By equivariance, this means that the distances between the images of consecutive corners on the fundamental region are less than $C\log T$. Define the map K on the edges of the fundamental region by requiring that the restriction of K to an edge be equal to the geodesic between the images of the corners, parametrized at constant speed.

Now choose a point q_1 in the middle of the fundamental region, and draw straight lines from q_1 to the edge. Choose $K(q_1)$ to be some point within a distance of $C\log T$ from $K(q)$. Then define the map K restricted to any straight line emanating from q_1, to be the geodesic going from $K(q_1)$ to the image of the point where the line meets the edge of the fundamental region, parametrized at constant speed. This serves to define the map K as restricted to the fundamental region. And, since the definitions at the edges are by geodesics parametrized at constant speed (with respect to metrics on the edges which come from metrics on X), the definitions at corresponding edges of the fundamental region will match. Thus we may extend our map K to an equivariant map from $\tilde{X}$ to $SU(2)\backslash S\ell(2)$.

The map K defined in this way is continuous but not everywhere differentiable. However, where the derivative is defined, we have $|dK| \leq C\log T$. This is because $SU(2)\backslash S\ell(2)$ is negatively curved, so geodesic triangles are thinner in the middle than in the euclidean case. Thus if an euclidean triangle is mapped to a geodesic triangle by making all of the lines extending from the top vertex become geodesics, then the norm of the derivative in the middle is smaller than the norm of the derivative along the bottom edge.

Now the map K may be replaced by a smooth equivariant map with the same bound on the energy. One can define a sequence of maps K_ϵ such that $K_\epsilon \to K$ in C^0 (on the fundamental region), and such that $dK_\epsilon \to dK$ in L^2 on the fundamental region. To make the smoothing, we may choose an appropriate open set in the target space, in which everything will take place, and embed this open set into euclidean space, so smoothing can be done by convolution. In order to make the smoothing equivariantly, first make the map smooth in a neighborhood of the point q. This determines a map which is smooth in neighborhoods of the translates of q. Take the resulting map and make it smooth on neighborhoods of the paths which generate the fundamental group. Again, this determines a map which is smooth on every translate of a neighborhood of the edge of the fundamental region. Finally, make the map smooth in the interior of the fundamental region.

Corollary 6 *Suppose* (N, ∇) *is a semisimple rank two determinant one system of ODE's, with monodromy representation bounded by* $|\rho(\gamma_i)| \leq T$. *Let* K *denote the harmonic metric on* N. *Let* $\nabla + d'' = \partial + \overline{\partial} + \theta + \overline{\theta}$ *be the decomposition of the connection described above. Then*

$$\mathcal{E}(K) = 8 \int_X |\theta|^2 \leq C(\log T)^2.$$

Proof: This follows from Propositions 4 and 5.

Rank 2 harmonic bundles

In this section we will make some calculations for the case of harmonic bundles of rank two. Something similar should work in the case of higher rank, but it might be more complicated. It is for this reason that the proof has been restricted to the case of bundles of rank two.

Suppose that (N, ∇) is a rank two system of ODE's, with $\delta : \bigwedge^2(N, \nabla) \cong (\mathcal{O}_X, d)$. Let V denote the underlying C^∞ bundle $V = N \otimes_{\mathcal{O}_X} \mathcal{C}^\infty$. Let d'' be the operator of type $(1,0)$ on V which gives the holomorphic structure of N. Choose a harmonic metric K with $det(K) = 1$, and let $\nabla + d'' = \partial + \overline{\partial} + \theta + \overline{\theta}$ decomposition described before. Note that $Tr(\theta) = 0$.

Let $a = det(\theta) \in H^0(Sym^2\Omega^1_X)$. Let $A = \|a\|_{L^1}$ (we will use this notation consistently below). *Make the assumption that* $A \geq C$ (the constant C may, as usual, be increased below). The set of systems V for which $A \leq C$ forms a compact subset of the moduli space [8], so we may safely make this assumption.

Since $|a| \leq |\theta|^2$ (see the calculations below) then in terms of our previous notation, we have $A \leq C(\log T)^2$. Also, by elliptic estimates for the fixed bundle $Sym^2\Omega^1_X$, we have $\sup_X |a| \leq CA$.

Let Y be the divisor of a, and let $|Y|$ denote the set-theoretic support of Y. We will call the points $y \in |Y|$ "turning points". One conjectures that these points are related to the turning points which appear in the singular perturbation theory of, for example, Langer [10].

Let $X' = X - |Y|$. Locally on X' we can choose $\sqrt{a}$. Write $V = V^+ \oplus V^-$, a direct sum decomposition of the holomorphic bundle $(V, \overline{\partial})$ into line bundles defined locally on X', such that θ acts by $\sqrt{a}$ on V^+ and $-\sqrt{a}$ on V^-. The line bundles have metrics K^+ and K^-, the restrictions of the harmonic metric K. Define a new metric on $V|_{X'}$,

$$J = K^+ \oplus K^-.$$

In other words, the restriction of J to V^+ or V^- is the same as the restriction of K, but $V^+ \perp V^-$ with respect to J.

Choose a smoothly varying K-orthonormal basis $\{v_1, v_2\}$ for V over an open subset of X', such that $v_1 \in V^+$. Choose a J-orthonormal basis $\{v^+, v^-\}$, with $v^+ = v_1$. Thus v^+ and v^- are unit vectors in V^+ and V^-.

Calculate now using the basis $\{v_1, v_2\}$. We can write $\theta(v_1) = \sqrt{a}v_1$ and $\theta(v_2) = \beta v_1 - \sqrt{a}v_2$ with β a $(1,0)$-form. In other words θ is given by the matrix

$$\theta = \begin{pmatrix} \sqrt{a} & \beta \\ 0 & -\sqrt{a} \end{pmatrix}.$$

The choices of v_1 and v_2 are unique up to scalars of norm one, so $|\beta|$ is well defined.

Let $\overline{\alpha}$ denote the $(0,1)$-form with coefficients in $End(V)$ which has eigenvalues $\pm\overline{\sqrt{a}}$, and eigenspaces $V^\pm$. In other words, $\overline{\alpha}(v^+) = \overline{\sqrt{a}}v^+$ and $\overline{\alpha}(v^-) = -\overline{\sqrt{a}}v^-$. We claim that $\overline{\alpha}$, written in terms of the basis $\{v_1, v_2\}$, is given by the matrix

$$\overline{\alpha} = \begin{pmatrix} \overline{\sqrt{a}} & \overline{\sqrt{a}}\beta/\sqrt{a} \\ 0 & -\overline{\sqrt{a}} \end{pmatrix}.$$

To prove this, note that

$$\theta = \begin{pmatrix} 1 & -\beta/2\sqrt{a} \\ 0 & 1 \end{pmatrix} \begin{pmatrix} \sqrt{a} & 0 \\ 0 & -\sqrt{a} \end{pmatrix} \begin{pmatrix} 1 & \beta/2\sqrt{a} \\ 0 & 1 \end{pmatrix}.$$

So the matrix $\overline{\alpha}$ (with the same eigenspaces), is

$$\overline{\alpha} = \begin{pmatrix} 1 & -\beta/2\sqrt{a} \\ 0 & 1 \end{pmatrix} \begin{pmatrix} \overline{\sqrt{a}} & 0 \\ 0 & -\overline{\sqrt{a}} \end{pmatrix} \begin{pmatrix} 1 & \beta/2\sqrt{a} \\ 0 & 1 \end{pmatrix} = \begin{pmatrix} \overline{\sqrt{a}} & \overline{\sqrt{a}}\beta/\sqrt{a} \\ 0 & -\overline{\sqrt{a}} \end{pmatrix}$$

as claimed.

Since $\{v_1, v_2\}$ is an orthonormal basis, the matrix for $\overline{\theta}$ is the complex conjugate of the transpose of the matrix for θ,

$$\overline{\theta} = \begin{pmatrix} \overline{\sqrt{a}} & 0 \\ \overline{\beta} & -\overline{\sqrt{a}} \end{pmatrix}.$$

So we can write $\overline{\theta} = \overline{\alpha} + q$ with

$$q = \begin{pmatrix} 0 & -\overline{\sqrt{a}\beta}/\sqrt{a} \\ \beta & 0 \end{pmatrix}.$$

Lemma 7 *For any section v of V over X',*

$$|v|_K \leq 2|v|_J$$

and

$$|v|_J \leq 2(1 + |a|^{-1}|\beta|^2)^{1/2}|v|_K.$$

Proof: From the previous discussion, the vector $v_2 - (\beta/2\sqrt{a})v_1$ is an eigenvector of θ with eigenvalue $-\sqrt{a}$; thus it is in V^-. Its norm is $(1+|\beta|^2/4|a|)^{1/2}$, so we may assume that

$$v^- = (1 + |\beta|^2/4|a|)^{-1/2}(v_2 - (\beta/2\sqrt{a})v_1).$$

Now write

$$\begin{aligned} v &= \mu^+ v^+ + \mu^- v^- \\ &= (\mu^+ - (\beta/2\sqrt{a})(1 + |\beta|^2/4|a|)^{-1/2}\mu^-)v_1 + \mu^- v_2, \end{aligned}$$

so

$$|v|_K^2 = |\mu^+ - (\beta/2\sqrt{a})(1 + |\beta|^2/4|a|)^{-1/2}\mu^-|^2 + |\mu^-|^2.$$

Note that $|\beta/2\sqrt{a}| \leq (1 + |\beta|^2/4|a|)^{1/2}$, so

$$|v|_K^2 \leq (|\mu^+| + |\mu^-|)^2 + |\mu^-|^2 \leq 3(|\mu^+|^2 + |\mu^-|^2) = 3|v|_J^2.$$

So the first inequality will hold.

For the second inequality, note that

$$v_2 = (1 + |\beta|^2/4|a|)^{1/2}v^- + (\beta/2\sqrt{a})v^+.$$

So if $v = \mu_1 v_1 + \mu_2 v_2$, then

$$v = (\mu_1 + (\beta/2\sqrt{a})\mu_2)v^+ + (1 + |\beta|^2/4|a|)^{1/2}\mu_2 v^-,$$

and

$$|v|_J^2 = |\mu_1 + (\beta/2\sqrt{a})\mu_2|^2 + (1 + |\beta|^2/4|a|)|\mu_2|^2.$$

This gives

$$|v|_J^2 \leq 3(1 + |\beta|^2/4|a|)(|\mu_1|^2 + |\mu_2|^2),$$

which gives the second claimed inequality.

We now discuss the curvatures of the various metrics. Let $F(\ldots)$ denote the curvature of a metric on a Higgs bundle, and let $R(\ldots)$ denote the curvature of a metric on a holomorphic vector bundle. Recall that for a Higgs bundle, $F = R + \theta\overline{\theta} + \overline{\theta}\theta$.

Recall that $\sqrt{-1}\omega$ is the volume form (equal to the Kähler form) of the metric on X. If R is a curvature form, then $\omega^{-1}R$ is a selfadjoint matrix-valued function on X. The condition of positive curvature are that its eigenvalues are positive.

Locally, V^+ and V^- are preserved by $\overline{\partial}$ and θ, so they form sub-Higgs bundles of $(V, \overline{\partial}, \theta)$. The principal that curvature decreases in subobjects holds for Higgs bundles as well as for holomorphic bundles. We have $F(V, \partial, \theta, K) = 0$ by the definition of harmonic metric. Hence, $\omega^{-1}F(V^+, \partial, \theta, K^+) \leq 0$ and $\omega^{-1}F(V^-, \partial, \theta, K^-) \leq 0$. On the other hand, $\theta\overline{\theta}_{K^+} + \overline{\theta}_{K^+}\theta = 0$, since V^+ is a bundle of rank one (and 1×1 matrices commute with each other). Hence $F(V^+, \overline{\partial}, \theta, K^+) = R(V^+, \overline{\partial}, K^+)$. The same is true for V^-, so we get

$$\omega^{-1}R(V^+, \overline{\partial}, K^+) \leq 0$$

$$\omega^{-1}R(V^-, \overline{\partial}, K^-) \leq 0.$$

The metric J on V is an orthogonal direct sum of K^+ and K^-, so $R(V, \overline{\partial}, J) = R(V^+, \overline{\partial}, K^+) \oplus R(V^-, \overline{\partial}, K^-)$. Thus

$$\omega^{-1}R(V, \overline{\partial}, J) \leq 0.$$

(This means that the eigenvalues of the hermitian matrix $\omega^{-1}R(V, \overline{\partial}, J)$ are negative.)

On $V|_{X'}$ define a new metric $J^* = det(J)^{-1}J$, the dual of J via the self duality δ.

Lemma 8 *We have*

$$\omega^{-1}R(V, \overline{\partial} + \overline{\alpha}, J^*) \geq 0$$

on X'. Furthermore, we have

$$|v|_{J^*} \leq C|v|_K$$

and

$$|v|_K \leq C(1 + |a|^{-1}|\beta|^2)^{1/2}|v|_{J^*}.$$

Proof: First of all, note that the $(0,1)$-form $\overline{\sqrt{a}}$ is antiholomorphic. Therefore, adding it to $\overline{\partial}$ doesn't change the curvature of the line bundles $V^{\pm}$. In particular, the curvature estimates obtained above hold also for $\overline{\partial} + \overline{\alpha}$. The isomorphism $\delta : \bigwedge^2(V) \cong \mathcal{O}_V$ gives $V \cong V^*$; and the metric K is self-dual. The dual of the metric J is $J^* = det(J)^{-1}J$. The estimates for J may be dualized to obtain the estimates for J^*.

Lemma 9 *There is a constant C depending only on X such that, first of all, $\int_X |\beta|^2 \leq C$. Second, $\sup_X |\beta|^2 \leq CA$. And third, let U denote the unit disc in $\mathbf{C}$ (and tU the disc of radius t). Suppose that $\phi : U \to X$ is a map such that $\sup_U |\phi^* a| \leq 1$. Then $\sup_{1/2U} |\phi^*\beta| \leq C$. Finally, the same estimates hold for $|q|_K$.*

Proof: From the above expressions for θ and $\overline{\theta}$ we get

$$[\theta, \overline{\theta}] = \begin{pmatrix} \beta\overline{\beta} & -2\beta\overline{\sqrt{a}} \\ -2\sqrt{a}\overline{\beta} & -\beta\overline{\beta} \end{pmatrix}.$$

Thus

$$|[\theta,\overline{\theta}]|^2 = 2|\beta|^4 + 8|a||\beta|^2.$$

The "distance decreasing property for variations of Hodge structure" translates in this case into the estimate

$$\omega^{-1}\overline{\partial}\partial \log|\theta|^2 \le -\frac{|[\theta,\overline{\theta}]|^2}{|\theta|^2} + C_X,$$

where C_X is a constant depending on the curvature of the (fixed) metric on X. This is a reflection of the fact that curvature decreases in holomorphic subbundles, applied to the subbundle of $End(V)$ spanned by $\theta \cdot TX$—see [13]. In terms of our calculations (and multiplying by -1) we get

$$\omega^{-1}\partial\overline{\partial}log(|\beta|^2 + 2|a|) \ge \frac{2|\beta|^4 + 8|a||\beta|^2}{|\beta|^2 + 2|a|} - C_X.$$

By throwing away four of the $|a||\beta|^2$ this simplifies to

$$\omega^{-1}\partial\overline{\partial}log(|\beta|^2 + 2|a|) \ge 2|\beta|^2 - C_X.$$

This estimate also holds distributionally on all of X, since the left hand side can gain only positive multiples of delta functions on $|Y|$. Recall that the integral over X of the Laplacian ($\omega^{-1}\partial\overline{\partial}$) of anything, is equal to zero. Hence we get

$$\int_X |\beta|^2 \le C_X vol(X)/2,$$

the first estimate. For the second estimate, we can write $u = (|\beta|^2 + 2|a|)$, and (assuming $A \ge C$) the inequality becomes

$$\omega^{-1}\partial\overline{\partial}\log u \ge u - C_1 A.$$

Then, at any point where $u > C_1 A$, we have $\omega^{-1}\partial\overline{\partial}\log u > 0$. In particular, $\log u$ cannot achieve its maximum at any point where $u > C_1 A$. Thus, $u \le C_1 A$ everywhere. This is the second estimate. The third estimate is done in a similar way on the disc—there, the global maximum argument is replaced by the full argument of Ahlfors for the disc [1].

Some partitions of unity

Recall that U denotes the unit disc in **C**. For any point $y \in X$ and any ϵ, choose a map $\phi = \phi_{y,\epsilon} : U \to B(y,\epsilon)$. We may assume that ϕ is holomorphic, and also if ϵ is small enough, that the derivative is bounded:

$$C^{-1}\epsilon \le |d\phi| \le C\epsilon.$$

For each $y \in X$, let τ_y be the largest number such that $|\phi^*_{y,\tau_y} a(x)| \le 1$ for all $x \in U$.

Lemma 10 *Suppose given $\delta > 0$. We may choose C such that if $A \geq C$, then for any point $y \in X$, we have $\tau_y \leq \delta$.*

Proof: Suppose not. Then for any A there are a_1 with $\|a_1\|_{L^1} = 1$ and $y_1 \in X$ such that $\sup_U |\phi^*_{y_1,\delta} A a_1| \leq 1$. Take the limit of a subsequence of (a_1, y_1) as $A \to \infty$; we get an a_2 and a point y_2 such that $\|a_2\|_{L^1} = 1$ but $\phi^*_{y_2,\delta} a_2 = 0$, which is not possible.

Lemma 11 *There is a δ and a constant C such that the following is true. Suppose $y \in X$, and $\epsilon \leq \delta$. Suppose a is a nonzero section of $Sym^2\Omega^1_X$. Let s_j be the zeros of $\phi^*_{y,\epsilon} a$ inside U, counted with multiplicity. There exists a number λ such that*

$$C^{-1} \leq \left| \frac{\lambda \prod (z - s_j) dz^2}{\phi^*_{y,\epsilon} a} \right| \leq C$$

for any $z \in 1/2U$.

Proof: Each a may be written as γa_1 where $\|a_1\|_{L^1} = 1$. The space of sections a is finite dimensional, so the set of such a_1 is compact. We can absorb the constant γ into the λ in the lemma, so it suffices to prove the lemma for the a_1. Fix a number δ smaller than the injectivity radius of X, and Consider first the case $\epsilon = \delta$. The set of $\phi_{y,\delta}$ is compact, so the set of functions $dz^{-2}\phi^*_{y,\delta} a_1$ on U is compact. We can cover this compact set by closed subsets where no zeros cross through the boundary of U (although some zeros may go to the boundary or appear there from the outside). On one such closed subset, the set of expressions of the form $\prod(z - s_j)$, the product taken over all interior zeros and some of the boundary zeros, is compact. On any such closed subset we get a compact collection of functions $\prod(z - s_j)dz^2/\phi^*_{y,\delta} a_1$ which don't vanish in the interior of U. Hence on $1/2U$ these functions are uniformly bounded above and away from 0. (The total number of zeros is bounded, so the question of whether zeros on the boundary are included or not doesn't hurt the estimate over $1/2U$.) This covers the case $\epsilon = \delta$. For smaller $\epsilon = t\delta$, first approximate $\phi^*_{y,\delta} a$ by a function of the form $\prod(z - s_j)dz^2$, and then note that $\mu(t, \{s_j\})\phi^*_{0,t} \prod(z - s_j)$ are contained in some compact family of nonzero functions, for suitable normalizations $\mu(t, \{s_j\})$. The same argument then gives an approximation for $\mu(t, \{s_j\})\phi^*_{0,t} \prod(z - s_j)$, and hence for $\phi^*_{y,\epsilon} a_1$.

Proposition 12 *There is a constant C—and let δ be as above. Suppose a is any section with $A \geq C$. We may choose points $y_i \in |Y|$ and numbers $\epsilon_i \leq \delta$ such that: the open balls $B_i = B(y_i, \epsilon_i)$ are disjoint; each $y \in |Y|$ is contained in one of the balls; we have*

$$\sup_U |\phi^*_{y_i, 2\epsilon_i} a| \leq 1;$$

and

$$\inf_{U - 1/2U} |\phi^*_{y_i, \epsilon_i} a| \geq C^{-1}.$$

Proof: We will first show that for each $y_i \in |Y|$, there is an ϵ_i so that the third and fourth conditions are satisfied. Let τ_i be chosen as above. Apply Lemma 11 to $\phi^*_{y_i,2\tau_i}(a)$ to obtain an approximation for $\phi^*_{y_i,\tau_i}(a)$. Since the supremum of $|\phi^*_{y_i,\tau_i}(a)|$ is 1, we may assume that $\lambda = 1$ (possibly after increasing the constants). Set $\epsilon_i = 8^{-\ell}\tau_i$ for some ℓ between 1 and $4g$, chosen so that there are no zeros of a in $B(y_i, 2\epsilon_i) - B(y_i, \epsilon_i/4)$. This is possible since the number of zeros of a is less than or equal to $4g-4$, the degree of $Sym^2\Omega^1_X$. Since $2\epsilon_i \leq \tau_i$, we get the third condition. From the approximation for $\phi^*_{y_i,\tau_i}(a)$, the infimum of $\tau_i^2|a(x)|$ for $x \in B(y_i, \epsilon_i) - B(y_i, \epsilon_i/2)$ is at least C^{-1}. Since ϵ_i is some bounded (away from zero) multiple of τ_i, we obtain the desired estimate of the last condition. We will also use below the condition that there is no point of $|Y|$ in $B(y_i, 2\epsilon_i) - B(y_i, \epsilon_i)$.

Next we will see how to choose some of the y_i so that the B_i are disjoint and cover $|Y|$. Choose $y_1 \in |Y|$, with the largest ϵ_1. Then choose $y_2 \in |Y|$ with y_2 not in $B(y_1, \epsilon_1)$, again with the largest ϵ_2 among those. Proceed inductively, choosing y_{k+1} in $|Y|$ but not in $\bigcup_{j=1}^k B(y_j, \epsilon_j)$, each time with the largest possible ϵ_{k+1}, until no further choices are possible. From our construction of ϵ_i, we know that if $j > i$ then in fact y_j is not in $B(y_i, 2\epsilon_i)$, so $d(y_i, y_j) \geq 2\epsilon_i$. But also, $\epsilon_j \leq \epsilon_i$. Thus $B_i \cap B_j = \emptyset$. By construction, the balls cover $|Y|$. (In fact the fourth condition implies that the $B(y_i, \epsilon_i/2)$ cover $|Y|$.)

Remark: From the third condition of the proposition, and from Lemma 9, we have

$$\sup |\phi^*_{y_i,\epsilon_i}\beta|_{\phi^*_{y_i,\epsilon_i}K} \leq C.$$

Lemma 13 *Let B_i be given by the above proposition. There is a constant C such that if $x \notin \bigcup_i B_i$, then $|a|^{-1}|\beta|^2(x) \leq C$.*

Proof: Let C_2 be a constant as specified in a moment. For any $y \notin \bigcup_i B_i$, let σ_y be the biggest number such that $\sup_U |\phi^*_{y,\sigma_y}a| \leq C_2^{-1}$. We claim that if C_2 is large enough, then the $B(y, \sigma_y)$ don't meet any $B(y_i, \epsilon_i/2)$. If $y \notin \bigcup_i B_i$ and $B(y, \sigma_y)$ meets some $B(y_i, \epsilon_i/2)$, then there is a point x such that

$$B(x, \epsilon_i/5) \subset B(y, \sigma_y) \cap (B(y_i, \epsilon_i) - B(y_i, \epsilon_i/2)).$$

In this case, $\inf_U |\phi^*_{x,\epsilon_i/5}a| \geq C^{-1}$, where the constant depends on the estimate of the fourth condition in Proposition 12. Note that $\sigma_y \geq \epsilon_i/2$. Thus, increasing the constant again, we get $\sup_U |\phi^*_{y,\sigma_y}a| \geq C^{-1}$. Let C_2 be a constant which is bigger than this last one—then we obtain a contradiction, which proves the claim.

Consequently, there are no zeros of $\phi^*_{y,\sigma_y}a$ in U. The approximation of Lemma 11 shows that

$$|\phi^*_{y,\sigma_y}a(0)| \geq C^{-1}.$$

On the other hand, the estimate of Lemma 9 shows that

$$|\phi^*_{y,\sigma_y}\beta(0)|^2 \leq C.$$

These combine to give the required estimate, $|\beta(y)|^2/|a(y)| \leq C$ (since the scaling factors for a and β^2 are the same).

We may choose positive functions ρ_i such that $\rho_i(x) = 1$ if $x \in B(y_i, 2\epsilon_i/3)$, $\rho_i(x) = 0$ if $x \notin B(y_i, \epsilon_i)$, and $\sup |d\rho_i| \leq C\epsilon_i^{-1}$. Make these choices so that $\phi^*_{y_i,\epsilon_i}\rho_i$ are some standard such functions on U. The supports of ρ_i are disjoint. Let $\rho = 1 - \sum_i \rho_i$. Note that $\rho = 0$ inside any $B(y_i, 2\epsilon_i/3)$. In particular, $\rho = 0$ in a neighborhood of any point of $|Y|$. (There will be no confusion between this use of the symbol ρ for partitions of unity, and the original use for monodromy representations.)

Choosing a line bundle

In this section we will see how to choose a line bundle L which will work in our solution of the $\overline{\partial} + \overline{\theta}$-problem, uniformly for all V.

Lemma 14 *There exist constants C, and k. Given a rank two harmonic bundle V, there exists a positive function r which satisfies the following properties (refering to notation established above which depends on V).*

1. r is constant in the disjoint balls B_i.

2. $r \geq \epsilon_i^{-2}$ on B_i.

3. If C_{22} is the constant which appears in Proposition 22 below, then $C_{22}|q|^2_K \leq r/4$.

4. $\int_X r \leq C$.

5. $\sup_X |r| \leq CA^k$.

Proof: First of all, note that $\epsilon_i \geq C^{-1}A^{-1/2}$. This is because we have the bound $\sup |a| \leq CA$. The numbers τ_i were defined to be the largest such that $|\phi^*_{y_i,\tau_i} a| \leq 1$. Since the size of a scales essentially by τ^{-2}, the τ_i are at least $C^{-1}A^{-1/2}$. But recall that ϵ_i was at least $8^{-4g}\tau_i$, giving the claimed lower bound on ϵ_i. Thus conditions 2 and 5 are compatible. Since the area of B_i is a multiple of ϵ_i^2, condition 2 is compatible with condition 4. By Lemma 9, we have $\sup |q|_K \leq CA^k$, so conditions 3 and 5 are compatible (even in view of condition 1). Furthermore, the estimate $\int_X |q|^2_K \leq C$ of Lemma 9 shows that condition 3 is compatible with condition 4—except possibly on the B_i (in view of condition 1). But on the B_i we have $\sup_U |\phi_i^* q| \leq C$, so $\sup_{B_i} |q|_K \leq C\epsilon_i^{-1}$. Hence the conditions 1, 3, and 4 are compatible too. The reader will verify that the constant C_{22} appearing in Proposition 22 below is independent of the function r—hence there is no circularity in that regard.

Lemma 15 *Suppose f is a function on X with $\int_X f = 0$, such that*

$$\int_X |f| \leq C_{15} \quad \text{and} \quad \sup_X |f| \leq B.$$

Let u be the function with $\int_X u = 0$ and $\Delta(u) = f$. Then there exist C and k depending only on C_{15} such that

$$\sup_X e^{|u|} \leq CB^k.$$

Proof: This is an Aubin-type result, the proof of which is based on the same idea as in [3]. Suppose X_1 is an open set isomorphic via coordinate z to the disc of radius $1/2$ in $\mathbf{C}$. (We can cover X by finitely many such open sets to get the estimate globally.) Let Δ_z denote the Laplacian in the euclidean coordinate z. A conformal change of metric on an open set in X has the result of multiplying the Laplacian by a function, so $\Delta = \alpha(z)\Delta_z$. Set

$$u_1(z) = C' \int_{X_1} \alpha(y)^{-1} f(y) \log|z-y| dy.$$

This is a solution of $\Delta_z(u_1) = \alpha^{-1} f$, or $\Delta(u_1) = f$. Thus $u - u_1$ is harmonic in X_1, and bounded in a way which is dependant only on the L^1 norm of f. So it suffices to show that

$$e^{|u_1|} \le CB^k,$$

after which, one would have to increase C and k to obtain the values sought in the statement of the lemma. Hölder's inequality says

$$|u_1(z)| \le C \left(\int_X |f|^p \right)^{1/p} \left(\int_{|y|\le 1/2} |\log|z-y||^q dy \right)^{1/q}$$

with $q = p/(p-1)$; the constant depends on the relationship between the euclidean metric and the original one. Let

$$D_q = \int_{|y|\le 1} |\log|y||^q dy.$$

We will give a bound for this below. Using our hypotheses on f we get a constant C_3 depending on C_{15} such that

$$|u_1(z)| \le C_3 B^{(p-1)/p} (D_q)^{1/q} = C_3 (BD_q)^{1/q}.$$

Hence, for any fixed k,

$$|u_1(z)|^n \le C_3^n B^k (D_{n/k})^k,$$

and

$$e^{|u_1(z)|} = \sum_n |u_1(z)|^n/n! \le B^k \sum_n (D_{n/k})^k (C_3)^n/n!.$$

We claim that, for given C_3, there exists k such that the right hand sum is finite. This will prove the lemma. We have

$$D_q = 2\pi \int_0^1 (-\log r)^q r dr$$

or, with the change of variable $r = e^{-v/2}$,

$$D_q = 2^{-q}\pi \int_0^\infty v^q e^{-v} dv = 2^{-q}\pi\Gamma(q+1).$$

By Stirling's approximation,

$$\sum_n (D_{n/k})^k (C_3)^n/n! \le \sum_n (C_4)^n k^{-n}.$$

If $k > C_4$, this sum is finite, giving the desired estimate.

On any line bundle, let S denote the standard metric with constant curvature. Let $h = \omega^{-1}R(\mathcal{O}_X(x), S)$. Then $\omega^{-1}R(\Omega^1_X, S) = (2g-2)h$. Normalize ω so that the volume of X is equal to one.

Lemma 16 *There exists a line bundle L''. For any rank two harmonic bundle V, and any choice of function r as prescribed in Lemma 14, there exists a metric I'' on L'', such that*

1. The curvature satisfies $\omega^{-1}R(L'', I'') \ge (2g-2)h + r$, and

2. I'' is bounded with respect to the standard metric in the following way:

$$C^{-1}A^{-k}|v|_S \le |v|_{I''} \le CA^k|v|_S.$$

Furthermore, if M is any line bundle of degree zero, then $(L'' \otimes M, I'' \otimes S)$ will work just as well.

Proof: Let n be an integer so that $nh \ge (2g-2)h + \int_X r$. Note that, by condition 4 of Lemma 14, the number n may be chosen independently of V. Let x be some point in X, and let $L'' = \mathcal{O}_X(nx)$. Let

$$f = (2g-2)h + r + c - nh,$$

where c is a positive constant which depends on r, chosen so that $\int_X f = 0$. Then f satisfies the hypotheses of Lemma 15, with $B = CA^k$ and C_{15} independent of r. By that lemma, we get a function u with $\omega^{-1}\overline{\partial}\partial(u) = f$ and $\sup_X e^{|u|} \le CA^k$. Here the constants C and k may have to be increased, but they still don't depend on V. Multiply the standard metric on L'' by e^u to get the metric $I'' = e^u S$. Then

$$\omega^{-1}R(L'', I'') = \omega^{-1}R(L'', S) + \omega^{-1}\overline{\partial}\partial \log e^u = (2g-2)h + r + c,$$

and the metric is bounded with respect to the standard metric, in the desired way.

Lemma 17 *We may choose a line bundle L', and for each harmonic bundle V, a metric I', such that (L', I') satisfies the two conditions of Lemma 16, and so that the following is true. For any harmonic bundle V, there is a line bundle L'' (whose degree is independent of V) with metric I'' (depending on V), satisfying the conditions of Lemma 16. There is a morphism $\iota : L'' \to L'$ (depending on V), such that $|v(x))|_{I''} \le C|\iota v(x)|_{I'}$ for $x \notin \bigcup_i B(y_i, 2\epsilon_i/3)$. And we have*

$$|\iota v|_{K\otimes I'} \le C|v|_{J^*\otimes I''}$$

on all of X'. Again, if M has degree zero then $(L' \otimes M, I' \otimes S)$ works just as well.

Proof: For each $s \in 1/2U$, define a positive function $f_s(z)$ such that $f_s(z) = 1$ for $|z| \geq 3/4$, and $f_s(z) = 5|z - s|$ for $|z| \leq 1/2$. We may do this in such a way that $0 \leq \omega^{-1}\overline{\partial}\partial \log f_s \leq C$ uniformly in s. Now, if V is a harmonic bundle, then look at the balls B_i constructed above. Define a function f on X in the following way. If x is not in any B_i then set $f(x) = 1$. Inside B_i, define f so that, if s_j are the zeros of $\phi_i^* a$ in $1/2U$ counted with multiplicity, we have $\phi_i^* f = \prod_j f_{s_j}$. We claim that there is a constant C so that $|a|^{-1}|\beta|^2 \leq Cf^{-1}$ on all of X. By Lemma 13, this works outside of the balls B_i. But, inside the balls B_i, we have $|\phi_i^*\beta|^2 \leq C$, and $|\phi_i^* a| \sim \lambda \prod_j |z - s_j|$. By the method for choosing τ_i and then ϵ_i, we can take $\lambda = 1$. Thus, $|\phi_i^* a|^{-1} \leq C \prod_j |z - s_j|^{-1}$. The scaling factors for ϕ_i^* acting on $|\beta|^2$ and $|a|$ are the same. Hence, $|a|^{-1}|\beta|^2 \leq Cf^{-1}$ as claimed.

We can use f to define a metric H on the line bundle $\mathcal{O}_X(Y)$, by setting $|1|_H = f$, where 1 is the unit section of $\mathcal{O}_X$. Note that $\omega^{-1}R(\mathcal{O}_X(Y), H) \geq 0$. Also, H is bounded with respect to the standard metric in the same way as required in condition 2 of Lemma 16. This is because the balls B_i have radius $\epsilon_i \geq C^{-1}A^{-k}$ (see Lemma 14), and because of the definition of the function f.

Choose the line bundle L' in such a way that $L'' = L' \otimes \mathcal{O}_X(-Y)$ satisfies the criteria of Lemma 16 (this only depends on the degree of L''—recall that the degree of Y is fixed). Then if V is any harmonic bundle, let $L'' = L' \otimes \mathcal{O}_X(-Y)$, with the natural morphism $\iota : L'' \to L'$. Choose a metric I'' on L'' as required by Lemma 16. Let $I' = I'' \otimes H$. If v is a section of L'' then

$$|\iota(v)|_{I'} = f|v|_{I''}.$$

The curvature is

$$\omega^{-1}R(L', I') = \omega^{-1}(R(L'', I'') + R(\mathcal{O}_X(Y), H)) \geq \omega^{-1}R(L'', I'').$$

In particular, if (L'', I'') satisfies condition 1 of Lemma 16, then so will (L', I'). Furthermore, if I'' satisfies condition 2 of Lemma 16, then since H is bounded with respect to the standard metric in a similar way, I' will satisfy condition 2.

Since the function f is uniformly bounded away from zero on the complements of $B(y_i, 2\epsilon_i/3)$, we have $|u(x)|_{I''} \leq C|\iota u(x)|_{I'}$ for x not in $\bigcup_i B(y_i, 2\epsilon_i/3)$. On the other hand, we have

$$(1 + |a|^{-1}|\beta|^2) \leq Cf^{-1}.$$

In view of the estimate of Lemma 8 for the metric J^*, this inequality proves the last condition of the lemma.

Lemma 18 *There is a line bundle L. For any rank two harmonic bundle V there is a metric I on L, with the following property. For any point $x \in X$, if we set $L' = L \otimes \mathcal{O}_X(-x)$ with metric $I' = I \otimes S$, then (L', I') satisfies the criteria of Lemma 16. Furthermore, the metric I satisfies condition 2 of Lemma 16 with respect to the standard metric on L.*

Proof: Let n_1 be the degree of the line bundle L' given above. By the remark at the end of the previous lemma, any line bundle L' of that degree will do. Let L be a fixed line bundle of degree n_1+1. Then for any $x \in X$, $L' = L \otimes \mathcal{O}_X(-x)$ is of degree n_1, so it works in the previous lemma. Given V, we get a metric I' for L', and we can let $I = I' \otimes S$.

Solution of the $\overline{\partial}+\overline{\alpha}$ and $\overline{\partial}+\overline{\theta}$ problems

In this section we will show how to solve the $\overline{\partial}+\overline{\alpha}$ problem in the bundle $V \otimes L$, for an appropriate line bundle L with metric. Then the solution of the $\overline{\partial}+\overline{\theta}$ problem follows by an inductive process.

Lemma 19 *Suppose Z is a compact Riemann surface with boundary, provided with a metric. Assume that the boundary is smooth and nontrivial. Suppose $(W, \overline{\partial}, H)$ is a holomorphic vector bundle with metric. Suppose r is a positive real function on Z, such that*

$$\omega^{-1}R(W, \overline{\partial}, H) \geq r + \omega^{-1}R(\Omega^1_Z).$$

Then if η is any $(0,1)$-form with coefficients in W, there is a section u of W such that $\overline{\partial}(u) = \eta$. We have the estimate

$$\int_Z |u|^2_H \leq \int_Z r^{-1}|\eta|^2_H.$$

Proof: Write $W = W' \otimes \Omega^1_Z$. Let H' be the metric on W' whose tensor product with the given metric on Ω^1_Z equals H. The hypothesis becomes $\omega^{-1}R(W', H') \geq r$. We may interpret our $(0,1)$ form η as a $(1,1)$-form η' with coefficients in W', and we may search for a $(1,0)$-form u' with coefficients in W', which will correspond to u. The equation is still $\overline{\partial}(u') = \eta'$, and the norms of these new sections are the same as of the old. Let ∂ be the operator on W' complex conjugate to $\overline{\partial}$, with respect to the metric H'. Then $\omega^{-1}\overline{\partial}\partial$ is a multiple of the ∂-Laplacian for sections of W. We can solve the Dirichlet problem on W (it will be clear from the estimate below that there are no harmonic functions satisfying the Dirichlet boundary conditions). Let v be a solution of $\omega^{-1}\overline{\partial}\partial(v) = \omega^{-1}\eta'$, in other words

$$\overline{\partial}\partial(v) = \eta'.$$

Now do some calculations, always in the metric H', and integrating two-forms over Z.

$$\begin{aligned}
\int_Z (\eta', v) &= \int_Z (\overline{\partial}\partial v, v) \\
&= \int_Z \overline{\partial}(\partial v, v) + \int_Z (\partial v, \partial v) \\
&= \int_Z d(\partial v, v) + \int_Z (\partial v, \partial v) \\
&= \int_{\partial Z} (\partial v, v) + \int_Z (\partial v, \partial v) \\
&= \int_Z (\partial v, \partial v).
\end{aligned}$$

The last step is from the Dirichlet conditions $v|_{\partial Z} = 0$. From this equation you can see that if $\omega^{-1}\overline{\partial}\partial(v) = 0$ then $\partial v = 0$; thus v is an antiholomorphic section of an antiholomorphic bundle which vanishes on some boundary pieces; thus $v = 0$.

By the same calculation, we get

$$\int_Z (\partial\overline{\partial}v, v) = \int_Z (\overline{\partial}v, \overline{\partial}v),$$

and adding the two together,

$$\int_Z (R(W', H')v, v) = \int_Z (\overline{\partial}v, \overline{\partial}v) + \int_Z (\partial v, \partial v).$$

The matrix $\omega^{-1}R(W', H')$ is self adjoint, with eigenvalues which are bigger than the function r. Recall that $\sqrt{-1}\omega$ is the volume form, so

$$\sqrt{-1}\int_Z (R(W', H')v, v) = \int_Z (\omega^{-1}R(W', H')v, v) \geq \int_Z r|v|^2.$$

On the other hand,

$$\sqrt{-1}\int_Z (\overline{\partial}v, \overline{\partial}v) = -\int |\overline{\partial}v|^2$$

whereas

$$\sqrt{-1}\int_Z (\partial v, \partial v) = \int |\partial v|^2.$$

Therefore we get

$$\int_Z r|v|^2 \leq \int_Z (|\partial v|^2 - |\overline{\partial}v|^2).$$

(To check the signs, note that this is consistent with the fact that a positively curved bundle may have holomorphic sections but has no antiholomorphic ones). Combining with the first calculation, we get

$$\int_Z r|v|^2 \leq \int_Z |(\eta', v)|,$$

and by Hölder's inequality,

$$\int_Z r|v|^2 \leq \left(\int_Z r|v|^2\right)^{1/2} \left(\int_Z r^{-1}|\eta'|^2\right)^{1/2}.$$

Dividing, we get

$$\left(\int_Z r|v|^2\right)^{1/2} \leq \left(\int_Z r^{-1}|\eta'|^2\right)^{1/2}.$$

Again by Hölder's inequality, we get

$$\int_Z |(\eta', v)| \leq \int_Z r^{-1}|\eta'|^2.$$

Thus,

$$\int_Z |\partial v|^2 \leq \int_Z r^{-1}|\eta'|^2.$$

Now set $u' = \partial v$. This is the desired solution, with the desired estimate.

Lemma 20 *Suppose $(L'', \overline{\partial}, I'')$ is a line bundle with metric on X, whose curvature satisfies $\omega^{-1}R(L'', I'') \geq (2g-2)h + r$. Suppose $(V, \overline{\partial} + \overline{\alpha}, J^*)$ is a vector bundle on X' whose curvature satisfies $R(V, \overline{\partial} + \overline{\alpha}, J^*) \geq 0$. Suppose η is a compactly supported $(0,1)$-form on X' with coefficients in $V \otimes L''$. Then there is a section u of $V \otimes L''$ on X', solving the equation*

$$(\overline{\partial} + \overline{\alpha})(u) = \eta,$$

with the estimate

$$\int_{X'} |u|^2_{J^* \otimes I''} \leq \int_{X'} r^{-1} |\eta|^2_{J^* \otimes I''}.$$

Proof: Choose a sequence of subsets $X_n \subset X$, such that each X_n is a Riemann surface with boundary, and such that the boundaries consist of circles around the points $y \in |Y|$. Make the boundary circles get smaller and smaller, so that $X' = \bigcup_n X_n$. Using a standard metric on X, we have $\omega^{-1}R(\Omega^1_{X_n}) = (2g-2)h$. Use the previous lemma to solve

$$(\overline{\partial} + \overline{\alpha})(u_n) = \eta|_{X_n},$$

with the estimate

$$\int_{X_n} |u_n|^2 \leq \int_{X_n} r^{-1} |\eta|^2.$$

Let $n \to \infty$, and take a weak local limit of the u_n: a subsequence converges to u on each relatively compact subset of X', with $\overline{\partial}(u) = \eta$. Elliptic estimates imply that this local convergence is strong (in whichever norm). Furthermore, we get an estimate for the L^2 norm of u' on any relatively compact open set. Since the L^2 norm of u' is the supremum of the norms over relatively compact open sets, we get

$$\int_{X'} |u|^2 \leq \int_{X'} r^{-1} |\eta|^2.$$

This proves the lemma.

Finally, we will collect our results in usable form. Let L L', and L'' refer to line bundles as constructed in the previous section. Next, suppose V is fixed, and let I, I', and I'' be the metrics constructed in the previous section, depending on V. Suppose η is a $(0,1)$-form with coefficients in $V \otimes L'$, supported on the complement of $\bigcup_i B(y_i, 2\epsilon_i/3)$. Then we may consider η to be a form with coefficients in $V \otimes L''$. By applying Lemma 20 to the line bundle (L'', I'') and vector bundle $(V, \overline{\partial} + \overline{\alpha}, J^*)$, we can find a section u of $V \otimes L''$ with

$$(\overline{\partial} + \overline{\alpha})(u) = \eta,$$

and

$$\int_{X'} |u|^2_{J^* \otimes I''} \leq C \int_{X'} r^{-1} |\eta|^2_{J^* \otimes I''}$$

(to verify the hypotheses, see Lemmas 8 and 16). Using the inclusion provided by Lemma 17, we may consider u as a section of $V \otimes L'$, with $|u|_{K \otimes I'} \leq C|u|_{J^* \otimes I''}$. Since the form η is supported away from $\bigcup_i B(y_i, 2\epsilon_i/3)$, we get $|\eta|_{J^*\otimes I''} \leq C|\eta|_{K\otimes I'}$ (see Lemmas 8 and 17). Thus our estimate becomes

$$\int_{X'} |u|^2_{K\otimes I'} \leq C \int_{X'} r^{-1}|\eta|^2_{K\otimes I'}.$$

Next we will make some local analysis of the situation near a turning point. This analysis will take place in one of the subsets $B_i = B(y_i, \epsilon_i)$ discussed above, but for convenience we will pull everything back by the map $\phi_i = \phi_{y_i,\epsilon_i} : U \to B_i$ and make estimates on U. Fix one of the B_i which occured in the construction of the partition of unity.

The pullback is a harmonic bundle $(\phi_i^*V, \phi_i^*(\overline{\partial} + \theta), \phi_i^*K)$ over U. By the condition of Proposition 12, and the estimate of Lemma 9, we have

$$\sup_U |\phi_i^*\theta|_{\phi_i^*K} \leq C.$$

Since the curvature is equal to $[\theta, \overline{\theta}]$, we have

$$\sup_U |\phi_i^* R(V, \overline{\partial}, K)| \leq C.$$

Lemma 21 *Let L' be a line bundle with metric I' satisfying the conditions of Lemma 16. Suppose η is a $(0,1)$-form with coefficients in $V \otimes L'$. Then there is a solution w_i of the equation $(\overline{\partial} + \overline{\alpha})(w_i) = \eta|_{B_i}$, defined in the ball B_i, and satisfying the estimate*

$$\int_U |\phi_i^* w_i|^2_{\phi_i^*K\otimes I'} \leq C \int_U \epsilon_i^{-2}\phi_i^* r^{-1}|\phi_i^*\eta|^2_{\phi_i^*K\otimes I'}.$$

Proof: We have bounds

$$\sup_U |\phi_i^* R(V, \overline{\partial}, K)| \leq C_5$$

$$\sup_U |\phi_i^*\overline{\alpha}| \leq C_6.$$

Choose a metric H for the trivial line bundle over U such that $\omega^{-1}R(\mathcal{O}_U, \overline{\partial}, H) \geq C_5 + 4C_6^2$. Let u_0 be a solution of $\phi_i^*\overline{\partial}(u_0) = \phi_i^*\eta$, and then let $u_1, u_2, \ldots$ be solutions of $\phi_i^*\overline{\partial}(u_{j+1}) = -\phi_i^*\overline{\alpha}u_j$. We have a bound on curvature,

$$\omega^{-1}R(\phi_i^*(V\otimes L'), \phi_i^*\overline{\partial}, \phi_i^*(K\otimes I')\otimes H) \geq C^{-1}\epsilon_i^2\phi_i^* r + 4C_6^2.$$

By Lemma 19, we have estimates

$$\int_U |u_0|^2 \leq C \int_U \epsilon_i^{-2}\phi_i^* r^{-1}|\phi_i^*\eta|^2$$

and

$$\begin{aligned} \int_U |u_{j+1}|^2 &\leq \int_U (2C_6)^{-2}|\phi_i^*(\overline{\alpha})u_j|^2 \\ &\leq 1/4 \int_U |u_j|^2. \end{aligned}$$

In particular,

$$\|u_j\|_{L^2} \le 2^{-j} C \left(\int_U \epsilon_i^{-2} \phi_i^* r^{-1} |\phi_i^* \eta|^2 \right)^{1/2}.$$

Therefore the sum $\phi_i^* w_i \stackrel{def}{=} \sum_{j=0}^{\infty} u_j$ converges in L^2, and hence locally in any norm. We have $(\overline{\partial} + \overline{\alpha})(w_i) = \eta$, and we get the estimate

$$\int_U |\phi_i^* w_i|^2 \le C \int_U \epsilon_i^{-2} \phi_i^* r^{-1} |\phi_i^* \eta|^2.$$

In all of the above, we have used the metric $\phi_i^*(K \otimes I'') \otimes H$. However, H is uniformly fixed, and hence uniformly comparable to the standard metric. Thus, after increasing the constants, we obtain the desired estimate without H.

Proposition 22 *Suppose L' is a line bundle chosen as described in Lemma 17. Suppose η is a smooth $(0,1)$-form on X with coefficients in $V \otimes L'$. Then there is a section u of $V \otimes L'$ on X', such that*

$$(\overline{\partial} + \overline{\alpha})(u) = \eta,$$

with the estimate

$$\int_X |u|^2_{K \otimes I'} \le C_{22} \int_X r^{-1} |\eta|^2_{K \otimes I'}.$$

The constant C_{22} is independent of the function r.

Proof: By Lemma 21, we may find sections w_i of $V \otimes L'$ defined over B_i, such that $(\overline{\partial} + \overline{\alpha})(w_i) = \eta|_{B_i}$, with the estimates

$$\int_U |\phi^* w_i|^2 \le C \int_U \epsilon_i^{-2} \phi_i^* r^{-1} |\phi_i^* \eta|^2.$$

In these estimates, the terms $\epsilon_i^{-2} \phi_i^* r^{-1}$ are less than one. Use our partition of unity to set

$$\xi = \rho\eta - \sum_i (\overline{\partial}\rho_i) w_i.$$

Then ξ is supported on the complement of the union of the $B(y_i, 2\epsilon_i/3)$. Therefore we may apply Lemma 20 and the subsequent discussion to find a solution w of $(\overline{\partial} + \overline{\alpha})(w) = \xi$. Using the norms $K \otimes I'$ we get the estimate

$$\begin{aligned}
\int_X |w|^2 &\le \int_X r^{-1} |\xi|^2 \\
&\le C \int_X r^{-1} |\eta|^2 + C \sum_i \int_{B_i} r^{-1} |(\overline{\partial}\rho_i) w_i|^2 \\
&\le C \int_X r^{-1} |\eta|^2 + C \sum_i \int_U \phi_i^* r^{-1} |\phi_i^* (\overline{\partial}\rho_i) w_i|^2 \\
&\le C \int_X r^{-1} |\eta|^2 + C \sum_i \int_U \phi_i^* r^{-1} |\phi_i^* w_i|^2 \\
&\le C \int_X r^{-1} |\eta|^2 + C \sum_i \int_U \phi_i^* r^{-1} |\phi_i^* \eta|^2 \\
&\le C \int_X r^{-1} |\eta|^2.
\end{aligned}$$

(The first line is from Lemma 20 and the discussion following it; the second is from the definition of ξ; the third is because there is no scale factor in changing variables for the integral of the square norm of a 1-form; the fourth line is because on U the partition of unity is standard; the fifth line is from the estimate for w_i–noting that r^{-1} is a constant there, and that the other term $\epsilon_i^{-2}\phi_i^* r^{-1}$ in the estimate for w_i is less than one; and the last is because there is no scaling for the square norm of the 1-form η, and the number of B_i is uniformly bounded.)

Finally, set $u = w + \sum_i \rho_i w_i$. Then

$$\begin{aligned}(\overline{\partial}+\overline{\alpha})(u) &= \xi + \sum_i (\overline{\partial}\rho_i) w_i + \sum_i \rho_i \eta|_{B_i} \\ &= (\rho + \sum_i \rho_i)\eta \\ &= \eta.\end{aligned}$$

And we have the estimate

$$\begin{aligned}\int_X |u|^2 &\leq C\int_X |w|^2 + C\sum_i \int_U \epsilon_i^2 |\phi_i^* w_i|^2 \\ &\leq C\int_X r^{-1}|\eta|^2 + C\sum_i \int_U \epsilon_i^2 \epsilon_i^{-2} \phi_i^* r^{-1} |\phi_i^*\eta|^2 \\ &\leq C\int_X r^{-1}|\eta|^2.\end{aligned}$$

This proves the proposition.

Proposition 23 *Let C_{22} be the constant of the last proposition. Suppose that the function r is chosen according to Lemma 14 (so that $C_{22}|q|_K^2 \leq r/4$), and suppose that the line bundle L' and metric I' are chosen accordingly as described in the previous section. If η is a smooth $(0,1)$-form on X with coefficients in $V \otimes L'$, then there is a section v of $V \otimes L'$, smooth over X', such that $(\overline{\partial}+\overline{\theta})(v) = \eta$. Furthermore, we have the estimate*

$$\int_{X'} |v|^2_{K\otimes I'} \leq 4C_{22}\int_X r^{-1}|\eta|^2_{K\otimes I'}.$$

Proof: Let v_0 be the solution of $(\overline{\partial}+\overline{\alpha})(v_0) = \eta$ given by the previous proposition. Then for each $k \geq 1$, let v_k be the solution of

$$(\overline{\partial}+\overline{\alpha})(v_k) = -qv_{k-1}.$$

We have estimates

$$\int_X |v_0|^2 \leq C_{22}\int_X r^{-1}|\eta|^2,$$

and for each $k \geq 1$,

$$\int_X |v_k|^2 \leq C_{22}\int_X r^{-1}|q|^2|v_{k-1}|^2 \leq 1/4\int_X |v_{k-1}|^2.$$

Hence by induction,

$$\|v_k\|_{L^2} \leq 2^{-k} \left(C_{22} \int_X r^{-1} |\eta|^2 \right)^{1/2}.$$

Set $v = \sum v_k$. Let $\tilde{C}$ denote constants which may depend on V, q, r, and even η; we have elliptic estimates (locally on X'):

$$\|v_0\|_{L^2_{s+1}} \leq \tilde{C} \|\eta\|_{L^2_s}$$

and

$$\|v_k\|_{L^2_{s+1}} \leq \tilde{C} \|v_{k-1}\|_{L^2_s}.$$

By induction on s, these imply that we have estimates of the form $\|v_k\|_{L^p_s} \leq \tilde{C} 2^{-k}$. Hence, the sum v converges in any norm, locally on X'. For the L^2 estimate on X', note that by Fatou's lemma, we have $\|v\|_{L^2} \leq \sum_k \|v_k\|_{L^2}$. Therefore

$$\int_{X'} |v|^2 \leq 4C_{22} \int_X r^{-1} |\eta|^2$$

as desired.

Corollary 24 *If the form η is smooth over X, then the solution v constructed above will be smooth over X. Thus $H^1(X, N \otimes L') = 0$.*

Proof: The bundle $(V \otimes L', \overline{\partial} + \overline{\theta}, K \otimes I')$ extends as a smooth holomorphic vector bundle with smooth metric on all of X. Let v' be a smooth solution of $(\overline{\partial} + \overline{\theta})(v') = \eta$ defined near a puncture $y \in |Y|$. Then $v - v'$ is a $(\overline{\partial} + \overline{\theta})$-holomorphic section on X', which is also in L^2. A section with a pole or an essential singularity would not be in L^2, so $v - v'$ is holomorphic across the puncture. In particular it is smooth, so v is smooth across the puncture.

Corollary 25 *There are constants k and C. We may choose a line bundle L with a standard metric S, which works in the following way for all V. Let $A = \|a\|_{L^1}$ as before. If $x \in X$, and η is a smooth $(0,1)$-form with coefficients in $V \otimes L(-x)$, then there is a section v of $V \otimes L(-x)$, smooth over X, with*

$$(\overline{\partial} + \overline{\theta})(v) = \eta$$

and

$$\int_X |v|^2_{K \otimes S_x} \leq C A^k \int_X |\eta|^2_{K \otimes S_x}.$$

Here S_x denotes the tensor product of the standard metric S with a standard metric on $\mathcal{O}_X(-x)$.

Proof: By Lemma 18 we may choose our line bundle L and metric S in such a way that $L' = L(-x)$ is a line bundle as required by Proposition 23. Recall that the relationship between the metric I' and the standard metric $S' = S_x$ on L' is

$$C^{-1} A^{-k} |e|_{S'} \leq |e|_{I'} \leq C A^k |e|_{S'}.$$

With this, Proposition 23 implies the desired estimate.

Proof of the theorem

In this section, we will complete the proof of the theorem. Let L be a fixed line bundle as provided by Corollary 25.

Proposition 26 *There are constants C and k. Suppose (N, ∇) is any semisimple system of ODE's of rank two with unit determinant. Let $(V, \overline{\partial}, \theta, K)$ be the corresponding harmonic bundle with harmonic metric of determinant one. Let $a = det(\theta)$ and $A = \|a\|_{L^1}$ (and assume $A \geq C$). Suppose $x \in X$, and suppose $v_x \in (V \otimes L)_x$ with $|v_x|_{K \otimes S} = 1$. Then there is a section v of $V \otimes L$ with $v(x) = v_x$, $(\overline{\partial} + \overline{\theta})(v) = 0$ (so v is a holomorphic section of $N \otimes L$), and $\|v\|_{L^2, K\otimes S} \leq CA^k$.*

Proof: Let $B = B(x, \tau_x)$, and let $\phi = \phi_{x,\tau_x} : U \to B$. Recall that τ_x was the largest number so that $\sup_U |\phi^* a| \leq 1$.

We claim that there is a frame $\{u_1, u_2\}$ for the bundle $\phi^* V$ over U, unitary with respect to the metric $\phi^* K$, such that if $\overline{\partial}_0$ is the constant operator with respect to this frame, then

$$\phi^*(\overline{\partial} + \overline{\theta}) = \overline{\partial}_0 + H^{0,1}$$

with $|H^{0,1}(x)| \leq C|x|$. And we may assume that $u_1(0) = \phi^*(v_x)$.

To prove the claim, note (as in the discussion before Lemma 21) that we have a curvature estimate for $\phi^* V$,

$$|R(\phi^* V, \phi^* \overline{\partial}, \phi^* K)| \leq C$$

on U. Choose a unitary frame for the fiber $\phi^* V_0$, with $u_1(0) = \phi^*(v_x)$. Extend the frame to all of U by parallel translating along the real axis, then parallel translating from each point on the real axis in the direction of the imaginary axis. Let $z = x + \sqrt{-1}\, y$ denote the usual coordinate system on U. We obtain a frame for $\phi^*(V)$ such that

$$\phi_i^*(\partial + \overline{\partial}) = \partial_0 + \overline{\partial}_0 + H(x, y)dx,$$

with $H(x, 0) = 0$. In this case the curvature is

$$\phi^* R(x, y) = -\frac{\partial H}{\partial y} dx dy.$$

Given the previous bound for the curvature, we see by integrating that $|H(x, y)| \leq C|y|$, which proves the claim.

The claim provides us with a section $\tilde{v} = \phi_*(u_1)$ of V over B, of unit length, such that $\tilde{v}(x) = v_x$, and such that

$$|(\overline{\partial} + \overline{\theta})(\tilde{v})(y)| \leq C\tau_x^{-2} d(x, y).$$

Let $\tilde{\rho}$ be a function such that $\phi^* \tilde{\rho}$ is a standard function equal to 1 in $1/4\,U$ and equal to zero in $1/2\,U$. Let $\eta = (\overline{\partial} + \overline{\theta})(\tilde{\rho}\tilde{v})$. We can consider η as a $(0, 1)$-form with coefficients in $V \otimes L \otimes \mathcal{O}_X(-x)$, with $\|\eta\|_{L^2(K \otimes S_x)} \leq C\tau_x^{-1}$. (To go from coefficients in

V to coefficients in $V \otimes L$, note that we can choose a holomorphic section of L which is uniformly bounded above and away from zero in the standard metric S.) Note that $\tau_x \geq C^{-1}A^{-k}$ (see the proof of Lemma 14). Hence by Corollary 25 we can solve $(\overline{\partial} + \overline{\theta})(u) = \eta$ for a section u of $V \otimes L \otimes \mathcal{O}_X(-x)$, with $\|u\|_{L^2(K\otimes S_x)} \leq CA^k$. Consider u as a section of $V \otimes L$ with $u(x) = 0$, and set $v = \tilde{\rho}\tilde{v} - u$. Then v is $\overline{\partial} + \overline{\theta}$-holomorphic, $v(x) = \tilde{\rho}\tilde{v}(x) = v_x$, and $\|v\|_{L^2(K\otimes S)} \leq CA^k$.

Corollary 27 *The set of rank two determinant one systems of ODE's is bounded.*

Proof: We have chosen a uniform line bundle L such that for any semisimple system of ODE's, the holomorphic bundle $N \otimes L$ is generated by global sections. Thus the set of vector bundles underlying semisimple systems of ODE's is bounded. For each bundle N, the space of connections is a finite dimensional affine space which varies constructibly with N, so the set of semisimple systems of ODE's is bounded. Finally, to cover the case of systems which are not semisimple, note that the set of rank one systems is bounded, and the set of extensions between two rank one systems is a finite dimensional space varying constructibly.

We now proceed with the proof of the theorem. Fix our line bundle L as described above, with standard metric S. Suppose (N, ∇, δ) is a semisimple system of ODE's of rank two and determinant one. Then we may choose a harmonic metric K of determinant one; let $(V, \overline{\partial}, \theta, K)$ be the associated harmonic bundle. Let $\nu_1, \ldots, \nu_p$ be a basis of sections of $H^0(N \otimes L)$ which is orthonormal with respect to the L^2 inner product made using the metric $K \otimes S$. This gives a point $\mathbf{z} = (N, \nabla, \delta, \nu)$ in our parameter space $\mathbf{Z}$. We must estimate

$$\psi_{\mathbf{Z}}(\mathbf{z}) = \psi_U(\mathbf{z}) + \psi_{\tilde{\Sigma}}(\mathbf{z}) + \psi_{Hom}(\mathbf{z}).$$

First of all,

$$\psi_U(\mathbf{z}) = \sup_{i<j} \|\delta(\nu_i \wedge \nu_j)\|_{L^1(S^{\otimes 2})} \leq (\sup_i \|\nu_i\|_{L^2(K\otimes S)})^2 = 1.$$

Next, suppose $x \in X$. We can choose $K \otimes S$-unit vectors $v_{1,x}$ and $v_{2,x}$ in $(V \otimes L)_x$ such that $|\delta(v_{1,x} \wedge v_{2,x})| = 1$. By Proposition 26 above, we may extend these to sections v_1 and v_2 of $N \otimes L$ such that $\|v_j\|_{L^2(K\otimes S)} \leq CA^k$. Write $v_j = \sum a^i_j \nu_i$. Then $|a^i_j| \leq CA^k$. Hence the equation

$$|\sum a^i_1 a^l_2 \delta(\nu_i \wedge \nu_l)(x)| = 1$$

implies that for some $i < l$ we have $|\delta(\nu_i \wedge \nu_l)(x)| \geq C^{-1}A^{-k}$ (after increasing the constants C and k by a fixed amount). This shows that $\psi_{\tilde{\Sigma}}(\mathbf{z}) \leq CA^k$.

Finally we consider ψ_{Hom}. Suppose z is a local coordinate and λ is a local generator for L^* on an open set X_0. Thought of as an isomorphism $\lambda : L \cong \mathcal{O}_X$, λ induces a connection $\lambda^{-1} \cdot d \cdot \lambda$ on $L|_{X_0}$. We can compare this with the metric connection d_L for the standard metric,

$$\lambda^{-1} \cdot d \cdot \lambda = d_L + \Lambda.$$

There is also a metric connection d_N for the metric K on the bundle N. In terms of the operators of the harmonic bundle,

$$d_N = \partial - \theta + \overline{\partial} + \overline{\theta}.$$

Thus, acting on holomorphic sections, $\nabla = d_N + 2\theta$, so

$$\nabla(\lambda(\nu_i)) = \lambda(d_{N\otimes L} + 2\theta + \Lambda)(\nu_i).$$

In particular,

$$\begin{aligned} \psi_{Hom}(\mathbf{z}) &= \sup_{i,j} \left(\|\lambda\delta(\nu_j \wedge \nabla_{\frac{\partial}{\partial z}}(\lambda(\nu_i)))\|_{L^1} + \|\lambda\delta(\nu_j \wedge \lambda(\nu_i))\|_{L^1} \right) \\ &\leq C \sup_{i,j} \|\nu_j\|_{L^2(K\otimes S)} \left(\sup(C + |\theta|_K + |\Lambda|_S) \|\nu_i\|_{L^2(K\otimes S)} + \|d_{N\otimes L}\nu_i\|_{L^2(K\otimes S)} \right) \\ &\leq C + C \sup |\theta|_K + C \sup_i \|d^{1,0}_{N\otimes L}\nu_i\|_{L^2(K\otimes S)}. \end{aligned}$$

Now the curvature of $N \otimes L$ with the metric $K \otimes S$ is the curvature of L plus the curvature of N, a constant plus $[\theta, \overline{\theta}]$. By the estimate of Lemma 9, $\sup |\theta|_K \leq CA^k$, so we have

$$|R(N \otimes L, K \otimes S)| \leq CA^k.$$

This implies that

$$\int_X \omega^{-1}(d^{1,0}\nu_i, d^{1,0}\nu_i) = \int_X (\omega^{-1} R(N \otimes L, K \otimes S)\nu_i, \nu_i) \leq CA^k.$$

Therefore,

$$\psi_{Hom}(\mathbf{z}) \leq CA^k.$$

Putting these together, we get $\psi_{\mathbf{Z}}(\mathbf{z}) \leq CA^k$. Since $A \leq C(\log T)^2$ (by Corollary 6 and the definition of A), we obtain the estimate of Theorem 1′, hence Theorem 1.

References

[1] L. Ahlfors, An extension of Schwarz's lemma. *Trans. Amer. Math. Soc.* **43** (1938), 359-364.

[2] M. Atiyah, Vector bundles on elliptic curves. *Proc. London Math. Soc.* **7** (1957), 424.

[3] T. Aubin, Sur la fonction exponentielle. *C. R. Acad. Sci. Paris* **270A** (1970), 1514-1516.

[4] K. Corlette, Flat G-bundles with canonical metrics. *J. Diff. Geom.* **28** (1988) 361-382.

[5] P. Deligne, Equations différentielles à points singuliers réguliers. *Lect. Notes in Math.* **163** Springer, N.Y. (1970).

[6] S. Donaldson, Twisted harmonic maps and self-duality equations. *Proc. London Math. Soc.* **55** (1987), 127-131.

[7] D. Gieseker, On the moduli of vector bundles on an algebraic surface. *Ann. of Math.* **106** (1977), 45-60.

[8] N. Hitchin, The self-duality equations on a Riemann surface. *Proc. London Math. Soc.* **55** (1987) 59-126.

[9] F. Kirwan, *Cohomology of Quotients in Symplectic and Algebraic Geometry.* Princeton Univ. Press, Princeton (1984).

[10] R. Langer, The asymptotic solutions of ordinary linear differential equations of the second order with special reference to a turning point. *Trans. Amer. Math. Soc.* **67** (1949), 461-490.

[11] M. Maruyama, On the boundedness of families of torsion free sheaves. *J. Math. Kyoto Univ.* **21**-4 (1981), 673-701.

[12] D. Mumford, *Geometric Invariant Theory.* Springer Verlag, New York (1965).

[13] C. Simpson, Harmonic bundles on noncompact curves. *J.A.M.S.* **3** (1990), 713-770.

[14] C. Simpson, Moduli of representations of the fundamental group of a smooth projective variety (preprint).

ABELIAN VARIETIES OF K3 TYPE AND ℓ-ADIC REPRESENTATIONS

Yuri G. Zarhin

Research Computing Center of the USSR Academy of Sciences
Pushchino, Moscow Region, 142292, USSR

In this paper we study the algebraic envelopes of one-dimensional ℓ-adic Lie algebras attached to the Galois actions on the Tate modules of Abelian varieties over finite fields. These envelopes are linear reductive commutative Lie algebras. We prove that these envelopes (after an extension of scalars) are generated by semisimple linear operators, whose spectrum coincides with the set of slopes of the Newton polygon of the Abelian variety. In addition, the multiplicity of each eigen value is equal to the length of the slope.

Recall that an Abelian variety A over a finite field is called *ordinary* if its set of slopes is $\{0,1\}$. If so, both slopes have length $\dim(A)$. The Abelian variety is called *supersingular* if it has only one slope 1/2 . We say that *an Abelian variety A over a finite field is of* K3 *type if either it is an ordinary elliptic curve or has the same Newton polygon as the product of an ordinary elliptic curve and a* $(\dim(A) - 1)$*-dimensional supersingular Abelian variety*. This means that the set of slopes is either $\{0,1\}$ or $\{0,1/2,1\}$ and the slopes 0 and 1 have length 1. A special case of a theorem of Lenstra and Oort [5] asserts that for each positive integer g and for each prime number p there exists an absolutely simple g-dimensional Abelian variety of K3 type defined over a certain finite field of characteristic p.

For Abelian varieties of K3 type we compute explicitly the algebraic envelopes and prove that they are "as large as possible". In fact, we consider the multiplicative group spanned by all eigen values of the Frobenius and prove that its rank is "as large as possible". This allows us to check that all Galois-invariant ℓ-adic cohomology classes are linear combinations of the products of 2-dimensional Galois-invariant classes. Since all Galois-invariant 2-dimensional ℓ-adic cohomology classes are linear combinations of the classes of divisors in the case of an arbitrary Abelian variety over a finite field (Tate [12]), we obtain that in the case of Abelian varieties of K3 type each Galois-invariant ℓ-adic cohomology class is a linear combination of the products of the classes of divisors and, therefore, is algebraic. This proves Tate's conjecture [11] on algebraicity of Galois-invariant cohomology classes in the case of Abelian variety of K3 type.

I am deeply grateful to F. Oort for helpful discussions and his interest to this paper. Part of this work was done during my stay in Tokyo in August-September of 1990 and I am very happy to be able to thank the Department of Mathematics of the Tokyo Metropolitan University for the hospitality. My special thanks go to Y. Miyaoka, whose efforts made possible my visit to Japan. The support of the Kajima Foundation is also gratefully acknowledged.

0. Generalities. We write $\mathbb{Z}$ for the ring of integers, $\mathbb{Q}$, $\mathbb{R}$ and $\mathbb{C}$ for the field of rational, real and complex numbers respectively. If ℓ is a prime then we write $\mathbb{Z}_\ell$ and $\mathbb{Q}_\ell$ for the ring of ℓ-adic integers and the field of ℓ-adic numbers respectively. We fix an algebraic closure $\bar{\mathbb{Q}}_\ell$ of $\mathbb{Q}_\ell$. We write $\bar{\mathbb{Z}}_\ell$ for the integral closure of $\mathbb{Z}_\ell$ in $\bar{\mathbb{Q}}_\ell$, log for the ℓ-adic logarithm map log: $\bar{\mathbb{Z}}_\ell^* \to \bar{\mathbb{Q}}_\ell$. It is well-known that the homomorphism log is surjective and its kernel coincides with the group of all roots of unity in $\bar{\mathbb{Q}}_\ell$ [4].

0.1. Let k be a commutative field, $k(a)$ its separable algebraic closure, $G(k) := \mathrm{Gal}(k(a)/k)$ the Galois group . If $k' \subset k(a)$ is a finite algebraic extension of k, then $k(a)$ is also a separable algebraic closure of k' and the Galois group $G(k') = \mathrm{Gal}(k(a)/k')$

is an open subgroup of finite index in $G(k)$.

If m is a positive integer prime to char k, then we write μ_m for the multiplicative group of all roots of 1 of power m in $k(a)$. Let ℓ be a prime different from char k. We write $\mathbb{Z}_\ell(1)$ for the projective limit of groups μ_m where m runs through all powers of ℓ and the transition map is raising to power ℓ It is well-known that $\mathbb{Z}_\ell(1)$ is a free $\mathbb{Z}_\ell$-module of rank 1. The group $G(k)$ operates on $\mathbb{Z}_\ell(1)$ in an obvious way and we write χ_ℓ for the corresponding *cyclotomic character* χ_ℓ: $G(k) \to \mathrm{Aut}\, \mathbb{Z}_\ell(1) = \mathbb{Z}_\ell^*$ defining the Galois action on $\mathbb{Z}_\ell(1)$ (i.e., on all the ℓ-power roots of 1). We also introduce a one-dimensional $\mathbb{Q}_\ell$-vector space $\mathbb{Q}_\ell(1) := \mathbb{Z}_\ell(1) \otimes_{\mathbb{Z}_\ell} \mathbb{Q}_\ell$. Clearly, $\mathbb{Z}_\ell(1)$ is a $\mathbb{Z}_\ell$-lattice in $\mathbb{Q}_\ell(1)$ and one may extend by $\mathbb{Q}_\ell$-linearity the Galois action to $\mathbb{Q}_\ell(1)$. This action is defined by the same cyclotomic character

$$\chi_\ell\colon G(k) \to \mathrm{Aut}\, \mathbb{Z}_\ell(1) = \mathbb{Z}_\ell^* \subset \mathbb{Q}_\ell^* = \mathrm{Aut}\, \mathbb{Q}_\ell(1).$$

We write $\mathbb{Q}_\ell(-1)$ for the dual one-dimensional $\mathbb{Q}_\ell$-vector space

$$\mathbb{Q}_\ell(-1) := \mathrm{Hom}_{\mathbb{Q}_\ell}(\mathbb{Q}_\ell(1), \mathbb{Q}_\ell)$$

with a natural structure of the dual $G(k)$-module. This means that the Galois structure on $\mathbb{Q}_\ell(-1)$ is defined by the character

$$\chi_\ell^{-1} : G(k) \to \mathbb{Q}_\ell^* = \mathrm{Aut}\, \mathbb{Q}_\ell(-1).$$

If i is an integer we define the one-dimensional $\mathbb{Q}_\ell$-vector space $\mathbb{Q}_\ell(i)$ as follows: $\mathbb{Q}_\ell(i) = \mathbb{Q}_\ell(1)^{\otimes i}$ if i is positive, $\mathbb{Q}_\ell(-1)^{\otimes(-i)}$ if i is negative and $\mathbb{Q}_\ell$ if $i = 0$. Clearly, the natural Galois action on $\mathbb{Q}_\ell(i)$ is defined by the character

$$\chi_\ell^{i} : G(k) \to \mathbb{Q}_\ell^* = \mathrm{Aut}\, \mathbb{Q}_\ell(i).$$

0.2. Let V be a finite-dimensional vector space over $\mathbb{Q}_\ell$. We write $\bar{V}$ for the corresponding $\bar{\mathbb{Q}}_\ell$-vector space defined as the tensor product of V with $\bar{\mathbb{Q}}_\ell$ over $\mathbb{Q}_\ell$. There are natural embeddings

$$V \subset \bar{V},\ \mathrm{End}(V) \subset \mathrm{End}(V) \otimes_{\mathbb{Q}_\ell} \bar{\mathbb{Q}}_\ell = \mathrm{End}(\bar{V}) \ .$$

Let $f \in \mathrm{End}(V)$ be a non-zero linear semisimple operator in V. This implies that f is a diagonalizable operator in $\bar{V}$. We write $\mathrm{spec}(f)$ for the set of all eigen values of f. Let $\mathbb{Q}(f)$ be the $\mathbb{Q}$-vector subspace of $\bar{\mathbb{Q}}_\ell$ spanned by $\mathrm{spec}(f)$. If the $\mathbb{Q}$-vector space $\mathbb{Q}(f)$ is one-dimensional then the linear $\bar{\mathbb{Q}}_\ell$-Lie subalgebra $\bar{\mathbb{Q}}_\ell f \subset \mathrm{End}(\bar{V})$ is *algebraic*. In general, let us consider its *algebraic envelope* $(\bar{\mathbb{Q}}_\ell f)^{\mathrm{al}}$: it is the smallest algebraic $\bar{\mathbb{Q}}_\ell$-Lie subalgebra of $\mathrm{End}(\bar{V})$ containing $\bar{\mathbb{Q}}_\ell f$. In order to describe the algebraic envelope explicitly let us recall the notion of *replica* [3]. To each $\mathbb{Q}$-linear map $\pi\colon \mathbb{Q}(f) \to \bar{\mathbb{Q}}_\ell$ corresponds a semisimple linear operator $f^{(\pi)}$ in $\bar{V}$ defined as follows. If x is an eigen vector of f then it is also an eigen vector of $f^{(\pi)}$. More precisely, if $fx = ax$ with $a \in \mathrm{spec}(f)$ then $f^{(\pi)}x = \pi(a)x$. The operator $f^{(\pi)}$ is called a *replica* of f with respect to π. Clearly, $\mathrm{spec}(f^{(\pi)}) = \pi(\mathrm{spec}(f))$ and if b is an eigen value of $f^{(\pi)}$ then its multiplicity is equal to the sum of the multiplicities of the eigen values a of f with $\pi(a) = b$. All the replicas constitute the algebraic envelope $(\bar{\mathbb{Q}}_\ell f)^{\mathrm{al}}$. In particular ,

$$\dim_{\mathbb{Q}}\mathbb{Q}(f) = \dim\,(\bar{\mathbb{Q}}_\ell f)^{\mathrm{al}}\ (\text{over } \bar{\mathbb{Q}}_\ell).$$

Notice, that if we fix a basis of $\bar{V}$ consisting of eigen vectors of $\bar{V}$, then all operators in $(\bar{\mathbb{Q}}_\ell f)^{\mathrm{al}}$ are *diagonal* with respect to this basis.

We write $(\mathbb{Q}_\ell f)^{\mathrm{al}}$ for the algebraic envelope of $\mathbb{Q}_\ell$-Lie subalgebra $\mathbb{Q}_\ell f \subset \mathrm{End}(V)$. By definition, it is the smallest algebraic $\mathbb{Q}_\ell$-Lie subalgebra of $\mathrm{End}(V)$ containing $\mathbb{Q}_\ell f$. Clearly,

$$(\mathbb{Q}_\ell f)^{\mathrm{al}} \otimes_{\mathbb{Q}_\ell} \bar{\mathbb{Q}}_\ell \subset \mathrm{End}(V) \otimes_{\mathbb{Q}_\ell} \bar{\mathbb{Q}}_\ell = \mathrm{End}(\bar{V})$$

is the smallest algebraic $\bar{\mathbb{Q}}_\ell$-Lie subalgebra of $\mathrm{End}(\bar{V})$ containing $\bar{\mathbb{Q}}_\ell f$. Using ([3], Ch.2), one may easily check that

$$(\bar{\mathbb{Q}}_\ell f)^{\mathrm{al}} = (\mathbb{Q}_\ell f)^{\mathrm{al}} \otimes_{\mathbb{Q}_\ell} \mathbb{Q}_\ell .$$

In particular, $\dim_{\mathbb{Q}_\ell}(\mathbb{Q}_\ell f)^{\mathrm{al}} = \dim_{\mathbb{Q}}\mathbb{Q}(f)$. It easily follows that if S is a set of $\mathbb{Q}$-linear maps $\pi\colon \mathbb{Q}(f) \to \mathbb{Q}$ such that the product-map

$$\mathbb{Q}(f) \to \mathbb{Q}^S,\ c \mapsto \{\pi(c)\}_{\pi \in S}$$

is an embedding, then the replicas $f^{(\pi)}$ spann $(\bar{\mathbb{Q}}_\ell f)^{\mathrm{al}}$ as a $\bar{\mathbb{Q}}_\ell$-vector subspace of $\mathrm{End}(\bar{V})$.

0.2.1. Theorem. *Let us assume that f enjoys the following properties:* 1) $(\bar{\mathbb{Q}}_\ell f)^{\mathrm{al}}$ *is generated as a $\bar{\mathbb{Q}}_\ell$-vector space by* *operators of rank* 2 (*e.g.*, *there exists a set* S *of* $\mathbb{Q}$*-linear maps* $\pi\colon \mathbb{Q}(f) \to \mathbb{Q}$ *such that the product-map* $\mathbb{Q}(f) \to \mathbb{Q}^S$, $c \mapsto \{\pi(c)\}_{\pi\in S}$ *is an embedding and each replica* $f^{(\pi)}$ *is an operator of rank* 2 ($\pi \in S$));2)*there exists a non-degenerate skew-symmetric bilinear form* $(\ ,\ )\colon V \times V \to \mathbb{Q}_\ell$ *such that the Lie algebra* $\mathbb{Q}_\ell f$ *preserves* $(\ ,\ )$, *i.e.*, $(fx,y) + (x,fy) = 0$ for all $x,y \in V$.

Let us extend the form $(\ ,\ )$ *by* $\bar{\mathbb{Q}}_\ell$*-linearity to a non-degenerate skew-symmetric bilinear form* $(\ ,\ )\colon \bar{V} \times \bar{V} \to \bar{\mathbb{Q}}_\ell$.

Then there is a canonical f*-invariant orthogonal splitting*

$$\bar{V} = \bar{V}^f \oplus f\bar{V} \qquad \text{with } \bar{V}^f := \{x \in \bar{V} \mid fx = 0\},$$

and the restrictions of $(\ ,\ )$ *to* $\bar{V}^f$ *and* $f\bar{V}$ *are non-degenerate.*

There is also a canonical f*-invariant orthogonal splitting*

$$\bar{V} = \bar{V}^f \oplus (\oplus W_i) \qquad (1 \le i \le (\dim(\bar{V}) - \dim(\bar{V}^f))/2\ )$$

where

$$f\bar{V} = \oplus W_i \qquad (1 \le i \le (\dim(\bar{V}) - \dim(\bar{V}^f))/2\ ),$$

and all W_i *are* 2*-dimensional vector subspaces such that the restrictions of* $(\ ,\ )$ *to* W_i *are non-degenerate*.. *In addition, there are semisimple operators* $f_i \in \mathrm{End}(W_i)$ *of rank* 2 *and with trace zero such that*

$$(\bar{\mathbb{Q}}_\ell f)^{\mathrm{al}} = \oplus\bar{\mathbb{Q}}_\ell f_i \quad (1 \le i \le (\dim(\bar{V}) - \dim(\bar{V}^f))/2\ ).$$

Here each f_i *is viewed as an operator in* $\bar{V}$ *killing* V^f *and all* W_j *with* $j \ne i$. *In other words*, $(\bar{\mathbb{Q}}_\ell f)^{\mathrm{al}} \subset \mathrm{End}(f\bar{V})$ *is a Cartan subalgebra of the Lie algebra* $\mathfrak{sp}(f\bar{V})$ *of the symplectic group attached to the restriction of* $(\ ,\ )$ *to* $f\bar{V}$.

Remark. Clearly, the Lie algebra $(\bar{\mathbb{Q}}_\ell f)^{\mathrm{al}}$ preserves the form $(\ ,\ )$, i.e. $(ux,y)+(x,uy) = 0$ for all $u \in (\bar{\mathbb{Q}}_\ell f)^{\mathrm{al}}$; $x,y \in \bar{V}$.

We will prove the Theorem 0.2.1 in Sect. 0.3.

We will use the following easy corollary of the Theorem 0.2.1 describing skew-symmetric polylinear forms preserved by the Lie algebras $\bar{\mathbb{Q}}_\ell f$ and $\mathbb{Q}_\ell f$.

Corollary to Theorem 0.2.1. 0) *There is a canonical orthogonal* $(\mathbb{Q}_\ell f)^{\mathrm{al}}$*-invariant splitting* $V = \mathrm{Ker}(f) \oplus fV$, *and the restrictions of* (,) *to* $\mathrm{Ker}(f)$ *and* fV *are non-degenerate. The Lie algebra* $(\mathbb{Q}_\ell f)^{\mathrm{al}}$ *kills* $\mathrm{Ker}(f)$. *If one view* $(\mathbb{Q}_\ell f)^{\mathrm{al}}$ *as a Lie subalgebra of* $\mathrm{End}(V)$, *then* $(\mathbb{Q}_\ell f)^{\mathrm{al}}$ *is a Cartan subalgebra of the Lie algebra* $\mathbf{sp}(fV)$ *of the symplectic group attached to the restriction of* (,) *to* fV . **1**) *All* $\bar{\mathbb{Q}}_\ell f$*-invariant skew-symmetric polylinear forms on* $\bar{V}$ *of even degree are linear combinations of the exterior products of* $\bar{\mathbb{Q}}_\ell f$*-invariant skew-symmetric bilinear forms on* $\bar{V}$. *More precisely, they are linear combinations of the products of elements of* $\mathrm{Hom}(\Lambda^2 W_i, \bar{\mathbb{Q}}_\ell)$ and (*or*) $\mathrm{Hom}(\Lambda^2 \bar{V}^f, \bar{\mathbb{Q}}_\ell)$. 2) *All* $\mathbb{Q}_\ell f$*-invariant skew-symmetric polylinear forms on* V *are linear combinations of the exterior products of* $\mathbb{Q}_\ell f$*-invariant skew-symmetric bilinear forms on* V .

Indeed, it is clear that these forms are preserved by the Lie algebras $(\bar{\mathbb{Q}}_\ell f)^{\mathrm{al}}$ and $(\mathbb{Q}_\ell f)^{\mathrm{al}}$ respectively and the second assertion of the Corollary follows immediately from the first one. As for the assertion 0 , one has only to notice that

$$\bar{V}^f = \mathrm{Ker}(f)\otimes_{\mathbb{Q}_\ell}\bar{\mathbb{Q}}_\ell,\quad f\bar{V} = (fV)\otimes_{\mathbb{Q}_\ell}\bar{\mathbb{Q}}_\ell,\quad \mathbf{sp}(f\bar{V}) = \mathbf{sp}(fV)\otimes_{\mathbb{Q}_\ell}\bar{\mathbb{Q}}_\ell,$$
$$(\bar{\mathbb{Q}}_\ell f)^{\mathrm{al}} = (\mathbb{Q}_\ell f)^{\mathrm{al}}\otimes_{\mathbb{Q}_\ell}\bar{\mathbb{Q}}_\ell.$$

0.3. Proof of the Theorem 0.2.1. Clearly ,

$$\bar{V}^f = \{x \in \bar{V} \mid ex = 0 \text{ for all } u \in (\bar{\mathbb{Q}}_\ell f)^{\mathrm{al}}\}$$

and the restriction of (,) is non-degenerate (because f is semisimple). Replacing $\bar{V}$ by the orthogonal complement to $\bar{V}^f$ we may and will assume that $\bar{V}^f = \{0\}$, i.e. $f\bar{V} = \bar{V}$.

Since $(\bar{\mathbb{Q}}_\ell f)^{\mathrm{al}}$ preserves (,) , it lies in the Lie algebra $\mathbf{sp}(\bar{V})$ of the corresponding symplectic group of $\bar{V}$. Since $(\bar{\mathbb{Q}}_\ell f)^{\mathrm{al}}$ is commutative and consists of semisimple operators, there exists a Cartan subalgebra $\mathfrak{h}$ in $\mathrm{sp}(\bar{V})$ containing $(\bar{\mathbb{Q}}_\ell f)^{\mathrm{al}}$. It follows from the explicit description of Cartan subalgebras in $\mathbf{sp}$ ([2], Ch. 8, § 13) that there is a canonical $\mathfrak{h}$-invariant orthogonal splitting

$$\bar{V} = \bar{V}^f \oplus (\oplus W_i) \qquad (1 \le i \le \dim(\bar{V})/2)$$

where all W_i are 2-dimensional vector subspaces such that the restrictions of (,) to W_i are non-degenerate. In addition, there are semisimple operators $f_i \in \mathrm{End}(W_i)$ of rank 2 and with trace zero such that

$$\mathfrak{h} = \oplus\bar{\mathbb{Q}}_\ell f_i \qquad (1 \le i \le \dim(\bar{V})/2)$$

Here each f_i is viewed as an operator in $\bar{V}$ killing all W_j with $j \ne i$. In particular,

$$\dim(\mathfrak{h}) = \dim(\bar{V})/2.$$

So, in order to prove the theorem it is enough to check that $(\bar{\mathbb{Q}}_\ell f)^{\mathrm{al}} = \mathfrak{h}$. Since $(\bar{\mathbb{Q}}_\ell f)^{\mathrm{al}} \subset \mathfrak{h}$,

$$m := \dim\, (\bar{\mathbb{Q}}_\ell f)^{\mathrm{al}} \le \dim(\mathfrak{h}) = \dim(\bar{V})/2$$

and we have only to prove that $m \ge \dim(\bar{V})/2$. In order to do that, let us choose a basis $e_1, e_2, \ldots, e_m$ of $(\bar{\mathbb{Q}}_\ell f)^{\mathrm{al}}$ consisting of operators of rank 2. We have

$$\Sigma_{i=1}^m e_i\bar{V} = (\bar{\mathbb{Q}}_\ell f)^{\mathrm{al}}\bar{V} := \Sigma\, u\bar{V} \quad (u \in (\bar{\mathbb{Q}}_\ell f)^{\mathrm{al}}).$$

In particular,

$$\dim((\bar{\mathbb{Q}}_\ell f)^{\mathrm{al}}\bar{V}) \le \Sigma_{i=1}^m \dim(e_i\bar{V}) = 2m.$$

Since $f \in (\bar{\mathbb{Q}}_\ell f)^{\mathrm{al}}$,

$$\dim(f\bar{V}) \le \dim((\bar{\mathbb{Q}}_\ell f)^{\mathrm{al}}\bar{V}).$$

But $f\bar{V} = \bar{V}$, i.e. $\dim(f\bar{V}) = \dim(\bar{V})$. It follows that

$$\dim(\bar{V}) \le \dim((\bar{\mathbb{Q}}_\ell f)^{\mathrm{al}}\bar{V}) \le 2m,$$

i.e., $m \ge \dim(\bar{V})/2$. **QED.**

0.4. Let us assume that there is a structure of continuous $\mathrm{G}(k)$-module on V, i.e., there is an ℓ-adic representation

$$\rho: \mathrm{G}(k) \to \mathrm{Aut}(V).$$

Then we write $V(i)$ for $V \otimes \mathbb{Q}_\ell(i)$ (the tensor product is taken over $\mathbb{Q}_\ell$). For example $[\mathbb{Q}_\ell(j)](i) = \mathbb{Q}_\ell(i+j)$.

1. Abelian varieties. Let A be an Abelian variety over k and $\mathrm{End}_k A$ be the ring of all k-endomorphisms of A. If m is a positive integer prime to char k then we write A_m for the group of elements $a \in A(k(\mathrm{a}))$ such that $ma = 0$. It is well-known that A_m is a free $\mathbb{Z}/m\mathbb{Z}$-module of rank 2 dim(A). Let ℓ be a prime number different from char k. The Tate module $T_\ell(A)$ is defined as the projective limit of groups A_m where m runs through all powers of ℓ and the transition map is multiplication by ℓ. It is well-known that $T_\ell(A)$ is a free $\mathbb{Z}_\ell$-module of rank 2 dim(A). The group G(k) operates on $T_\ell(A)$ and there is a natural embedding

$$\mathrm{End}_k A\otimes\mathbb{Z}_\ell \to \mathrm{End}_{G(k)} T_\ell(A).$$

1.0. Let us put $V_\ell(A) := T_\ell(A)\otimes_{\mathbb{Z}_\ell}\mathbb{Q}_\ell$. Clearly, $V_\ell(A)$ is a $\mathbb{Q}_\ell$-vector space of dimension 2 dim(A) and $T_\ell(A)$ is a $\mathbb{Z}_\ell$-lattice in $V_\ell(A)$. It is also clear that $\mathrm{End}\, T_\ell(A) := \mathrm{End}_{\mathbb{Z}_\ell} T_\ell(A)$ is a $\mathbb{Z}_\ell$-lattice in $\mathrm{End}\, V_\ell(A) := \mathrm{End}_{\mathbb{Q}_\ell} V_\ell(A)$ .We also have a natural embedding

$$\mathrm{Aut}\, T_\ell(A) := \mathrm{Aut}_{\mathbb{Z}_\ell} T_\ell(A) \subset \mathrm{Aut}\, V_\ell(A) := \mathrm{Aut}_{\mathbb{Q}_\ell} V_\ell(A),$$

Let us denote by N the order of the finite general linear group $\mathrm{GL}(2\dim(A), \mathbb{Z}/\ell^2\mathbb{Z}) \approx \mathrm{Aut}(T_\ell(A)/\ell^2 T_\ell(A))$. One may easily check that

$$u^N \in \mathrm{id} + \ell^2\, \mathrm{End}\, T_\ell(A) \text{ for all } u \in \mathrm{Aut}\, T_\ell(A)$$

Clearly, $\mathrm{Aut}\, T_\ell(A)$ is a compact ℓ-adic Lie group and its Lie algebra coincides with $\mathrm{End}\, V_\ell(A)$. Therefore, we have everywhere defined logarithm map $\mathit{log}\colon \mathrm{Aut}\, T_\ell(A) \to \mathrm{End}\, V_\ell(A)$ attached to the compact ℓ-adic Lie group $\mathrm{Aut}\, T_\ell(A)$ [1]. We have

$$\mathit{log}(u^n) = n\, \mathit{log}(u) \text{ for all integers } n ,$$

$$\mathit{log}(a\, \mathrm{id}) = \log(a)\, \mathrm{id},\ \mathit{log}(au) = \log(a)\mathrm{id} + \mathit{log}(u)$$

for all $a \in \mathbb{Z}_\ell^*$. Here $\mathrm{id}\colon V_\ell(A) \to V_\ell(A)$ is the identity map. If $u \in \mathrm{id} + \ell^2\, \mathrm{End}\, T_\ell(A)$ then $u = \mathit{exp}(\mathit{log}(u))$ where exp is the ℓ-adic exponent. It follows that

$$u^N = (\mathit{exp}(N\, \mathit{log}(u))) \text{ for all } u \in \mathrm{Aut}\, T_\ell(A) .$$

This equality implies that if H is a bilinear form on $V_\ell(A)$ then

$$H(log(u)x,y) + H(x,log(u)y) = 0 \text{ for all } x,y \in V_\ell(A)$$

if and only if

$$H(u^N x,u^N y) = H(x,y) \text{ for all } x,y \in V_\ell(A) .$$

If $u \in \mathrm{Aut}\, T_\ell(A) \subset \mathrm{Aut}\, V_\ell(A)$ is a semisimple operator, then $log(u) \in End\, V_l(A)$ is also semisimple and its eigen values are just the ℓ-adic logarithms of the eigen values of u . For example,

$$log(a \text{ id}) = \log(a) \text{ id for all } a \in \mathbb{Z}_\ell^*.$$

More precisely, let us consider the 2 dim(A)-dimensional $\bar{\mathbb{Q}}_\ell$-vector space $\bar{V}_\ell(A) := V_\ell(A) \otimes_{\mathbb{Q}_\ell} \bar{\mathbb{Q}}_\ell$. One may view u and $log(u)$ as $\bar{\mathbb{Q}}_\ell$-linear operators in $\bar{V}_\ell(A)$. Let

$$\mathcal{P}_u(t) := \det(t \text{ id} - u,\ V_\ell(A)) \in \mathbb{Z}_\ell[t]$$

be the characteristic polynomial of u. Let $R(u) \subset \bar{\mathbb{Q}}_\ell$ be the set of roots of $\mathcal{P}_u$, i.e., the set of eigen values of u in $\bar{V}_\ell(A)$. Clearly, $R(u) \subset \bar{\mathbb{Z}}_\ell^*$. Then $R(log(u)) = \log(R(u))$ and the multiplicity of an eigen value b of $log(u)$ is equal to the sum of the multiplicities of all eigen values a of u with $\log(a)=b$.

One may easily check that the kernel of log consists of roots of id . More precisely, if $log(u)=0$ then $u^N = \text{id}$.

Now, let us give an explicit description of the kernel of $log(u)$ for semisimple $u \in \mathrm{Aut}\, T_\ell(A) \subset \mathrm{Aut}\, V_\ell(A)$ First, notice that if an eigen value α of a linear operator in $V_\ell(A)$ is a root of 1 then $\alpha^N = 1$. Let us put

$$V_\ell(A)^{(u)} := \{x \in V_\ell(A) \mid u^N x = x\} \subset V_\ell(A) .$$

Then one may easily check, using the description of the spectrum of $log(u)$ given above, that $V_\ell(A)^{(u)}$ is just the kernel of $log(u)$. There is a canonical u-invariant splitting

$$V_\ell(A) = V_\ell(A)^{(u)} \oplus V_\ell(A)_u \text{ where } V_\ell(A)_u := log(u)(V_\ell(A)) .$$

Clearly, $\log(u)(V_\ell(A)_u) = V_\ell(A)_u$.

1.1. Let $\Gamma(u)$ be the multiplicative subgroup in $\bar{\mathbb{Z}}_\ell^* \subset \bar{\mathbb{Q}}_\ell^*$

spanned by $R(u)$. It is a finitely generated commutative group. By definition, $R(u) \subset \Gamma(u)$. Since

$$\log(R(u)) = R(log(u)) \subset \mathbb{Q}(log(u)),$$

we obtain that $\log(\Gamma(u)) \subset \mathbb{Q}(log(u))$. Extending the map $\log: \Gamma(u) \to \mathbb{Q}(log(u))$ by $\mathbb{Q}$-linearity to $\Gamma(u)\otimes\mathbb{Q}$ we obtain the $\mathbb{Q}$-linear map $\log_{\mathbb{Q}}: \Gamma(u)\otimes\mathbb{Q} \to \mathbb{Q}(log(u))$ .Clearly, it is surjective and, since the kernel of the ℓ-adic logarithm consists of the roots of unity, we obtain easily that $\log_{\mathbb{Q}}$ is injective. It follows that $\log_{\mathbb{Q}}$ is an isomorphism of the $\mathbb{Q}$-vector spaces. This allows us to attach to any homomorphism $\phi: \Gamma(u) \to \bar{\mathbb{Q}}_\ell$ a replica $log(u)_\phi :=$ $(log(u))^{(\phi_{\mathbb{Q}} \log_{\mathbb{Q}}^{-1})}$ attached to the $\mathbb{Q}$-linear map

$$\phi \log_{\mathbb{Q}}^{-1}: \mathbb{Q}(log(u)) \to \Gamma(u)\otimes\mathbb{Q} \to \bar{\mathbb{Q}}_\ell .$$

Here we denote by $\phi_{\mathbb{Q}}$ the extension of ϕ by $\mathbb{Q}$-linearity to $\Gamma(u)\otimes\mathbb{Q}$. Clearly, $\operatorname{spec}(log(u)_\phi) = \phi(R(u))$ and the multiplicity of each eigen value b of the operator $log(u)_\phi$ is equal to the sum of multiplicities of $a \in R(u)$ with $\phi(a)=b$. This allows us to describe the algebraic envelope $(\mathbb{Q}_\ell log(u))^{\mathrm{al}} \subset \operatorname{End} V_\ell(A)$ in terms of eigen values of u . For example , the following assertion easily follows from constructions of Sect. 0.2.

1.1.1. Lemma. *Let us assume that there exist a finite set* $B \subset \mathbb{Q}$, *a positive integral-valued function* $\mathrm{mult}: B \to \mathbb{Z}_+$ *and a set* S *of homomorphisms* $\phi: \Gamma(u) \to \mathbb{Q}$, *enjoying the following properties*: 1) *the kernel of the product-map*

$$\Gamma(u) \to \mathbb{Q}^S,\ c \mapsto \{\phi(c)\}_{\pi\in S}$$

consists of roots of 1; 2) $\phi(R(u)) = B$ *and* $\mathrm{mult}(a)$ *is equal to the sum of multiplicities of roots* α *of* $\mathcal{P}_u$ *with* $\phi(u) = a$ *for all* $a \in B$ *and* $\phi \in S$.

Then: 1) $(\mathbb{Q}_\ell log(u))^{\mathrm{al}}\otimes_{\mathbb{Q}_\ell}\bar{\mathbb{Q}}_\ell$ is spanned as a $\bar{\mathbb{Q}}_\ell$-vector subspace of $\operatorname{End}(V_\ell(A))\otimes_{\mathbb{Q}_\ell}\bar{\mathbb{Q}}_\ell = \operatorname{End}(\bar{V}_\ell(A))$ by $u_\phi :=$ replicas of $log(u)$ attached to $\phi_{\mathbb{Q}}\ log_{\mathbb{Q}}^{-1}$; 2) $\operatorname{spec}(u_\phi) = B$ and the multiplicity of each eigen value a of u_ϕ *is equal to the sum of multiplicities of all roots* α *of* $\mathcal{P}_u$ *with* $\phi(u) = a$.

The map log is a local diffeomorphism in an open neighborhood of id . In particular, the image of exp contains an open neighborhood of the zero operator. Here exp is the ℓ-adic exponential map [1].

The action of $G(K)$ on $T_\ell(A)$ gives us a canonical ℓ-adic representation [8] ρ_ℓ: $G(k) \to \mathrm{Aut}\, T_\ell(A) \subset \mathrm{Aut}\, V_\ell(A)$ attached to A. Since $G(k)$ is compact, its image $\mathrm{Im}\,\rho_\ell$ is also compact and, therefore is a closed subgroup of $\mathrm{Aut}\, T_\ell(A)$. Therefore, it is a Lie subgroup of $\mathrm{Aut}\, T_\ell(A)$ and one may define the Lie algebra $\mathfrak{g}_\ell$ of $\mathrm{Im}\,\rho_\ell$ which is a $\mathbb{Q}_\ell$-Lie subalgebra of $\mathrm{End}\, V_\ell(A)$. It is well-known that $log(\mathrm{Im}\,\rho_\ell) \subset \mathfrak{g}_\ell$; in addition, $log(\mathrm{Im}\,\rho_\ell)$ contains an open neighborhood of zero in $\mathfrak{g}_\ell$. One may also easily check using elementary properties of the ℓ-adic exponential map that if $f \in \mathfrak{g}_\ell$ then there exists a sufficiently large positive integer M such that $\ell^M f \in log(\mathrm{Im}\,\rho_\ell)$.

1.2. Let us fix an ample invertible sheaf $\mathcal{L}$ on A. Then one may attach to $\mathcal{L}$ a non-degenerate skew-symmetric Galois-equivariant bilinear form [7] $H_{\mathcal{L}} : V_\ell(A) \times V_\ell(A) \to \mathbb{Q}_\ell(1)$. Here Galois-equivariance means that $H_{\mathcal{L}}(\sigma x, \sigma y) = \sigma(H_{\mathcal{L}}(x,y)) = \chi_\ell(\sigma) H_{\mathcal{L}}(x,y)$ for all $\sigma \in G(k)$; $x,y \in V_\ell(A)$ .This form is called the Riemann form attached to $\mathcal{L}$. Let us fix a non-canonical isomorphism of 1-dimensional $\mathbb{Q}_\ell$-vector spaces $\mathbb{Q}_\ell$ and $\mathbb{Q}_\ell(1)$. Then $H_{\mathcal{L}}$ becomes a non-degenerate skew-symmetric bilinear form

$$H_{\mathcal{L}}' : V_\ell(A) \times V_\ell(A) \to \mathbb{Q}_\ell$$

such that

$$H_{\mathcal{L}}'(\sigma x, \sigma y) = \chi_\ell(\sigma)\, H_{\mathcal{L}}'(x,y) \quad \text{for all } \sigma \in G(k);\ x,y \in V_\ell(A) .$$

Since $\sigma x = \rho_\ell(\sigma) x$ by the definition of ρ_ℓ, we obtain the existence of the continuous homomorphism $\chi_\ell' : \mathrm{Im}\,\rho_\ell \to \mathbb{Z}_\ell^*$ such that

$$\chi_\ell'(\rho_\ell(\sigma)) = \chi_\ell'(\sigma) \qquad \text{for all } \sigma \in G(k)$$

and

$$H_{\mathcal{L}}'(ux, uy) = \chi_\ell'(u)\, H_{\mathcal{L}}'(x,y) \quad \text{for all } u \in \mathrm{Im}\,\rho_\ell;\ x,y \in V_\ell(A) .$$

In particular, each $u \in \operatorname{Im} \rho_\ell$ is a symplectic similitude. It follows that if α is an eigen value of u then $\chi_\ell'(u)/\alpha$ is also an eigen value of u. In particular, if u is semisimple, then

$$\mathbb{Z}_\ell^* \supset \chi_\ell'(u) = \alpha\,(\chi_\ell'(u)/\alpha) \in \Gamma(u).$$

1.2.0. Let us put

$$u^\circ := \chi_\ell'(u)^{-1} u^2 \in \operatorname{Aut} T_\ell(A) \subset \operatorname{Aut} V_\ell(A)$$

One may easily check that

$$H_{\mathcal{L}}'(u^\circ x, u^\circ y) = H_{\mathcal{L}}'(x,y) \qquad \text{for all } u \in \operatorname{Im} \rho_\ell;\ x,y \in V_\ell(A).$$

It follows that

$$H_{\mathcal{L}}'(log(u^\circ)x, y) + H_{\mathcal{L}}'(x, log(u^\circ)y) = 0,$$

i.e., $log(u^\circ)$ lies in the Lie algebra $\mathfrak{sp}_{\mathcal{L}}$ of the symplectic group attached to $H_{\mathcal{L}}'$. Since $\mathfrak{sp}_{\mathcal{L}}$ is an algebraic Lie subalgebra of $\operatorname{End} V_\ell(A)$,

$$log(u^\circ) \in (\mathbb{Q}_\ell log(u^\circ))^{al} \subset \mathfrak{sp}_{\mathcal{L}} \subset \operatorname{End} V_\ell(A).$$

It follows from the definition of u° that

$$log(u^\circ) = 2\, log(u) - (\log(\chi_\ell'(u))\mathrm{id} \in \mathfrak{g}_\ell + \mathbb{Q}_\ell \mathrm{id}.$$

Clearly, $(u^i)^\circ = (u^\circ)^i$ for all integers i. It follows that $log((u^i)^\circ) = i\, log(u^\circ)$. One may also easily check that if H is a bilinear form on $V_\ell(A)$ then:

$$H(u^\circ x, u^\circ y) = H(x,y) \quad \text{for all } x,y \in V_\ell(A) \quad \text{if and only if}$$
$$H(u^2 x, u^2 y) = \chi_\ell'(u^2)\, H(x,y) \qquad \text{for all } x,y \in V_\ell(A).$$

Now, let us assume that u is semisimple. Then u° and $log(u^\circ)$ are also semisimple, and one may easily check that the restrictions of $H_{\mathcal{L}}'$ to $V_\ell(A)_{(u^\circ)} = \operatorname{Im}(log(u^\circ)$ and $V_\ell(A)^{(u^\circ)} = \operatorname{Ker}(log(u^\circ))$ are non-degenerate, and $V_\ell(A)_{(u^\circ)}$ is orthogonal to $V_\ell(A)^{(u^\circ)}$. We write $\mathfrak{sp}_{\mathcal{L},u} \subset \operatorname{End}(V_\ell(A)_{(u^\circ)})$ for the Lie algebra of the symplectic group attached to the restriction of $H_{\mathcal{L}}'$ on $V_\ell(A)_{(u^\circ)}$. If we view $\operatorname{End}(V_\ell(A)_{(u^\circ)})$ as a subalgebra of $\operatorname{End} V_\ell(A)$ killing $V_\ell(A)^{(u^\circ)}$, then we have embeddings

$$log(u^\circ) \in \mathfrak{sp}_{\mathcal{L},u} \subset \operatorname{End}(V_\ell(A)_{(u^\circ)}) \subset \operatorname{End} V_\ell(A).$$

Since $\mathfrak{sp}_{\mathcal{L},u}$ is an algebraic Lie subalgebra of $\operatorname{End} V_\ell(A)$, we obtain embeddings

$$(\mathbb{Q}_\ell log(u^\circ))^{al} \subset sp_{\mathcal{L},u} \subset End(V_\ell(A)_{(u^\circ)}) \subset End\ V_\ell(A) .$$

1.2.0.1. In order to describe the kernel and the image of $log(u^\circ)$ explicitly let us split the characteristic polynomial $\mathcal{P}_u$ into the product $\mathcal{P}_u = \mathcal{P}_{u,ss}\mathcal{P}_{u,0}$ of reduced polynomials $\mathcal{P}_{u,ss}$, $\mathcal{P}_{u,0} \in \mathbb{Z}_\ell[t]$ such that the set $R(\mathcal{P}_{u,ss}) \subset R(u)$ of roots of $\mathcal{P}_{u,ss}$ enjoys the following properties: a root $\alpha \in R(u)$ belongs to $R(\mathcal{P}_{u,ss})$ if and only if $\alpha^2/\chi_\ell'(u)$ is a root of 1 ; if $\beta \in R(u)$ is a root of $\mathcal{P}_{u,0}$ then $\beta^2/\chi_\ell'(u)$ is not a root of 1. There is a canonical u-invariant splitting

$$V_\ell(A) = V_\ell(A)^{ss} \oplus V_\ell(A)^{u,0}$$

such that the characteristic polynomials of the restrictions of u to $V_\ell(A)^{ss}$ and $V_\ell(A)^{u,0}$ coincide with $\mathcal{P}_{u,ss}$ and $\mathcal{P}_{u,0}$ respectively. Then (see 1.0)

$$V_\ell(A)^{(u^\circ)} = Ker(log(u^\circ)) = V_\ell(A)^{ss},$$
$$V_\ell(A)_{(u^\circ)} = Im\ (log(u^\circ)) = V_\ell(A)^{u,0}.$$

1.2.1. Theorem. *Let us assume that u is semisimple and the algebraic envelope* $(\mathbb{Q}_\ell log(u^\circ))^{al} \otimes_{\mathbb{Q}_\ell} \bar{\mathbb{Q}}_\ell$ is spanned as a $\bar{\mathbb{Q}}_\ell$-vector subspace of $End(V_\ell(A)) \otimes_{\mathbb{Q}_\ell} \bar{\mathbb{Q}}_\ell = End(\bar{V}_\ell(A))$ *by operators of rank* 2. *Then:* 1) *the algebraic envelope* $(\mathbb{Q}_\ell log(u^\circ))^{al}$ *of the Lie algebra* $\mathbb{Q}_\ell log(u^\circ)$ *is a Cartan subalgebra of* $sp_{\mathcal{L},u}$; 2) *All* $\mathbb{Q}_\ell log(u^\circ)$*-invariant skew-symmetric polylinear forms of even degree on* $V_\ell(A)$ *are linear combinations of the exterior products of* $\mathbb{Q}_\ell log(u^\circ)$*-invariant skew-symmetric bilinear forms on* $V_\ell(A)$.

Proof. One has only to apply the Corollary of the Theorem 0.2.1 to $V = V_\ell(A)$ and $f = log(u^\circ)$. **QED.**

Clearly, $R(u^\circ) = \{\chi_\ell'(u)^{-1}\alpha^2 | \alpha \in R(u)\} \subset \Gamma(u)$. In particular, $\Gamma(u^\circ) \subset \Gamma(u)$. It follows that if $\phi : \Gamma(u) \to \mathbb{Q}$ is a homomorphism , then its restriction to $\Gamma(u^\circ)$ defines a homomorphism $\phi^\circ : \Gamma(u^\circ) \to \mathbb{Q}$ such that
$\phi^\circ(R(u^\circ)) = 2\phi(R(u)) - \log(\chi_\ell'(u)) := \{2a - \log(\chi_\ell'(u)) | a \in \phi(R(u)\}$,

and {the sum of multiplicities of roots α of $\mathcal{P}_u$ with $\phi(\alpha) = b$} =
{the sum of multiplicities of roots α° of $\mathcal{P}_{u^\circ}$ with $\phi^\circ(\alpha) = 2b - \log(\chi_\ell'(u))$}.

1.2.2. Lemma. *Let us assume that there exist a finite set $B \subset \mathbb{Q}$, a positive integral-valued function* mult: $B \to \mathbb{Z}_+$, *a set S of homomorphisms ϕ: $\Gamma(u) \to \mathbb{Q}$ and a rational number $c \in \mathbb{Q}$, enjoying the following properties:* 0) $\phi(\chi_\ell'(u)) = c$ *for all* $\phi \in S$; 1) *the kernel of the product-map* $\Gamma(u) \to \mathbb{Q}^S$, $c \mapsto \{\phi(c)\}_{\pi\in S}$ *consists of roots of* 1; 2) $\phi(R(u)) = B$ *and* mult(a) *is equal to the sum of multiplicities of roots α of $\mathcal{P}_u$ with $\phi(u) = a$ for all $a \in B$ and $\phi \in S$.*

Then: 1) $(\mathbb{Q}_\ell log(u^\circ))^{\mathrm{al}} \otimes_{\mathbb{Q}_\ell} \bar{\mathbb{Q}}_\ell$ *is spanned as a $\bar{\mathbb{Q}}_\ell$-vector subspace of* $\mathrm{End}(V_\ell(A)) \otimes_{\mathbb{Q}_\ell} \bar{\mathbb{Q}}_\ell = \mathrm{End}(\bar{V}_\ell(A))$ *by* $u^\circ_\phi :=$ *replicas of* $log(u^\circ)$ *attached to* $\phi^\circ_{\mathbb{Q}}\, log_{\mathbb{Q}}^{-1}$; 2) $\mathrm{spec}(u^\circ_\phi) = B^\circ := 2B - c = \{2b - c \mid b \in B\}$ *and the multiplicity of each eigen value* $2b-c$ of u°_ϕ *is equal to the sum of multiplicities of all roots α of $\mathcal{P}_u$ with $\phi(u) = b$.*

Proof. One has only to apply the Lemma 1.2.1 to u° and the set S° of homomorphisms $\phi^\circ : \Gamma(u^\circ) \to \mathbb{Q}$ $(\phi \in S)$.

1.2.3. Corollary. *Let us assume that there exist a set S of homomorphisms ϕ: $\Gamma(u) \to \mathbb{Q}$ and a non-zero rational number $c \in \mathbb{Q}$, enjoying the following properties:* 0) $\phi(\chi_\ell'(u)) = c$ *for all* $\phi \in S$; 1) *the kernel of the product-map* $\Gamma(u) \to \mathbb{Q}^S$, $c \mapsto \{\phi(c)\}_{\pi\in S}$ *consists of roots of* 1; 2) *if $\phi \in S$ then $\phi(R(u)) = \{0,c\}$ or $\{0,c/2, c\}$ and in the both cases the following conditions hold:* a) *there exists precisely one root α of $\mathcal{P}_u$ such that $\phi(u) = 0$; in addition, the multiplicity of α is equal to* 1; b) *there exists precisely one root β of $\mathcal{P}_u$ such that $\phi(u) = c$; in addition, the multiplicity of β is equal to* 1.

Then: 0)*the algebraic envelope* $(\mathbb{Q}_\ell log(u^\circ))^{al} \otimes_{\mathbb{Q}_\ell} \bar{\mathbb{Q}}_\ell$ *is spanned as a* $\bar{\mathbb{Q}}_\ell$*-vector subspace of* $\mathrm{End}(V_\ell(A)) \otimes_{\mathbb{Q}_\ell} \bar{\mathbb{Q}}_\ell = \mathrm{End}(\bar{V}_\ell(A))$ *by operators of rank* 2; 1) *the algebraic envelope* $(\mathbb{Q}_\ell log(u^\circ))^{al}$ *of the Lie algebra* $\mathbb{Q}_\ell log(u^\circ)$ *is a Cartan subalgebra of* $\mathbf{sp}_{\mathcal{L},u}$; 2) *All* $\mathbb{Q}_\ell log(u^\circ)$*-invariant skew-symmetric polylinear forms on* $V_\ell(A)$ *are linear combinations of the exterior products of* $\mathbb{Q}_\ell log(u^\circ)$*-invariant skew-symmetric bilinear forms* H *on* $V_\ell(A)$ *such that*

$$H((u^\circ)^N x, (u^\circ)^N y) = H(x,y), \qquad H(u^{2N}x, u^{2N}y) = \chi_\ell'(u^{2N})H(x,y)$$

$$\text{for all} \qquad x, y \in V_\ell(A).$$

Here N *is the order of* $\mathrm{GL}(2\dim(A), \mathbb{Z}/\ell^2\mathbb{Z})$.

Proof. Applying the Lemma 1.2.2, we obtain that

$$\mathrm{spec}(u^\circ_\phi) = \{-c, c\} \text{ or } \{0, -c, c\}$$

and the both eigen values c and $(-c)$ have multiplicity 1 for all the replicas u°_ϕ $(\phi \in S)$. This implies that all the u°_ϕ are operators of rank 2 and, therefore, $(\mathbb{Q}_\ell log(u^\circ))^{al} \otimes_{\mathbb{Q}_\ell} \bar{\mathbb{Q}}_\ell$ is spanned by operators of rank 2 (by lemma 1.2.2). One has only to apply the Theorem 1.2.1 and recall that if a bilinear form H is killed by $\mathbb{Q}_\ell log(u^\circ)$ then it is $(u^\circ)^N = (u^N)^\circ$-invariant (1.0), and if it is $(u^N)^\circ$-invariant, then it is $(u^N)^2 = u^{2N}$-semiinvariant with coefficient $\chi_\ell(u^{2N})$ (see 1.2.0.).

1.3. Let us put $A_a := A \times k(a)$. There are canonical isomorphisms of $G(k)$-modules (compatible with "the" products)

$$H^j(A_a, \mathbb{Q}_\ell) = \mathrm{Hom}(\Lambda^j V_\ell(A), \mathbb{Q}_\ell)$$

identifying the jth ℓ-adic étale cohomology groups $H^j(A_a, \mathbb{Q}_\ell)$ of A_a and the spaces $\mathrm{Hom}(\Lambda^j V_\ell(A), \mathbb{Q}_\ell)$ of skew-symmetric j-linear forms on $V_\ell(A)$ with values in $\mathbb{Q}_\ell$. We will mainly interested in the twisted ℓ-adic cohomology groups

$$H^{2i}(A_a, \mathbb{Q}_\ell)(i) = \mathrm{Hom}(\Lambda^{2i} V_\ell(A), \mathbb{Q}_\ell)(i) = \mathrm{Hom}(\Lambda^{2i} V_\ell(A), \mathbb{Q}_\ell(i)).$$

Let us fix a non-canonical isomorphism $\mathbb{Q}_\ell(1) \approx \mathbb{Q}_\ell$. Then one may identify $H^{2i}(A_a, \mathbb{Q}_\ell)(i)$ with the space $\mathrm{Hom}(\Lambda^{2i} V_\ell(A), \mathbb{Q}_\ell)$ of all skew-symmetric $2i$-linear forms E on $V_\ell(A)$ taking values in $\mathbb{Q}_\ell$, and the Galois action $E \mapsto \sigma E$ is defined by the formula

$$(\sigma E)(x_1, \ldots, x_{2i}) := \chi_\ell(\sigma)^i E(\rho_\ell(\sigma)^{-1} x_1, \ldots, \rho_\ell(\sigma)^{-1} x_{2i})$$
$$\text{for all } \sigma \in G(k);\ x_1, \ldots, x_{2i} \in V_\ell(A).$$

Under this identification the Galois-invariant part $H^{2i}(A_a, \mathbb{Q}_\ell)(i)^{G(k)}$ admits the following description. The subspace

$$H^{2i}(A_a, \mathbb{Q}_\ell)(i)^{G(k)} \subset H^{2i}(A_a, \mathbb{Q}_\ell)(i) = \mathrm{Hom}(\Lambda^{2i} V_\ell(A), \mathbb{Q}_\ell)$$

coincides with the subspace of all skew-symmetric $2i$-linear forms E on $V_\ell(A)$ taking values in $\mathbb{Q}_\ell$ and such that

$$E(\rho_\ell(\sigma)x_1, \ldots, \rho_\ell(\sigma)x_{2i}) = \chi_\ell(\sigma)^i E(x_1, \ldots, x_{2i})$$
$$\text{for all } \sigma \in G(k);\ x_1, \ldots, x_{2i} \in V_\ell(A).$$

Here is an equivalent description:

$$H^{2i}(A_a, \mathbb{Q}_\ell)(i)^{G(k)} \subset \mathrm{Hom}(\Lambda^{2i} V_\ell(A), \mathbb{Q}_\ell)$$

is the subspace of all skew-symmetric $2i$-linear forms E on $V_\ell(A)$ such that

$$E(ux_1, \ldots, ux_{2i}) = \chi_\ell'(u)^i E(x_1, \ldots, x_{2i})$$
$$\text{for all } u \in \mathrm{Im}\, \rho_\ell;\ x_1, \ldots, x_{2i} \in V_\ell(A).$$

Example. One may easily check that the bilinear form $H_{\mathcal{L}}' \in \mathrm{Hom}(\Lambda^2 V_\ell(A), \mathbb{Q}_\ell)$ belongs to $H^2(A_a, \mathbb{Q}_\ell)(1)^{G(k)}$.

It easily follows that if

$$E \in H^{2i}(A_a, \mathbb{Q}_\ell)(i)^{G(k)} \subset \mathrm{Hom}(\Lambda^{2i} V_\ell(A), \mathbb{Q}_\ell)$$

then

$$E(u^\circ x_1, \ldots, u^\circ x_{2i}) = E(x_1, \ldots, x_{2i})$$
$$\text{for all } u \in \mathrm{Im}\, \rho_\ell;\ x_1, \ldots, x_{2i} \in V_\ell(A).$$

This implies that the Lie algebra $\mathbb{Q}_\ell \log(u) \subset \mathrm{End}\, V_\ell(A)$ and, therefore, its algebraic envelope $(\mathbb{Q}_\ell \log(u^\circ))^{al}$ kills all the Galois-invariant forms E.

1.3.1. Theorem. *Let us assume that u is semisimple and the algebraic envelope $(\mathbb{Q}_\ell \log(u^\circ))^{al} \otimes_{\mathbb{Q}_\ell} \bar{\mathbb{Q}}_\ell$ is spanned as a $\bar{\mathbb{Q}}_\ell$-vector subspace of* $\mathrm{End}(V_\ell(A)) \otimes_{\mathbb{Q}_\ell} \bar{\mathbb{Q}}_\ell = \mathrm{End}(\bar{V}_\ell(A))$ *by operators of rank* 2. *Then:* 1) *the algebraic envelope $(\mathbb{Q}_\ell \log(u^\circ))^{al}$ of the Lie algebra*

$\mathbb{Q}_\ell log(u^\circ)$ is a Cartan subalgebra of $sp_{\mathcal{L},u}$; 2) All G(k)-invariant skew-symmetric polylinear forms

$$E \in H^{2i}(A_a, \mathbb{Q}_\ell)(i)^{G(k)} \subset \mathrm{Hom}(\Lambda^{2i} V_\ell(A), \mathbb{Q}_\ell)$$

of even degree on $V_\ell(A)$ are linear combinations of the products of $\mathbb{Q}_\ell log(u^\circ)$-invariant skew-symmetric bilinear forms

$$H \in H^2(A_a, \mathbb{Q}_\ell)(1) = \mathrm{Hom}(\Lambda^2 V_\ell(A), \mathbb{Q}_\ell)$$

on $V_\ell(A)$ such that

$$H((u^\circ)^N x, (u^\circ)^N y) = H(x,y), \qquad H(u^{2N}x, u^{2N}y) = \chi_\ell'(u^{2N})H(x,y)$$

for all $x, y \in V_\ell(A)$.

Here N is the order of $GL(2\dim(A), \mathbb{Z}/\ell^2\mathbb{Z})$.

Proof. Let us recall that $\mathbb{Q}_\ell log(u^\circ)$ preserves E and apply the Theorem 1.2.1 . Now, one has only to recall that if a bilinear form H is killed by $\mathbb{Q}_\ell log(u^\circ)$, then it is $(u^\circ)^N = (u^N)^\circ$-invariant (1.0), and, therefore, is $(u^N)^2 = u^{2N}$-semiinvariant with coefficient $\chi_\ell(u^{2N})$ (1.2.0.).

2. Abelian varieties over finite fields. Let k be be a finite field of characteristic p. We write q for the number of elements of k; clearly, q is a power of the prime p. It is well-known that the Galois group G(k) is procyclic and its generator is the Frobenius automorphism $\sigma_k: k(a) \to k(a), x \mapsto x^q$. One may easily check that

$$\chi_\ell(\sigma_k) = q \in \mathbb{Z}_\ell^*.$$

If k' is a finite algebraic extension of k then k' is also finite and

$$\sigma_{k'} = \sigma_k^{[k':k]} \in G(k') \subset G(k) .$$

Let A be an Abelian variety over k,

$$\mathrm{Fr} := \rho_\ell(\sigma_k) \in \mathrm{Im}\, \rho_\ell \subset \mathrm{Aut}\, T_\ell(A) \subset \mathrm{Aut}\, V_\ell(A) \quad (\ell \neq p).$$

One may easily check that $log(\mathrm{Im}\, \rho_\ell) = \mathbb{Z}_\ell\, log(\mathrm{Fr})$, because the set $\{\sigma_k^n | n \in \mathbb{Z}\}$ is everywhere dense in G(k) . It follows that

$$\mathfrak{g}_\ell = \mathbb{Q}_\ell\, log(\mathrm{Fr}) \subset \mathrm{End}\, V_\ell(A) .$$

Since $\chi_\ell(\sigma_k) = q$ and $\mathrm{Fr} = \rho_\ell(\sigma_k)$, $\chi_\ell'(\mathrm{Fr}) = q$ In particular, $\mathrm{Fr}^\circ = \chi_\ell'(\mathrm{Fr})^{-1}\mathrm{Fr}^2 = q^{-1}\mathrm{Fr}^2$. Let

$$\mathcal{P}_{\mathrm{Fr}}(t) = t^{2\dim(A)} + a_1 t^{2\dim(A)-1} + \ldots + a_{2\dim(A)} \in \mathbb{Z}_\ell[t]$$

be the characteristic polynomial of Fr . By a well-known theorem of A. Weil [7] Fr is semisimple, $\mathcal{P}_{\mathrm{Fr}}$ lies in $\mathbb{Z}[t]$ and does not depend on the choice of ℓ. For example, $a_{2\dim(A)} = q^{2\dim(A)}$.
The set $R(\mathrm{Fr})$ of the roots of $\mathcal{P}_{\mathrm{Fr}}$ consists of *algebraic integers* and if we denote by L the decomposition field of $\mathcal{P}_{\mathrm{Fr}}$, i.e., the field obtained by adjoining to $\mathbb{Q}$ the set $R(\mathrm{Fr})$, then the following conditions hold (the continuation of the theorem of Weil):

. $R(\mathrm{Fr}) \subset L^*$ and the map $\alpha \mapsto q/\alpha$ is a permutation of $R(\mathrm{Fr})$. In addition, the multiplicity of roots α and q/α of $\mathcal{P}_{\mathrm{Fr}}$ coincide.

We have $|\alpha| = q^{1/2}$ for all embeddings of L into $\mathbb{C}$ and for all $\alpha \in R(\mathrm{Fr})$. In other words, all archimedean valuations on L take the exactly one value (namely, $q^{1/2}$) on the set of all roots of $\mathcal{P}_{\mathrm{Fr}}$.

Notice, that $\Gamma(\mathrm{Fr})$ becomes a multiplicative subgroup in L^*. It follows that if $\gamma \in \Gamma(\mathrm{Fr})$ then $|\gamma|$ does not depend on the choice of an archimedean valuation $|\ |$ on L , because $\Gamma(\mathrm{Fr})$ is , by definition, the multiplicative subgroup generated by $R(\mathrm{Fr})$.

Let O_L be the ring of all algebraic integers in L. Clearly, $R(\mathrm{Fr}) \subset O_L$. Therefore, we have $\alpha \in O_L$ and $q/\alpha \in O_L$ for all $\alpha \in R(\mathrm{Fr})$. Since q is a power of p, we obtain that if $\mathfrak{B}$ is a maximal ideal in O_L not lying over p, then

$$\mathrm{ord}_{\mathfrak{B}}(\alpha) = 0 \text{ for all } \alpha \in R(\mathrm{Fr})$$

where ord: $L^* \to \mathbb{Z}$ is a discrete valuation attached to $\mathfrak{B}$. It follows that $\mathrm{ord}_{\mathfrak{B}}(\gamma) = 0$ for all $\gamma \in \Gamma(\mathrm{Fr})$, because $\Gamma(\mathrm{Fr})$ is , by definition, the multiplicative subgroup generated by $R(\mathrm{Fr})$.

We denote by S the set of of prime ideals $\mathfrak{P}$ in O_L, which lie over p. If $\mathfrak{P} \in S$ then we write

$$\mathrm{ord}_{\mathfrak{P}} : L^* \to \mathbb{Q}$$

for the discrete valuation attached to $\mathfrak{P}$ and normalized by the condition

$$\mathrm{ord}_{\mathfrak{P}}(q) = 1.$$

2.1. Proposition. *Let us define a homomorphism*

$$ord : \Gamma(\mathrm{Fr}) \to \mathbb{Q}^S$$

by the formula $\quad ord(\gamma) = \{\mathrm{ord}_{\mathfrak{P}}(\gamma)\}_{\mathfrak{P}\in S}$.

Then the kernel of ord consists of the roots of 1.

Proof. If γ lies in the kernel of *ord* then γ is a $\mathfrak{B}$-adic unit for *all* maximal ideals $\mathfrak{B}$ in O_L. This means that γ is an unit, i.e. $\gamma \in O_L^*$. Recall that $|\gamma|$ does not depend on the choice of an archimedean valuation $|\ |$ on L. Therefore, the product formula for L implies that $|\gamma| = 1$ for *all* archimedean valuations $|\ |$ on L. This shows that γ is a root of 1 (cf. for instance, Th. 8 of Chap. IV, §4 in [14]). (Compare with [16].)

2.1.1. Remark. Let us choose an archimedean valuation $|\ |$ on L. Since $|\alpha| = q^{1/2}$ for each $\alpha \in R(\mathrm{Fr})$ and $R(\mathrm{Fr})$ generates the multiplicative group $\Gamma(\mathrm{Fr})$, there is a canonical homomorphism $\pi : \Gamma(\mathrm{Fr}) \to \frac{1}{2}\mathbb{Z} \subset \mathbb{Q}$ defined by the formula

$$|\gamma| = q^{\pi(\gamma)} \quad \text{for all } \gamma \in \Gamma(\mathrm{Fr}) .$$

By the definition $\pi(R(\mathrm{Fr})) = \{1/2\}$ and, therefore, the corresponding replica of $log(\mathrm{Fr})$ attached to the homomorphism

$$\pi_{\mathbb{Q}}\log_{\mathbb{Q}}^{-1} : \mathbb{Q}(log(\mathrm{Fr})) \to \Gamma(\mathrm{Fr})\otimes\mathbb{Q} \to \mathbb{Q}$$

is equal to $(1/2)\mathrm{id}$. This implies that the algebraic envelope $(\mathbb{Q}_\ell log(\mathrm{Fr}))^{\mathrm{al}}\otimes_{\mathbb{Q}_\ell}\bar{\mathbb{Q}}_\ell$ contains $\bar{\mathbb{Q}}_\ell\mathrm{id}$. It follows that $(\mathbb{Q}_\ell log(\mathrm{Fr}))^{\mathrm{al}}$ contains $\mathbb{Q}_\ell\mathrm{id}$ (Deligne[9]).

2.2. Now, we are ready to recall the definition of the Newton polygon of A. It consists of a finite set $\mathrm{Slp}_A \subset \mathbb{Q}$ named the set of slopes and a positive integral-valued function $\mathrm{length}_A : \mathrm{Slp}_A \to \mathbb{Z}_+$ which assigns to each slope its length. If $\mathfrak{P} \in S$ then we write $\mathrm{Slp}_A := \mathrm{ord}_{\mathfrak{P}}(R(\mathrm{Fr})) \subset \mathbb{Q}$ and for each $c \in \mathrm{Slp}_A$ $\mathrm{length}_A(c) = \{$the number of roots α of $\mathcal{P}_{\mathrm{Fr}}$ (with multiplicities) such that $\mathrm{ord}_{\mathfrak{P}}(\alpha) = c$ $\}$.

Since L is a Galois extension of $\mathbb{Q}$, the Galois group $\mathrm{Gal}(L/\mathbb{Q})$ of this extension acts transitively on S . This implies that the definition of Slp_A and length_A does not depend on the choice of $\mathfrak{P}$.

because $R(\mathrm{Fr})$ is the set of roots of $\mathcal{P}_{\mathrm{Fr}}$ and the polynomial $\mathcal{P}_{\mathrm{Fr}}$ has rational coefficients (i.e., $R(\mathrm{Fr})$ is $\mathrm{Gal}(L/\mathbb{Q})$-invariant). The geometric definition of the Newton polygon used the "visible part" of the convex hull of the set

$$\{(i, \mathrm{ord}_{\mathfrak{P}}(a_i)) \mid 1 \le i \le 2\dim(A),\ a_i \ne 0\}.$$

(See [4], [6]). This "visible part" is the graph of the real-valued continuous piece-wise linear function on $[0, 2\dim(A)]$ which takes the value 0 at 0 and whose derivatives (slopes) constitute the set Slp_A

Since all elements of $R(u)$ are algebraic integers, all slopes are non-negative rational numbers. Recall that if $\alpha \in R(\mathrm{Fr})$ then $q/\alpha \in R(\mathrm{Fr})$. Since $\mathrm{ord}_{\mathfrak{P}}(q) = 1$ we obtain that if $b \in \mathrm{Slp}_A$ then

$$0 \le b \le 1\ ,\ 1-b \in \mathrm{Slp}_A \text{ and } \mathrm{length}_A(b) = \mathrm{length}_A(1-b)\ .$$

If k' is a finite algebraic extension of k and $A' = A \times k'$ then $\mathrm{Slp}_{A'} = \mathrm{Slp}_A$ and $\mathrm{length}_{A'} = \mathrm{length}_A$. If X is an Abelian variety isogenous to A then $\mathrm{Slp}_X = \mathrm{Slp}_A$ and $\mathrm{length}_X = \mathrm{length}_A$. If $A = A_1 \times A_2$ is a product of Abelian varieties A_1 and A_2 then $\mathrm{Slp}_A = \mathrm{Slp}_{A_1} \cup \mathrm{Slp}_{A_2}$ and $\mathrm{length}_A(b) = \mathrm{length}_{A_1}(b) + \mathrm{length}_{A_2}(b)$ (here we assume $\mathrm{length}_{A_1}(b) = 0$ if $b \notin \mathrm{Slp}_{A_1}$, $\mathrm{length}_{A_2}(b) = 0$ if $b \notin \mathrm{Slp}_{A_2}$).

Applying Lemma 1.1.1 to $u = \mathrm{Fr}$, $B = \mathrm{Slp}_A$ and mult $= \mathrm{length}_A$ and using Proposition 2.1 and the equality $\mathfrak{g}_\ell = \mathbb{Q}_\ell \log(\mathrm{Fr})$, we obtain the following assertion (compare with [10]).

2.3. Theorem. 1) *The algebraic envelope*

$$\bar{\mathfrak{g}}_\ell^{\mathrm{al}} := \mathfrak{g}_\ell^{\mathrm{al}} \otimes_{\mathbb{Q}_\ell} \bar{\mathbb{Q}}_\ell = (\mathbb{Q}_\ell \log(\mathrm{Fr}))^{\mathrm{al}} \otimes_{\mathbb{Q}_\ell} \bar{\mathbb{Q}}_\ell \text{ of } \bar{\mathfrak{g}}_\ell := (\mathfrak{g}_\ell \otimes_{\mathbb{Q}_\ell} \bar{\mathbb{Q}}_\ell)$$

is spanned as a $\bar{\mathbb{Q}}_\ell$*-vector subspace of*

$$\mathrm{End}(V_\ell(A)) \otimes_{\mathbb{Q}_\ell} \bar{\mathbb{Q}}_\ell = \mathrm{End}(\bar{V}_\ell(A))$$

by $\mathrm{fr}_{\mathfrak{P}}$:= *replicas of* $\log(\mathrm{Fr})$ *attached to* $(\mathrm{ord}_{\mathfrak{P}})_{\mathbb{Q}} \log_{\mathbb{Q}}^{-1}$ $(\mathfrak{P} \in S)$;

2) $\mathrm{spec}(\mathrm{fr}_{\mathfrak{P}}) = \mathrm{Slp}_A$ *and the multiplicity of each eigen value* b *of* $\mathrm{fr}_{\mathfrak{P}}$ *is equal to* $\mathrm{length}_A(b)$.

2.3.1. Recall (2.1.1) that $\mathfrak{g}_\ell^{\mathrm{al}} = (\mathbb{Q}_\ell \log(\mathrm{Fr}))^{\mathrm{al}}$ contains homotheties $\mathbb{Q}_\ell \mathrm{id}$. It follows that

$$log(\mathrm{Fr}^\circ) \in \mathfrak{g}_\ell + \mathbb{Q}_\ell \mathrm{id} \subset \mathfrak{g}_\ell^{\mathrm{al}} + \mathbb{Q}_\ell \mathrm{id} = \mathfrak{g}_\ell^{\mathrm{al}}.$$

On the other hand,

$$log(\mathrm{Fr}) = (log(\mathrm{Fr}^\circ) + \log(\chi_\ell'(\mathrm{Fr}))\mathrm{id})/2 \in \mathbb{Q}_\ell log(\mathrm{Fr}^\circ) + \mathbb{Q}_\ell \mathrm{id} .$$

It follows that $\mathfrak{g}_\ell^{\mathrm{al}} = \mathbb{Q}_\ell \mathrm{id} + (\mathbb{Q}_\ell log(\mathrm{Fr}^\circ))^{\mathrm{al}}$.
Since $(\mathbb{Q}_\ell log(\mathrm{Fr}^\circ))^{\mathrm{al}}$ lies in the Lie algebra $\mathfrak{sp}_{\mathcal{E}}$ of the symplectic group of $V_\ell(A)$ (Remark 1.2.3),

$$\mathfrak{g}_\ell^{\mathrm{al}} = \mathbb{Q}_\ell \mathrm{id} \oplus (\mathbb{Q}_\ell log(\mathrm{Fr}^\circ))^{\mathrm{al}} .$$

2.4. In order to obtain an analog of the Theorem 2.3 for $(\mathbb{Q}_\ell log(\mathrm{Fr}^\circ))^{\mathrm{al}}$, notice, that

$$c := \mathrm{ord}_{\mathfrak{P}}(\chi_\ell'(\mathrm{Fr})) = \mathrm{ord}_{\mathfrak{P}}(q) = 1 \text{ for all } \mathfrak{P} \in S .$$

Now, we are ready to apply Lemma 1.2.1 to $u = \mathrm{Fr}$, $B = \mathrm{Slp}_A$, $c = 1$ and mult = length_A. Using Proposition 2.1, we obtain the following assertion.

2.5. Theorem. 1) *The algebraic envelope*

$$(\mathbb{Q}_\ell log(\mathrm{Fr}^\circ))^{\mathrm{al}} \otimes_{\mathbb{Q}_\ell} \bar{\mathbb{Q}}_\ell \text{ of } (\mathbb{Q}_\ell log(\mathrm{Fr}^\circ)) \otimes_{\mathbb{Q}_\ell} \bar{\mathbb{Q}}_\ell$$

is spanned as a $\bar{\mathbb{Q}}_\ell$*-vector subspace of*

$$\mathrm{End}(V_\ell(A)) \otimes_{\mathbb{Q}_\ell} \bar{\mathbb{Q}}_\ell = \mathrm{End}(\bar{V}_\ell(A))$$

by $\mathrm{fr}_{\mathfrak{P}}^\circ$:= *replicas of* $log(\mathrm{Fr}^\circ)$ *attached to* $((\mathrm{ord}_{\mathfrak{P}})^\circ)_{\mathbb{Q}}\ log_{\mathbb{Q}}^{-1}$ ($\mathfrak{P} \in S$);

2) $\mathrm{spec}(\mathrm{fr}_{\mathfrak{P}}^\circ) = 2\ \mathrm{Slp}_A - 1 = \{2b-1 \mid b \in \mathrm{Slp}_A\}$ *and the multiplicity of each eigen value* $2b-1$ of $\mathrm{fr}_{\mathfrak{P}}^\circ$ *is equal to* $\mathrm{length}_A(b)$.

2.6. Definition. *An Abelian variety* A *is called supersingular if* $\mathrm{Slp}_A = \{1/2\}$.

It follows from the Proposition 2.1 that A is supersingular if and only if α^2/q is a root of 1 for all $\alpha \in R(\mathrm{Fr})$. It follows from the Theorem 2.5 that A is supersingular if and only if $log(\mathrm{Fr}^\circ) = 0$.

2.6.1. Lemma. *Let us assume that* $A = A^{\mathrm{ss}} \times A_0$ *is a product of a supersingular Abelian variety* A^{ss} *and an Abelian variety* A_0 *such that* A_0 *does not contain non-zero supersingular Abelian subvarieties. (Each Abelian variety over a finite field is isogenous to such a product.) Then* $V_\ell(A) = V_\ell(A^{\mathrm{ss}}) \oplus V_\ell(A_0)$,

$$\mathrm{Ker}(log(\mathrm{Fr}^\circ)) = \mathrm{V}_\ell(A^{ss}) \text{ and } \mathrm{V}_\ell(A_0) = log(\mathrm{Fr}^\circ)\mathrm{V}_\ell(A) .$$

In particular, if A is a simple non-supersingular Abelian variety then $log(\mathrm{Fr}^\circ)$ is an automorphism of $\mathrm{V}_\ell(A)$.

Proof. Clearly, $\mathrm{Ker}(log(\mathrm{Fr}^\circ)) \subset \mathrm{V}_\ell(A_0)$ and we have only to check that if α is an eigen value of Frobenius acting on $\mathrm{V}_\ell(A_0)$ then α^2/q is not a root of 1 . (Recall that $q = \chi_\ell'(\mathrm{Fr})$ and use (1.2.0.1).) If α^2/q is a root of 1 then there exists a simple supersingular Abelian variety X over k such that α is an eigen value of Frobenius acting on $\mathrm{V}_\ell(\mathrm{X})$ [13]. So, the characteristic polynomials of Frobenius acting on $\mathrm{V}_\ell(\mathrm{X})$ and $\mathrm{V}_\ell(A_0)$ have non-trivial common factors. Now, a corollary of Tate's theorem on homomorphisms [12,§ 3, Th.1] implies an existence of non-zero homomorphism $X \to A_0$, whose image is a supersingular Abelian variety , because X is supersingular . Contradiction .

2.7. Definition. *An Abelian variety A is of* K3 *type if* $\mathrm{Slp}_A = \{0,1\}$ or $\{0, 1/2, 1\}$ *and in both cases*

$$\mathrm{length}_A(0) = \mathrm{length}_A(1) = 1 .$$

2.7.0. Remarks. 1)If k' is a finite algebraic extension of k and $A' = A \times k'$ then A is of K3 type if and only if A' is of K3 type. 2)if an Abelian variety X is isogenous to A then X is of K3 type if and only if A is of K3 type. 3)let $A = A^{\mathrm{ss}} \times A_0$ is a product of a supersingular Abelian variety A^{ss} and an Abelian variety A_0 such that A_0 does not contain non-zero supersingular Abelian subvarieties. Then A is of K3 type if and only if A_0 is an absolutely simple Abelian variety of K3 type. Indeed, if A is of K3 type then $\mathrm{Slp}_{A_0} \subset \{0,1/2,1\}$ and $\mathrm{length}_A(0)=\mathrm{length}_A(1) = 1$. If A_0 is not absolutely simple, then there exist a finite algebraic extension k' of k and non-zero non-supersingular Abelian varieties A_1 and A_2 over k' such that $A_1 \times A_2$ is isogenous to $A_0 \times k'$. It follows that

$$\mathrm{Slp}_{A_1} \cup \mathrm{Slp}_{A_2} \subset \{0, 1/2, 1\} , \ \mathrm{Slp}_{A_1} \neq \{1/2\}, \ \mathrm{Slp}_{A_2} \neq \{1/2\} .$$

This implies that $\{0,1\} \subset \mathrm{Slp}_{A_1}$, $\{0,1\} \subset \mathrm{Slp}_{A_2}$ and

$$1 = \mathrm{length}_{A_0}(0) = \mathrm{length}_{A_1}(0) + \mathrm{length}_{A_2}(0) \geq 1+1=2.$$

Contradiction.

2.7.1. Remarks. 1)If A is of K3 type then all the replicas $\mathrm{fr}_{\mathfrak{p}}^{\circ}$ (see 2.5) are operators of rank 2 ! Then it follows from Theorem 2.5. that $(\mathbb{Q}_{\ell}\log(\mathrm{Fr}^{\circ})^{\mathrm{al}}\otimes_{\mathbb{Q}_{\ell}}\bar{\mathbb{Q}}_{\ell}$ is generated by operators of rank 2.More precisely, $\mathrm{spec}(\mathrm{fr}_{\mathfrak{p}}^{\circ}) = \{-1,1\}$ or $\{-1.0,1\}$ and eigen values 1 and (-1) have multiplicity 1. 2)assume, in addition , that A is simple. Then $\mathrm{End}^{\circ}A := \mathrm{End}_k A\otimes\mathbb{Q}$ is a commutative field. Indeed, we have the embedding $\mathrm{End}_k A\otimes\mathbb{Z}_{\ell} \to \mathrm{End}_{G(k)}\mathrm{T}_{\ell}(A)$ which can be extend by $\mathbb{Q}_{\ell}$-linearity to embeddings [7]

$$\mathrm{End}^{\circ}A = \mathrm{End}_k A\otimes\mathbb{Q} \subset (\mathrm{End}_k A\otimes\mathbb{Q})\otimes_{\mathbb{Q}}\mathbb{Q}_{\ell} = \mathrm{End}_k A\otimes\mathbb{Q}_{\ell} \to \mathrm{End}_{G(k)}\mathrm{V}_{\ell}(A) \subset \mathrm{End}_{\mathfrak{g}_{\ell}}\mathrm{V}_{\ell}(A) \subset \mathrm{End}_{\bar{\mathfrak{g}}_{\ell}}\bar{\mathrm{V}}_{\ell}(A) .$$

Notice that the identity automorphism $1 \in \mathrm{End}^{\circ}A$ of A goes to id . It is known [7] that $\mathrm{End}^{\circ}A$ is a finite-dimensional division algebra over $\mathbb{Q}$ if A is simple. Clearly, $\mathrm{End}^{\circ}A$ leaves invariant all eigen subspaces of all $\mathrm{fr}_{\mathfrak{p}}^{\circ}$. But they have 1-dimensional eigen $\bar{\mathbb{Q}}_{\ell}$-vector subspaces and we obtain non-zero homomorphisms (1 goes to id) $\mathrm{End}^{\circ}A \to \bar{\mathbb{Q}}_{\ell}$. Since $\mathrm{End}^{\circ}A$ is a division algebra, this homomorphism is an embedding and, therefore, $\mathrm{End}^{\circ}A$ is commutative. 2bis)it follows from the Tate's theorem on homomorphisms [12] that $\mathrm{End}^{\circ}A$ is an imaginary quadratic extension of a totally real number field and $\dim_{\mathbb{Q}}\mathrm{End}^{\circ}A = 2\dim(A)$ .It is also well-known that $\mathrm{Fr} \in \mathrm{End}^{\circ}A = \mathbb{Q}[\mathrm{Fr}]$ (here we identify $\mathrm{End}^{\circ}A$ with its image in $\mathrm{End}\,\mathrm{V}_{\ell}(A)$). It follows that

$$\log(\mathrm{Fr}),\ \log(\mathrm{Fr}^{\circ}) \in \mathrm{End}^{\circ}A\otimes_{\mathbb{Q}}\mathbb{Q}_{\ell} \subset \mathrm{End}\,\mathrm{V}_{\ell}(A) .$$

Let $\tau\colon \mathrm{End}^{\circ}A \to \mathrm{End}^{\circ}A$ be the "complex conjugation" on $\mathrm{End}^{\circ}A$. One may extend τ by $\mathbb{Q}_{\ell}$-linearity to the automorphism of $\mathrm{End}^{\circ}A\otimes_{\mathbb{Q}}\mathbb{Q}_{\ell}$. It is well-known [7] that all elements of $\mathrm{End}^{\circ}A\otimes_{\mathbb{Q}}\mathbb{Q}_{\ell}$ are semisimple operators in $\mathrm{V}_{\ell}(A)$ and

$$H_{\mathcal{L}}'(ex,y) = H_{\mathcal{L}}'(x,(\tau e)y) \text{ for all } e \in \mathrm{End}^{\circ}A\otimes_{\mathbb{Q}}\mathbb{Q}_{\ell};\ x,y \in \mathrm{V}_{\ell}(A) .$$

It follows that $\tau(\log(Fr^{\circ})) = -\log(Fr^{\circ})$. One may also easily check that $\{e \in \mathrm{End}^{\circ}A\otimes_{\mathbb{Q}}\mathbb{Q}_{\ell} \mid \tau e = -e\}$ lies in the Lie algebra $\mathfrak{sp}_{\mathcal{L}}$ of the symplectic group of $\mathrm{V}_{\ell}(A)$ attached to $H_{\mathcal{L}}'$. Dimension arguments imply that $\{e \in \mathrm{End}^\circ A \otimes_{\mathbb{Q}} \mathbb{Q}_\ell \mid \tau e = -e\}$ is a Cartan subalgebra of $\mathbf{sp}_{\mathcal{L}}$; in particular, it is a commutative algebraic Lie subalgebra of $\mathrm{End}\, V_\ell(A)$ containing $log(\mathrm{Fr}^\circ)$.

2.7.2. Theorem. *If A is a simple Abelian variety of K3 type then $(\mathbb{Q}_\ell log(\mathrm{Fr}^\circ))^{\mathrm{al}}$ is a Cartan subalgebra of the Lie algebra $\mathbf{sp}_{\mathcal{L}}$ of the symplectic group of $V_\ell(A)$ attached to $H_{\mathcal{L}}'$. More precisely,*

$$(\mathbb{Q}_\ell log(\mathrm{Fr}^\circ))^{\mathrm{al}} = \{e \in \mathrm{End}^\circ A \otimes_{\mathbb{Q}} \mathbb{Q}_\ell \mid \tau e = -e\}.$$

Proof. By the Lemma 2.6.1 $log(\mathrm{Fr}^\circ)$ is an automorphism of $V_\ell(A)$, i.e., $log(\mathrm{Fr}^\circ) V_\ell(A) = V_\ell(A)$. If we apply the Theorem 1.3.1 to $u = \mathrm{Fr}$ then we obtain that $(\mathbb{Q}_\ell log(\mathrm{Fr}^\circ))^{\mathrm{al}}$ is a Cartan subalgebra of $\mathbf{sp}_{\mathcal{L}}$. Since $log(\mathrm{Fr}^\circ) \subset \{e \in \mathrm{End}^\circ A \otimes_{\mathbb{Q}} \mathbb{Q}_\ell \mid \tau e = -e\}$, we obtain that $(\mathbb{Q}_\ell log(\mathrm{Fr}^\circ))^{\mathrm{al}} \subset \{e \in \mathrm{End}^\circ A \otimes_{\mathbb{Q}} \mathbb{Q}_\ell \mid \tau e = -e\}$, because the latter is algebraic. One has only to recall that all the Cartan subalgebras of $\mathbf{sp}_{\mathcal{L}}$ have the same dimension; in particular, $\dim_{\mathbb{Q}_\ell} (\mathbb{Q}_\ell log(\mathrm{Fr}^\circ))^{\mathrm{al}} = \dim_{\mathbb{Q}_\ell} \{e \in \mathrm{End}^\circ A \otimes_{\mathbb{Q}} \mathbb{Q}_\ell \mid \tau e = -e\}$. **QED.**

Now we are ready to prove Tate's conjecture on algebraicity of Galois-invariant classes for Abelian varieties of K3 type.

2.7.3. Theorem. *Let A be an Abelian variety of K3 type. Then all elements of $\mathrm{H}^{2i}(A_{\mathrm{a}}, \mathbb{Q}_\ell)(i)^{\mathrm{G}(k)}$ are linear combinations of the products of classes of divisors on A_{a} and, therefore, are algebraic.*

Proof. Applying Theorem 1.3.1 to $u = \mathrm{Fr}$ we obtain that each element

$$E \in \mathrm{H}^{2i}(A_{\mathrm{a}}, \mathbb{Q}_\ell)(i)^{\mathrm{G}(k)} \subset \mathrm{Hom}(\Lambda^{2i} V_\ell(A), \mathbb{Q}_\ell)$$

is a linear combination of the exterior products $H_1 \wedge \ldots \wedge H_i$ of skew-symmetric bilinear forms H_j on $V_\ell(A)$ such that

$$\mathrm{H}_j(\mathrm{Fr}^{2N} x, \mathrm{Fr}^{2N} y) = \chi_\ell'(\mathrm{Fr}^{2N}) H_j(x,y) \text{ for all } x,y \in V_\ell(A) \ (1 \le j \le i).$$

In order to prove the theorem, it is enough to check that each H_j is a linear combination of the divisor classes on A_{a}.

Let k' be a finite algebraic extension of k of degree $2N$. Then $\sigma_{k'} = \sigma_k^{2N}$ and $\mathrm{Fr}^{2N} = \rho_\ell(\sigma_{k'})$. Since $\sigma_{k'}$ generates $\mathrm{G}(k')$, Fr^{2N} generates $\rho_\ell(\mathrm{G}(k'))$ and, therefore,

$$\mathrm{H}_j(ux, uy) = \chi_\ell'(u) H_j(x,y) \text{ for all } u \in \rho_\ell(\mathrm{G}(k')); x,y \in V_\ell(A) \ (1 \le j \le i).$$

This means that

$H_j(\sigma x, \sigma y) = \chi_\ell(\sigma) H_j(x,y)$ for all $\sigma \in G(k')$; $x,y \in V_\ell(A)$ $(1 \le j \le i)$,
i.e. all

$$H_j \in H^2(A_a, \mathbb{Q}_\ell)(1)^{G(k')}.$$

It follows from a theorem of Tate ([12], Th.4) that each H_j is a linear combination of divisor classes on $A \times k'$. **QED.**

Concluding Remark. Using ideas of this paper, I can also prove the Tate's conjecture for powers of Abelian varieties of K3 type and for powers of ordinary K3 surfaces over finite fields. (For ordinary K3 surfaces the Tate's conjecture was proven by Nygaard [15].)

REFERENCES

1. N. Bourbaki, Groupes et algébres de Lie, Chapitres 2-3, Hermann. Paris, 1972
2. N. Bourbaki, Groupes et algébres de Lie, Chapitres 7-8, Hermann, Paris, 1975.
3. C. Chevalley, Theorie de groupes de Lie. Groupes Algébriques, Hermann, Paris, 1951.
4. N. Koblitz, p-adic numbers, p-adic analysis and zeta functions, second edition, Springer-Verlag, 1984.
5. H. W. Lenstra Jr., F. Oort, Simple Abelian varieties having a prescribed formal isogeny type, J. Pure Appl. Algebra 4 (1974), 47-53.
6. B. Mazur, Eigenvalues of Frobenius acting on algebraic varieties over finite fields, AMS Proc. Symp. Pure Math. 29(1975), 231-261.
7. D. Mumford, Abelian varieties, second edition, Oxford University Press, 1974.
8. J.-P. Serre, Abelian ℓ-adic representations and elliptic curves, second edition, Addison-Wesley, 1989.
9. J.-P. Serre, Representations ℓ-adiques, Kyoto Symposium on Algebraic Number Theory (1976), p.177-193, Japan Society for the Promotion of Science, Tokyo, 1977.
10. J.-P. Serre, Resumé de cours de 1984-85. Ann. Collège de France, Paris 1985.
11. J. Tate, Algebraic cycles and poles of zeta functions, Arithmetical Algebraic Geometry, p. 93 - 110, Harper and Row, New York, 1965.
12. J. Tate, Endomorphisms of Abelian varieties over finite fields, Invent. Math. 2(1966), 134-144.
13. J. Tate, Classes d'isogénie des variétes abéliennes sur un corps fini(d'apres T. Honda). Sém. Bourbaki 352 (1968/69), p. 95-110. Springer Lecture Notes in Math. 179(1971).
14. A. Weil, Basic Number Theory, Springer-Verlag, 1967.
15. N. O. Nygaard, The Tate conjecture for ordinary K3 surfaces over finite fields, Inv. Math. 74(1983), 213 - 237.
16. Yu. G. Zarhin, Abelian varieties, ℓ-adic representations and Lie algebras. Rank independence on ℓ, Invent. Math. 55(1979), 165 - 176.

List of Talks

Talks in the Morning Sessions

Beilinson, A. A. : Polylogarithms and regulators.
Bloch, S. : Motives, cycles and K-theory(some conjectures).
Kollár, J. : Extremal rays on 3-folds.
Kronheimer, P. : Embedded 2-manifolds in complex surfaces.
Lazarsfeld, R. : Syzygies of algebraic varieties.
Mazur, B. : Introduction to families of modular eigenforms.
Simpson, C. : Moduli spaces of representations of π_1.
Siu, Y. -T. : Some rigidity problems in complex and Riemannian geometry.
Soulé, C. : Arakelov Geometry.
Varchenko, A. : Hypergeometric functions and conformal field theory.

Talks in the Afternoon Sessions

Baily, W. L, Jr. : Exceptional modular forms.
Berthelot, P. : Introduction to arithmetic theory of $\mathcal{D}$-module.
Bogomolov, F. : The Grothendieck and Bloch conjectures.
Breen, L. : Non-abelian cohomology.
Carlson, J. : Harmonic mappings and Hodge theory.
Catanese, F. : Fibrations of Kähler manifolds and classification of irregular manifolds.
Chang, M. C. : Some examples of obstructed manifolds with very ample canonical bundle.
Coates, J. : p-adic L-functions attached to motives.
Damon, J. : The vanishing topology of discriminants.
Dlousky, G. : Separate analyticity and singularities of analytic mappings.
Durfee, A. : Critical points of real polynomials of two variables.
Ein, L. : Vanishing theorems and syzygies.
Esnault, H. : Hodge type of subvarieties of $\mathbf{P}^n$ of small degree.
Falk, M. : The homotopy classification of hyperplane complements.
Flenner, H. : Moduli problem in supergeometry.
Fujiwara, K. : Arithmetic compactification of modular varieties.
Gomez-Mont, X. : Prescribing singularities of meromorphic vector fields.
Greuel, G. M. : Global moduli spaces for modules on curve singularities.
Gurjar, R. V. : A conjecture of C.T.C.Wall on 2-dimensional linear algebraic group quotients of affine spaces.
Harris, M. : Petersson norms of quaternionic modular forms.
Holme, A. : Codimension two subvarieties of $\mathbf{P}^N$.
Hulek, K. : On the geometry of certain moduli spaces of abelian surfaces.
Ionescu, P. : Mori theory and ample divisors.
Ishii, S. : Simultaneous canonical modifications of deformations of isolated singularities.
Iskovskikh, V. A. : Birational automorphisms of algebraic varieties.
Jurkiewicz, J. : Linearizations and toric varieties.
Kobayashi, R. : Ricci flat Kaehler metrics on affine algebraic manifolds and a geometric picture of degeneration of Kaehler-Einstein $K3$ surfaces.
Kollár, J. : Example of algebraic surfaces with π_1 not residually finite (after Toledo).
Kosarew, S. : On some problems in complex hyperbolic geometry.
Kurke, H. : Explicit construction of twistor spaces of self-dual metrics on $\#_n\mathbf{P}^2(\mathbf{C})$ and their deformation theory.

Laumon, G. : Cohomology with compact support of Drinfeld modular varieties.
Lê Dũng Tráng : Spaces with maximal topological depth.
Mehta, V : Grauert-Riemenschneider in char. p.
Mok, N. : Kähler geometry on arithmetic varieties.
Mukai, S. : Fano 3-folds of genus 12 and new compactification of the affine 3-space.
Mulase, M. : Super Grassmannians and Jacobians of super curves.
Murre, J. P. : Motives and algebraic cycles.
Murty, V. K. : The Albanese of unitary Shimura varieties.
Nakayama, N. : Elliptic fibration over a surface.
Narasimhan, M. S. : Generalized theta divisors on moduli spaces of vector bundles on curves.
Nikulin, V. V. : Linear systems of Weil divisors on singular $K3$ surfaces.
Ohsawa, T. : Cheeger-Goresky-Macpherson conjecture on varieties with isolated singularities.
Pasarescu, O. : Existence of curves in projective n-space.
Persson, U. : A survey on geography of surfaces
Peternell, T. : Manifolds whose tangent bundles are numerically effective.
Ran, Z. : Deformation invariance and unobstructed deformations.
Reid, M. : Problems of surface geometry.
Riemenschneider, O. : Deformations of rational surface singularities.
Saito, S. : Cycle map for Chow groups of surfaces with $p_g = 0$.
Saito, T. : Ramification in higher dimension.
Schneider, P. : p-adic L-functions.
Seshadri, C. S. : Study of moduli of vector bundles associated with a curve.
Shastri, A. R. : Equisingularity criterion for 2-dimensional Jacobian problem.
Shepherd-Barron, N. : Unstable vector bundles in char. p, and applications.
Shioda, T. : Algebraic surfaces and sphere packing.
Shokurov, V. V. : Log-terminal models of 3-folds.
Siersma, D. : Monodromy, vanishing cycles and the Zariski examples.
Slodowy, P. : Elliptic genera of homogeneous spaces.
Srinivas, V. : Gysin maps and cycle classes for Hodge cohomology.
Steenbrink, J. H. M. : Semicontinuity of singular spectrum.
Todorov, A. : Moduli of Calabi-Yau manifolds.
Trautmann, G. : On compactification of moduli spaces of instantons.
Tsuji, H. : Zariski decomposition of canonical divisors.
Tyurin, A. N. : Moduli of vector bundles on regular surfaces.
Viehweg, E. : Moduli of polarized varieties.
Wall, C. T. C. : Discriminant matrices and topological triviality.
Watanabe, K. : Distribution formula for terminal singularities on the minimal resolution of a simple $K3$ singularity.
Wilson, P. M. H. : Some questions of Calabi-Yau manifolds.
Zarhin, Yu. G. : ℓ-adic representations and semisimple Lie algebras.

List of Participants

Participants from Overseas

Andreatta, M. (Univ. di Milano)
Avritzer, D. (Univ. di Minas Gerais)
Baily, W. L., Jr. (Univ. of Chicago)
Ballico, E. (Univ. Trento)
Beilinson, A. A. (Landau Inst.)
Berthelot, P. (Univ. de Rennes)
Blasius, D. (UCLA)
Bloch, S. (Univ. of Chicago)
Bogomolov, F. A. (Steklov Inst.)
Breen, L. (Univ. de Paris-Nord)
Brundu, M. (Univ. Trieste)
Burchard, P. (Univ. of Utah)
Carlson, J. A. (Univ. of Utah)
Catanese, F. (Univ. di Pisa)
Chang, M. -C. (Univ. of California)
Ciliberto, C. (Univ. di Roma)
Coates, J. (Univ. of Cambridge)
Damon, J. (Univ. of North Carolina)
Diamond, F. (Boston Univ.)
Dloussky, G. (Univ. de Provence)
Durfee, A. H. (Mount Holyoke Coll.)
Ein, L. (Univ. of Illinois at Chicago)
Esnault, H. (Univ. Essen)
Falk, M. (Northern Arizona Univ.)
Flenner, H. (Univ. Göttingen)
Friedland, S. (Univ. of Illinois at Chicago)
Gomez-Mont, X. (Univ. de Mexico)
Graham, J. (Univ. Sydney)
Greuel, G. -M. (Univ. Kaiserslautern)
Gurjar, R. V. (Tata Inst.)
Hanamura, M. (MIT)
Harris, M. (Johns Hopkins Univ.)
Hoang, L. M. (Intl. C. for Th. Phys.)
Höfer, T. (Kyoto Univ.)
Holme, A. (Univ. of Bergen)
Hulek, K. (Univ. Bayreuth)
Im, J. (Ohio State Univ.)
Ionescu, P. (Univ. of Bucharest)
Ionescu, P. C. (INCREST)
Iskovskikh, V. A. (Moscow State Univ.)
Jambu, M. (Univ. de Nantes)
Jurkiewicz, J. (Warsaw Univ.)
Keum, J. (Univ of Utah)
Kobayashi, S. (Univ. of California)
Kollár, J. (Univ. of Utah)
Kosarew, I. (Univ. Göttingen)
Kosarew, S. (MPI, Bonn)
Kronheimer, P. (Univ. of Oxford)
Kurke, H. (Humboldt Univ.)
Lakshmibai, V. (Northeastern Univ.)
Laumon, G. H. (Univ. de Paris-Sud)
Lazarsfeld, R. (UCLA)
Lê Dũng Tráng (Univ. Paris VII)
Lorenzini, D. J. (Yale Univ.)
Luo, T. (Univ. of Utah)
Marathe, K. B. (City Univ. of New York)
Martucci, G. (Univ. di Firenze)
Matsuki, K. (Brandeis Univ.)
Matsusaka, T. (Brandeis Univ.)
Mazur, B. (Harvard Univ.)
Mehta, V. B. (Tata Inst.)
Mestrano, N. (Univ. de Toulouse)
Mok, N. (Univ. de Paris-Sud)
Mulase, M. (Univ. of California)
Murre, J. P. (Univ. of Leiden)
Murty, V. K. (Univ. of Chicago)
Narasimhan, M. S. (Tata Inst.)
Navarro-Aznar, V. (Univ. Barcelona)
Nikulin, V. V. (Steklov Inst.)
Paranjape, K. H. (Tata Inst.)
Pardini, R. (Univ. Pisa)
Pasarescu, O. F. (INCREST)
Persson, U. (Chalmers Univ.)
Peternell, T. (Univ. Bayreuth)
Ramanan, S. (Tata Inst.)
Ran, Z. (Univ. of California)
Randell, R. (Univ. of Iowa)
Reid, M. (Univ. of Warwick)
Riemenschneider, O. W. (Univ. Hamburg)
Schneider, P. (Univ. Köln)
Seshadri, C. S. (SPIC Sc. Found.)
Shastri, A. R. (Ind. Inst. of Tech.)
Shepherd-Barron, N. I. (Univ. of Illinois)
Shiota, T. (Brandeis Univ.)
Shokurov, V. V. (Yaroslavle St. Ped. Inst.)
Shukla, P. (Suffolk Univ.)
Siersma, D. (Univ. of Utrecht)
Simpson, C. T. (Princeton Univ.)
Siu, Y. -T. (Harvard Univ.)
Slodowy, P. (Univ. Stuttgart)
Soulé, C. (I.H.E.S.)
Srinivas, V. (Tata Inst.)
Steenbrink, J. H. M. (Univ. Nijmegen)
Todorov, A. (MPI, Bonn)
Trautmann, G. (Univ. Kaiserslautern)
Tyurin, A. N. (Steklov Inst.)
Van de Ven, A. (Univ. of Leiden)
Varchenko, A. N. (Moscow Inst. of Gas & Oil)
Viehweg, E. (Univ. Essen)
Voloch, J. F. (IMPA)
Wall, C. T. C. (Univ. of Liverpool)
Wilson, P. M. H. (Univ. of Cambridge)
Yui, N. (Queen's Univ.)
Zarhin, Y. G.(Res. Comp. C. of USSR Acad. of Sci.)

Participants from inside Japan

Ando, T. (Chiba Univ.)
Asanuma, T. (Toyama Univ.)
Ashikaga, T. (Tohoku Gakuin Univ.)
Bando, S. (Tohoku Univ.)
Cho, K. (Kyushu Univ.)
Ebihara, M. (Gakushuin Univ.)
Endo, S. (Tokyo Metropolitan Univ.)
Enoki, I. (Osaka Univ.)
Fujiki, A. (Kyoto Univ.)
Fujimoto, H. (Kanazawa Univ.)
Fujisaki, G. (Univ. of Tokyo)
Fujisawa, T. (Tokyo Inst. of Tech.)
Fujita, T. (Tokyo Inst. of Tech.)
Fujiwara, K. (Univ. of Tokyo)
Fukui, T. (Nagano Tech. Coll.)
Furushima, M. (Ryukyu Univ.)
Furuta, H. (Tohoku Univ.)
Hamahata, Y. (Univ. of Tokyo)
Hara, N. (Waseda Univ.)
Hashimoto, N. (Tokyo Inst. of Tech.)
Hayashida, T. (Ochanomizu Univ.)
Hazama, F. (Tokyo Denki Univ.)
Hidaka, H. (Senshu Univ. Hokkaido Jun. Coll.)
Hino, T. (Tohoku Univ.)
Hino, Y. (Tokyo Metropolitan Univ.)
Hirano, A. (Saitama Univ.)
Hironaka, Y. (Shinshu Univ.)
Homma, M. (Ryukyu Univ.)
Homma, Y. (Ryukyu Univ.)
Hosoh, T. (Sci. Univ. of Tokyo)
Ichikawa, T. (Kyushu Univ.)
Ichimura, F. (Yokohama City Univ.)
Iitaka, S. (Gakushuin Univ.)
Ishida, M. (Tohoku Univ.)
Ishii, A. (Kanazawa Inst. of Tech.)
Ishii, S. (Tokyo Inst. of Tech.)
Ito, H. (Osaka Univ.)
Ito, H. (Univ. of Tokyo)
Iwasawa, K.
Iyanaga, S.
Izumi, S. (Kinki Univ.)
Kaga, T. (Niigata Univ.)
Kagesawa, M. (Tokyo Metropolitan Univ.)
Kaji, H. (Yokohama City Univ.)
Kaneko, M. (Kyoto Inst. of Tech.)
Kaneko, Y. (Univ. of Tsukuba)
Kaneyuki, S. (Sophia Univ.)
Katano, S. (Rikkyo Univ.)
Kato, K. (Univ. of Tokyo)
Kato, M. (Sophia Univ.)
Katsura, T. (Ochanomizu Univ.)
Kawamata, Y. (Univ. of Tokyo)
Kobayashi, M. (Univ. of Tokyo)
Kobayashi, R. (Univ. of Tokyo)
Koizumi, S. (Kogakuin Univ.)
Komatsu, K. (Tokyo Univ. of Agr. and Tech.)
Komori, Y. (RIMS, Kyoto)
Kondo, S. (Saitama Univ.)
Kondo, S. (Waseda Univ.)
Konno, H. (Univ. of Tokyo)
Konno, K. (Kyushu Univ.)
Kumo, T. (Univ. of Tokyo)
Kuramoto, Y. (Kagoshima Univ.)
Kurihara, A. (Nihon Women's Univ.)
Kurihara, M. (Tokyo Metropolitan Univ.)
Mabuchi, T. (Osaka Univ.)
Maeda, H. (Tokyo Univ. of Agr. and Tech.)
Maeda, H. (Waseda Univ.)
Maeda, T. (Ryukyu Univ.)
Maehara, K. (Tokyo Inst. of Polytech.)
Maehashi, T. (Kumamoto Univ.)
Majima, H. (Ochanomizu Univ.)
Masuda, T. (Univ. of Tsukuba)
Matsuda, S. (Univ. of Tokyo)
Matsuoka, S. (Hokkaido Univ.)
Matsuoka, T. (Saitama Univ.)
Miyajima, K. (Kagoshima Univ.)
Miyanishi, M. (Osaka Univ.)
Miyaoka, Y. (Tokyo Metropolitan Univ.)
Mizuno, H. (Univ. of Electro-Commun.)
Momose, F. (Chuo Univ.)
Mori, S. (RIMS, Kyoto)
Morimoto, T. (Waseda Univ.)
Morita, Y. (Tohoku Univ.)
Moriwaki, A. (Kyoto Univ.)
Mukai, S. (Nagoya Univ.)
Murata, Y. (Nagasaki Univ.)
Nakagawa, N. (Hiroshima Univ.)
Nakagawa, Y. (Osaka Univ.)
Nakagoshi, T. (Saitama Univ.)
Nakai, I. (Hokkaido Univ.)
Nakajima, S. (Univ. of Tokyo)
Nakajima, T. (Kyoto Univ.)
Nakajima, Y. (Univ. of Tokyo)
Nakamura, H. (Univ. of Tokyo)
Nakamura, I. (Hokkaido Univ.)
Nakano, T. (Setsunan Univ.)
Nakano, T. (Tokyo Denki Univ.)
Nakayama, N. (Univ. of Tokyo)
Namba, M. (Osaka Univ.)
Namikawa, Y. (Nagoya Univ.)
Namikawa, Y. (Sophia Univ.)
Nishiyama, K. (Saitama Univ.)
Nitta, T. (Osaka Univ.)
Noguchi, J. (Tokyo Inst. of Tech.)
Oda, T. (Tohoku Univ.)

Ogata, S. (Tohoku Univ.)
Ogiwara, T. (Waseda Univ.)
Oguiso, K. (Univ. of Tokyo)
Ohbuchi, A. (Yamaguchi Univ.)
Ohhira, M. (Tokyo Metropolitan Univ.)
Ohmori, Z. (Hokkaido Univ.)
Ohnita, Y. (Tokyo Metropolitan Univ.)
Ohno, K. (Univ. of Tokyo)
Ohno, M. (Waseda Univ.)
Ohsawa, T. (RIMS, Kyoto)
Ohta, H. (Univ. of Tokyo)
Ohta, K. (Tsuda Coll.)
Ohta, M. (Tokai Univ.)
Oka, M. (Tokyo Inst. of Tech.)
Okai, T. (Hiroshima Univ.)
Ooe, T. (Univ. of Tokyo)
Park, H. S. (Tohoku Univ.)
Saito, H. (Kanazawa Univ.)
Saito, H. (Kyoto Univ.)
Saito, K. (RIMS, Kyoto)
Saito, T. (Univ. of Tokyo)
Sakagawa, H. (Univ. of Tokyo)
Sakai, F. (Saitama Univ.)
Sasaki, T. (Kyoto Univ.)
Sasakura, N. (Tokyo Metropolitan Univ.)
Satake, I. (RIMS, Kyoto)
Satake, I. (Tokhoku Univ.)
Sato, A. (Tohoku Univ.)
Sato, E. (Kyushu Univ.)
Seki, T. (Waseda Univ.)
Sekiguchi, T. (Chuo Univ.)
Shiga, H. (Chiba Univ.)
Shimada, I. (Univ. of Tokyo)
Shimizu, H. (Univ. of Tokyo)
Shimizu, Y. (Tohoku Univ.)
Shioda, T. (Rikkyo Univ.)
Shiota, J. (Saitama Univ.)
Shiratani, K. (Kyushu Univ.)
Somekawa, M. (Univ. of Tokyo)
Sudo, M. (Seikei Univ.)
Sueyoshi, Y. (Kyushu Univ.)
Sugie, T. (Kyoto Univ.)
Sugihara, T. (Waseda Univ.)
Sugimoto, M. (Univ. of Tokyo)
Sugiyama, K. (Chiba Univ.)
Sumihiro, H. (Hiroshima Univ.)
Suwa, N. (Tokyo Denki Univ.)
Suwa, T. (Hokkaido Univ.)
Suzuki, S. (Teikyo Univ.)
Taguchi, Y. (Univ. of Tokyo)
Takahashi, T. (Tohoku Univ.)
Takayama, S. (Tokyo Metropolitan Univ.)
Takeda, Y. (Nara Women's Univ.)
Takegoshi, K. (Osaka Univ.)
Takeuchi, T. (Nagoya Univ.)
Taya, H. (Waseda Univ.)
Terao, H. (ICU)
Terao, T. (Univ. of Tokyo)
Toki, K. (Yokohama Nat. Univ.)
Tokunaga, H. (Kochi Univ.)
Tomari, M. (Kanazawa Univ.)
Tomaru, T. (Coll. of Med. Care & Tech. Gunma Univ.)
Tominaga, T. (Tsuda Coll.)
Tomiyama, Y. (Univ. of Tokyo)
Tsuboi, S. (Kagoshima Univ.)
Tsuji, H. (Tokyo Metropolitan Univ.)
Tsuru, H. (Saitama Univ.)
Tsushima, R. (Meiji Univ.)
Tsuyumine, S. (Mie Univ.)
Tsuzuki, N. (Univ. of Tokyo)
Umemura, H. (Kumamoto Univ.)
Umezu, Y. (Toho Univ.)
Urabe, T. (Tokyo Metropolitan Univ.)
Urakawa, H. (Tohoku Univ.)
Usui, S. (Osaka Univ.)
Wakabayashi, I. (Tokyo Univ. of Agr. and Tech.)
Watanabe, K. (Tohoku Univ.)
Watanabe, K. (Univ. of Tsukuba)
Watanabe, M. (Hokkaido Univ.)
Yamagishi, M. (Univ. of Tokyo)
Yamamoto, S. (Nihon Univ.)
Yamawaki, N. (Waseda Univ)
Yokogawa, K. (Kyoto Univ.)
Yokoyama, T. (Tohoku Univ.)
Yonemura, T. (Univ. of Tsukuba)
Yoshihara, H. (Niigata Univ.)
Yoshikawa, K. (RIMS, Kyoto)
Yoshioka, M. (Tohoku Univ.)
Yusa, T. (Kochi Univ.)
Zhang, D. -Q. (Osaka Univ.)